Robin Matheson

English for AGRIBUSINESS AND AGRICULTURE in Higher Education Studies

Teacher's Book

Series editor: Terry Phillips

esap
English for Specific Academic Purposes

Garnet EDUCATION

Published by
Garnet Publishing Ltd.
8 Southern Court
South Street
Reading RG1 4QS, UK

First published 2014.
Reprinted 2016.

ISBN 978-1-85964-451-5

British Cataloguing-in-Publication Data
A catalogue record for this book is available from the British Library.

Production

Series editor:	Terry Phillips
Project management:	Claire Forsyth, Matthew George
Editorial team:	Kirsten Campbell, Karen Kinnair-Pugh, Matthew George
Specialist review:	Alison Bailey
Design:	Mike Hinks, Bob House
Illustration:	Doug Nash
Photography:	Alamy, Corbis, Getty Images, Shutterstock

Audio recorded at Silver Street Studios.

Acknowledgements
Page 23, graph, reproduced with kind permission of the Soil Association, Organic market report 2013, www.soilassociation.org.
Page 51, Figure 1, self-sufficiency graph, taken from the website of the Cabinet Office Strategy Unit.
Page 84, Slide 1, graph, taken from the website of the Dutch Soft Drinks Association.
Page 133, quotation, reproduced with kind permission of The Guardian newspaper. Copyright © Guardian News & Media Ltd 2013.

Printed and bound in Lebanon by International Press:
interpress@int-press.com

Contents

Book map

Unit	Topics
1 Agriculture and business **Listening · Speaking**	• defining agribusiness • ethics • the origins of agribusiness
2 Foundations of modern agriculture **Reading · Writing**	• agricultural advances through the ages • key agriculturalists
3 Soil, land and the environment **Listening · Speaking**	• food security, self-sufficiency and land use • soil quality • organic farming methods
4 Technology and agribusiness **Reading · Writing**	• technological change: positive and negative consequences • using IT for research
5 Agribusiness marketing **Listening · Speaking**	• definitions of marketing • types of market • market research
6 Products and strategies **Reading · Writing**	• product life cycles • the marketing mix
7 The effective production manager **Listening · Speaking**	• job description • the qualities needed by a production manager • the importance of scheduling
8 Rural economies and financial management **Reading · Writing**	• agribusiness in the 'free market' • concepts in agri-economics • adaptive management techniques
9 Managing financial accounts **Listening · Speaking**	• documentation: profit and loss accounts and balance sheets • management accountancy methods • cash flow statements
10 Food safety **Reading · Writing**	• food hygiene • food safety regulations • case study: a proactive approach to food safety
11 The future of farming **Listening · Speaking**	• EU agricultural policy • agribusiness strategy in the 21st century • the impact of environmental issues
12 The good agribusiness **Reading · Writing**	• company performance: SWOT analysis • managing change • case study: responding to external factors

1 AGRICULTURE AND BUSINESS

This introductory unit explores what we understand by the term 'agribusiness'. Students hear a lecture in which they are introduced to the idea of simple and complex businesses. The lecturer presents features that are common to all businesses, and concepts which are more specific to the food and agriculture sector. Students also listen to a series of mini-lectures that outline the history of agriculture and agribusiness, introduce some key current concepts and predict future trends. The content of the mini-lectures will be explored in more detail in subsequent units.

Skills focus

Listening

- preparing for a lecture
- predicting lecture content from the introduction
- understanding lecture organization
- choosing an appropriate form of notes
- making lecture notes

Speaking

- speaking from notes

Vocabulary focus

- words from general English with a special meaning in agribusiness and agriculture
- prefixes and suffixes

Key vocabulary

agribusiness	free-range	output (n)
agriculture	gene	producer
agri-food	genetic	raw materials
agri-tourism	growth	research and development (R&D)
alternative	hand	resource (n)
battery (adj)	head	sector
cereal	holistic	service industry
consumer	human resources	shareholder
cooperative	industrial (adj)	stakeholder
crops	input (n)	stewardship
direct	intensive	supply chain
diversify	livestock	sustainability
ethical	manufacturing	variant
extractive industry	market (n)	variety
factory (adj)	marketing	
finance (n)	organic	
fossil fuels	operations	

1.1 Vocabulary

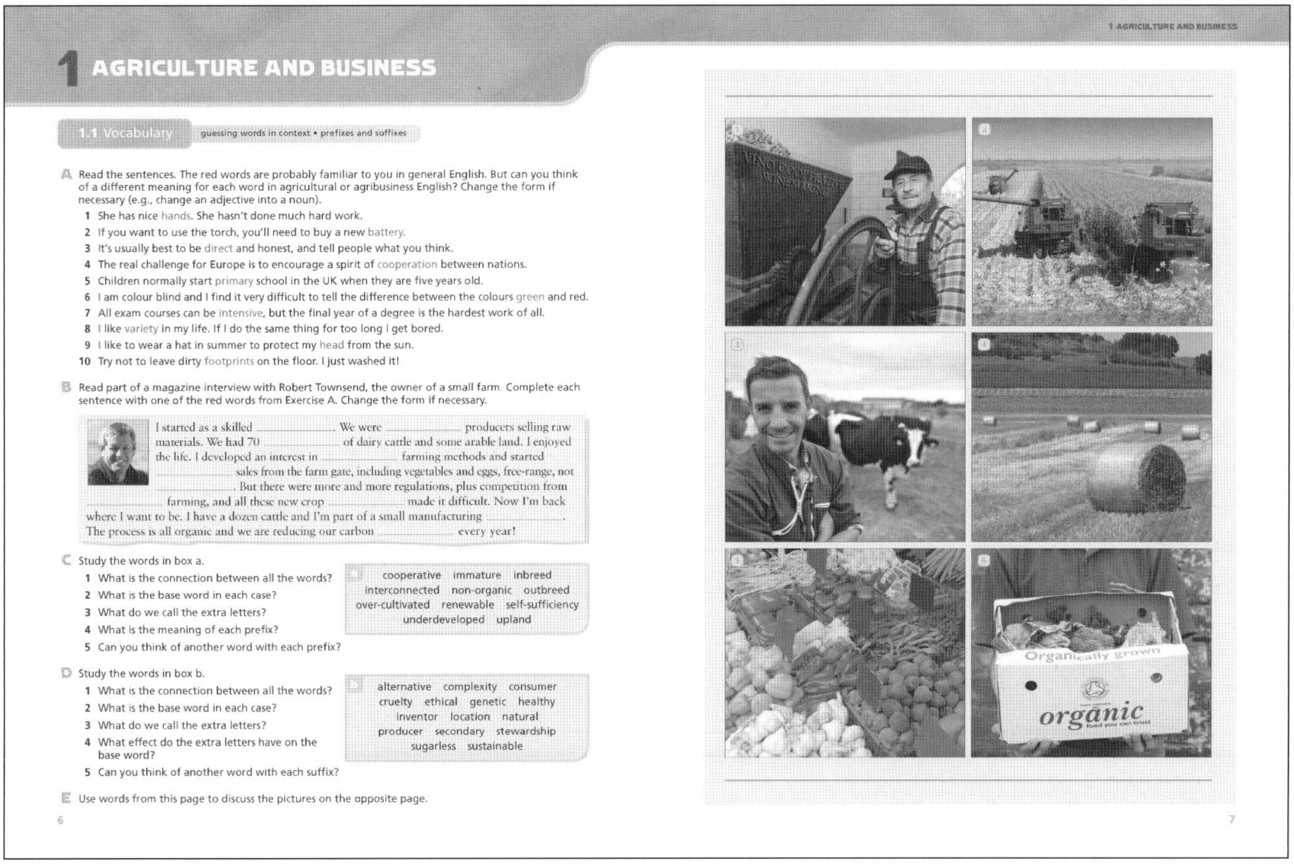

1 AGRICULTURE AND BUSINESS

1 AGRICULTURE AND BUSINESS

1.1 Vocabulary guessing words in context • prefixes and suffixes

A Read the sentences. The red words are probably familiar to you in general English. But can you think of a different meaning for each word in agricultural or agribusiness English? Change the form if necessary (e.g., change an adjective into a noun).

1 She has nice hands. She hasn't done much hard work.
2 If you want to use the torch, you'll need to buy a new battery.
3 It's usually best to be direct and honest, and tell people what you think.
4 The real challenge for Europe is to encourage a spirit of cooperation between nations.
5 Children normally start primary school in the UK when they are five years old.
6 I am colour blind and I find it very difficult to tell the difference between the colours green and red.
7 All exam courses can be intensive, but the final year of a degree is the hardest work of all.
8 I like variety in my life. If I do the same thing for too long I get bored.
9 I like to wear a hat in summer to protect my head from the sun.
10 Try not to leave dirty footprints on the floor. I just washed it!

B Read part of a magazine interview with Robert Townsend, the owner of a small farm. Complete each sentence with one of the red words from Exercise A. Change the form if necessary.

I started as a skilled ______. We were ______ producers selling raw materials. We had 70 ______ of dairy cattle and some arable land. I enjoyed the life. I developed an interest in ______ farming methods and started ______ sales from the farm gate, including vegetables and eggs, free-range, not ______. But there were more and more regulations, plus competition from ______ farming, and all these new crop ______ made it difficult. Now I'm back where I want to be. I have a dozen cattle and I'm part of a small manufacturing ______. The process is all organic and we are reducing our carbon ______ every year!

C Study the words in box a.

1 What is the connection between all the words?
2 What is the base word in each case?
3 What do we call the extra letters?
4 What is the meaning of each prefix?
5 Can you think of another word with each prefix?

a: cooperative immature inbreed interconnected non-organic outbreed over-cultivated renewable self-sufficiency underdeveloped upland

D Study the words in box b.

1 What is the connection between all the words?
2 What is the base word in each case?
3 What do we call the extra letters?
4 What effect do the extra letters have on the base word?
5 Can you think of another word with each suffix?

b: alternative complexity consumer cruelty ethical genetic healthy inventor location natural producer secondary stewardship sugarless sustainable

E Use words from this page to discuss the pictures on the opposite page.

6

7

General note

Read the *Vocabulary bank* at the end of the Course Book unit. Decide when, if at all, to refer students to it. The best time is probably at the very end of the lesson or the beginning of the next lesson, as a summary/revision.

Lesson aims

- identify words for the discipline in context, including words which contain affixes
- gain fluency in the target vocabulary

Introduction

Write the following sentences on the board:

1 *A good farm is a good home for wildlife.*
2 *My family has farmed this land for generations.*
3 *Farming is his life.*

Ask students the meaning of *farm* in the three sentences:

- In sentence 1, *farm* is a countable noun and refers to an area of cultivated land.
- In sentence 2, *farmed* is a transitive verb (present perfect) and refers to a way of life working on the land.
- In sentence 3, *farming* is an uncountable noun and it refers to a career or life choice, cf. *medicine is her life,* meaning 'she is a doctor and loves being one'.

Ask students to think of some phrases with *agri-* or *agro-* in them. You can ask them to use their imaginations and make up words and then check in their dictionaries to see if they exist. You can also take a real word and a made-up one and discuss why one works and the other does not. Here are some examples with the made-up words in bold:

agri-food
agro-chemical
agribusiness
agri-sausage
agriculture
agronomy
agro-tractor

Notice that the pronunciation of *~business* – /bɪznəs/ – is not the same as *~busyness* /bɪzɪnɪs/ (being busy, having a lot of things to do). Make sure students can pronounce the difference.

Note: Don't spend too long on this.

Exercise A

Set for individual work and pairwork checking. Point out that this is a text which introduces some important basic vocabulary related to agribusiness and agriculture – although it may not seem like that, at first glance. Do the first one as an example, e.g., *In general English, a hand refers to a part of the body. In agribusiness and agriculture, it refers to either a unit of measurement for the height of a horse (e.g., the horse is 16 hands at the shoulder) or a manual labourer (e.g., John works as a farmhand).*

Point out that there is often a relationship with the general English meaning, and if you know the general English meaning it can help to guess the agribusiness meaning (as in the case of *intensive*). Remind students to change the form if necessary, e.g., from verb to noun.

Answers

Model answers.

Check students understand grammar or other changes.

Feed back, putting the agribusiness English meanings in a table on the board.

Tell students to use these structures where possible:

- *a(n) X is (a(n))* … to define a noun
- *to X is to Y* to define a verb

Make sure students can say the words correctly. For example:

- vowel /e/ in *processed*
- long vowel /əʊ/ in *cooperation*
- schwa /ə/ in *cooperation*

Students may not be familiar with all meanings in general English, e.g., *intense* = extreme, profound; *cooperation* = the experience of working together towards a common goal.

Word	Meaning	Comments
hand	an unskilled farm worker	used in compound nouns such as *farmhand*
battery	raised indoors at high density	used in compound nouns such as *battery farming* and *battery chicken*
direct	from the producer to the customer without intermediary	the Internet is enabling more and more manufacturers to sell direct to the customer
cooperative	a group of, e.g., farmers who work together to, e.g., market their produce	cooperative can be a countable noun or an adjective change from noun to adjective to mean working together through choice: *Teaching should be a cooperative activity.*
primary	an economic sector like agriculture or coal mining producing raw materials	
green	describes a type of farming that is environmentally friendly	also used as abstract noun, as in the slogan 'green is better'
intensive	designed to produce the highest possible outputs	cf. *intense*, also an adjective but used, e.g., to describe an experience or flavour
variety	a type of (plant)	From same origin as *various* and linked to *variegated* (adj): *Variegated holly has more than one colour on its leaves.*
head	single or multiple units of domestic, herd animals like cattle, bison, deer: *100 head of deer or 1 head on its own …*	'head' is used for larger farmed animals that tend to be counted, e.g., through gates
footprint	an abstract noun meaning 'impact'	most common with carbon (*carbon footprint* = the amount of carbon an activity/person generates)

Language note

The prefixes *agri-* and *agro-* refer to land management or cultivation. They can be used, for example, to denote business and scientific activities relating to agriculture such as *agrochemical* and *agri-tourism*. The compound noun *agribusiness* seems to have come into use in the 1950s, initially in the United States. Ask students if they have an equivalent word to *agribusiness* in their language. You can also ask why they think it was necessary to invent the word.

Exercise B

Set for individual work and pairwork checking. Do the first sentence as an example.

Feed back with the whole class. Ask students for any other words they know which have a special meaning in agribusiness and agriculture.

Answers

I started as a skilled hand. We were primary producers selling raw materials. We had 70 head of dairy cattle and some arable land. I enjoyed the life. I developed an interest in green farming methods and started direct sales from the farm gate, including vegetables and eggs, free-range, not battery. But there were more and more regulations, plus competition from intensive farming, and all these new crop varieties made it difficult. Now I'm back where I want to be. I have a dozen cattle and I'm part of a small manufacturing cooperative. The process is all organic and we are reducing our carbon footprint every year!

Examples of other possible words from general English in agribusiness and agriculture:

combine (n) – short for combine harvester

rust (n) – a disease that affects plants with a powder-like dust

cake (n) – compound feed made of different ingredients

holding (n) – land and building owned freehold or as part of a tenancy

rustle (v) – to steal livestock

Exercise C

Set the first question for pairwork. See which pair can work out the answer first.

Set the remainder for pairwork. Feed back, building up the table in the Answers section on the board.

Answers

Model answers:

1 They all have a base word + extra letters at the beginning/prefixes.
2 See table below.
3 Prefix.
4 See table.
5 See table.

Language note

English is a lexemic language. In other words, the whole meaning of a word is usually contained within the word itself, rather than coming from a root meaning plus prefixes or suffixes (affixes). In most texts, written or spoken, there will only be a tiny number of words with affixes. However, these often add to a base meaning in a predictable way and it is important that students learn to detach affixes from a new word and see if they can find a recognizable base word.

Some words beginning with letters from prefixes are NOT in fact base + prefix, e.g., *refuse*. In other cases, the base word does not exist anymore in English and therefore will not help students, e.g., *transfer*, *transit*, although even in these cases the root meaning of the prefix may be a guide to the meaning of the whole word.

Exercise D

Repeat the procedure from Exercise C.

Answers

Model answers:

1 They all have a base word + extra letters at the end/suffixes.
2 See table on next page.
3 Suffix.
4 See table.
5 See table.

Prefix	Base word	Meaning of prefix	Another word
co~	operative	jointly or together	co-financed
im~	mature	not	impossible
in~	breed	within	ingoing
inter~	connected	between	international
non-	organic	not	non-productive
out~	breed	outside	outgoing
over~	cultivated	excessive	overgrown
re~	newable	again	replace
self-	sufficiency	by itself or oneself	self-motivated
under~	developed	insufficient	underperform
up~	land	higher or above	upstairs

Base word	Suffix	Effect/meaning of suffix	Another word
alternat(e)	~ive	verb → adjective	progressive
complex	~ity	adjective → noun	accountability
consum(e)	~er	verb → noun	provider
cruel	~ty	adjective → noun	activity
ethic	~al	noun → adjective	natural
gene	~tic	noun → adjective	aquatic
health	~y	noun → adjective	woody
invent	~or	verb → noun	originator
locat(e)	~ion	verb → noun	progression
natur(e)	~al	noun → adjective	ethical
produc(e)	~er	person that does something	grower
second	~ary	noun → adjective	revolutionary
steward	~ship	condition of being a …	statesmanship
sugar	~less	without	tasteless
sustain	~able	verb → adjective/capable of being	obtainable

Language note

Note that with prefixes, we rarely change the form of the base word. However, with suffixes, there are often changes to the base word, so students must:

- take off the suffix
- try to reconstruct the base word

Exercise E

Set for pairwork. Try to elicit more than just the words from this lesson. Students should describe the pictures as fully as they can at this stage.

Students may use the following words in their discussion of each picture:

1 **cooperative**
2 **intensive**, industrial, factory
3 unskilled, **hand, head**
4 **renewable, sustainable**
5 **produce**
6 **green**

Closure

If you have not done so already, refer students to the *Vocabulary bank* at the end of Unit 1. Tell students to explain how this lesson can help them deal with new words in context. If you wish, make three groups. Group A looks at the first section, *Using related words*. Group B looks at the second section, *Removing prefixes*. Group C looks at the third section, *Removing suffixes*. Then make sets of three with an ABC in each to explain the concepts to each other.

1.2 Listening

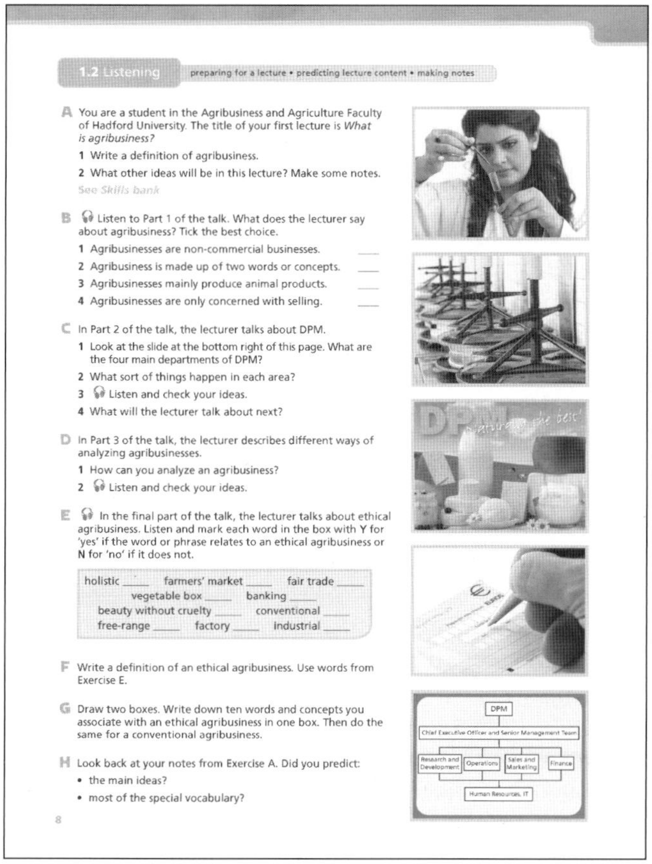

1.2 Listening — preparing for a lecture • predicting lecture content • making notes

A You are a student in the Agribusiness and Agriculture Faculty of Hadford University. The title of your first lecture is *What is agribusiness?*

1 Write a definition of agribusiness.

2 What other ideas will be in this lecture? Make some notes.

See *Skills bank*

B Listen to Part 1 of the talk. What does the lecturer say about agribusiness? Tick the best choice.

1 Agribusinesses are non-commercial businesses. ___

2 Agribusiness is made up of two words or concepts. ___

3 Agribusinesses mainly produce animal products. ___

4 Agribusinesses are only concerned with selling. ___

C In Part 2 of the talk, the lecturer talks about DPM.

1 Look at the slide at the bottom right of this page. What are the four main departments of DPM?

2 What sort of things happen in each area?

3 Listen and check your ideas.

4 What will the lecturer talk about next?

D In Part 3 of the talk, the lecturer describes different ways of analyzing agribusinesses.

1 How can you analyze an agribusiness?

2 Listen and check your ideas.

E In the final part of the talk, the lecturer talks about ethical agribusiness. Listen and mark each word in the box with Y for 'yes' if the word or phrase relates to an ethical agribusiness or N for 'no' if it does not.

holistic ___ farmers' market ___ fair trade ___ vegetable box ___ banking ___ beauty without cruelty ___ conventional ___ free-range ___ factory ___ industrial ___

F Write a definition of an ethical agribusiness. Use words from Exercise E.

G Draw two boxes. Write down ten words and concepts you associate with an ethical agribusiness in one box. Then do the same for a conventional agribusiness.

H Look back at your notes from Exercise A. Did you predict:

- the main ideas?
- most of the special vocabulary?

DPM

Chief Executive Officer and Senior Management Team

Research and Development | Operations | Sales and Marketing | Finance

Human Resources, IT

8

General note

The recording should only be played once, since this reflects what happens in a real lecture. Students should be encouraged to listen for the important points, since this is what a native speaker would take from the text. However, students can be referred to the transcript at the end of the lesson to check their detailed understanding and word recognition, or to try to discover reasons for failing to comprehend.

Read the *Skills bank* at the end of the Course Book unit. Decide when, if at all, to refer students to it. The best time is probably at the very end of the lesson or the beginning of the next lesson, as a summary/revision.

Lesson aims

- prepare for a lecture
- predict lecture content
- make notes

Introduction

1 Show students flashcards of some or all of the words from Lesson 1.1. Tell students to say the words correctly and quickly as you flash them. Give out each word to one of the students. Say the words again. The student with the word must hold it up. Repeat the process saying the words in context.

2 Refer students to the photos. Briefly elicit ideas of what they depict. (They will look at the different aspects of agribusiness activity in more detail in Exercise E.)

Exercise A

1 Set for pair or group work. Feed back, but do not confirm or correct at this time.

2 Set for pairwork. Elicit some ideas but do not confirm or correct.

Methodology note

You may want to refer students to the *Skills bank – Making the most of lectures* at this point. Set the following for individual work and pairwork checking. Tell students to cover the points and try to remember what was under each of the Ps – Plan, Prepare, Predict, Produce. Then tell students to work through the points to make sure they are prepared for the lecture they are about to hear.

Exercise B

Give students time to read the choices. Point out that they are only going to hear the introduction once, as in an authentic lecture situation. Play Part 1. Feed back. If students' answers differ, discuss which is the best answer and why.

Answers

2 Agribusiness is made up of two words or concepts.

Transcript 1.1

Part 1

Good morning, everyone, and welcome to the Faculty of Agribusiness and Agriculture. I'd like to start today by asking the question: What do we mean by the term 'agribusiness'? Well, let's start with the name.

OK. Agribusiness is a combination of two words or 'concepts': 'agriculture' and 'business'. If you look it up, the dictionary will define a business as 'an organization that makes and sells goods or services to people' or something similar to those words. And the same dictionary will define agriculture as 'the making and selling of food and goods by farming'. Both definitions mention 'selling' and we are going to focus on commercial agribusinesses on this course.

Exercise C

1/2 Set for pairwork discussion before listening. Tell students to make notes.

3/4 Play Part 2 for students to check their ideas and listen for question 4. Feed back, and focus on the complex agribusiness model, building up a diagram on the board and explaining that this is a matrix diagram. You can use the diagram to highlight differences between 'complex' and 'simple'. Finally, check the answer to question 4.

Answers

Model answers:

1/2

DPM

Research and Development	Operations	Sales and Marketing	Finance
testing new products	ensuring efficient production methods	customer liaison	preparing budgets
inventing new products	quality standards	market research	controlling costs
researching new raw materials	health and safety/food hygiene	product definition	providing financial data
	packaging	advertising	
		sales	

4 The lecturer is now going to answer the question: 'What does an agribusiness do?'

Transcript 1.2

Part 2

We often hear the term 'agribusiness' used to describe large-scale or industrial farming operations. And words like 'large scale' and 'industrial' are often associated with negative trends in farming, for example overuse of chemical fertilizers and the damage it causes to the environment.

So, yes, it is important to be aware of these views. However, on this course we will use 'agribusiness' as a neutral term. When we talk about agribusiness, we will be talking about a food- or agriculture-related business run on conventional lines. In other words, a business which is managed by a team of specialist managers.

Now, let's take a typical agribusiness, a dairy product manufacturer, to help us define some of the features of an agribusiness. And let's call it DPM, an acronym for Dairy Product Manufacturer.

First things first, DPM, our manufacturing agribusiness, employs 100 people and the company is run on conventional lines. What does that mean? Well, one important point is that DPM is a limited company. That means that the owners of the company are protected from the risk of losing their personal assets – their property or money – if the business fails.

So DPM is a conventional agribusiness with a conventional structure. It has four main departments or business functions, and each department is run by specialist managers.

DPM has four departments and three main product lines. First, they make cheeses. Second, they make butter and spreads. And finally, they also make desserts, or what we sometimes call 'puddings'. Each of the products has its own management team. So, for example, the team responsible for making cheese has a quality expert, and the quality expert also belongs to the marketing department. We call this type of organization a 'matrix' organization. I'll talk more about different types of business organizations later in the course.

The first of the four departments is the Research and Development department. The R and D department, as it's known, is responsible for testing products, for inventing new products and for researching new and better types of raw materials.

The second department is Operations. The key responsibility for the department is making sure that the production process is efficient and the product meets quality standards. Operations includes health and safety and food hygiene, as well as packaging.

Sales and Marketing is the third department. Their responsibilities include liaising with the customers to ensure DPM makes products that customers and consumers want. They work closely with R and D to provide market research and help define new products. They look after advertising and they include a team of specialist sales people.

And fourthly, to ensure that the company has the money it needs to operate, there is the Finance department. They take charge of making budgets, controlling costs and providing financial data so that the managing director, or MD, can make good decisions.

There is a strategic or senior management team made up of department heads and an MD. Sometimes they are called the Chief Executive Officer, or CEO, meaning the person with the authority to make high-level decisions. Finally, the departments are supported by specialist functions such as IT, HR and Admin.

You'll have noticed, I'm sure, that I've been talking about what people *do* in agribusiness, and this gives us an idea about how to answer our original question: What do we mean by the term 'agribusiness'? So now perhaps we should ask a slightly different question: What does an agribusiness do? In other words, we are asking: What are the activities that an agribusiness organization engages in?

Exercise D

You can set this exercise for pair work. Give students, say, ten minutes to brainstorm. Asking them for ways to *analyze* might be unclear. Encourage them to come up with different and inventive ways to *define* an agribusiness. Write ideas on the board. Make the connection with the different types/definitions and ways to analyze. Suggest they use real agribusinesses.

Answers

Possible answers:

- it's big → analyze by size → number of employees
- it makes drinks → analyze by what it makes
- it's organic → analyze by its principles → ideas → business methods
- it sells direct to the public → analyze by business type → wholesaler, manufacturer, retailer, etc.
- it's very profitable → analyze by financial performance

Transcript 1.3

Part 3

A farm supplies DPM with milk. The farm is, of course, an agribusiness too, like DPM. But this agribusiness has a simple organizational structure by comparison with DPM. It has one major function and that is production. The farm owner is the manager. He has responsibility for strategy. He also handles negotiations with the bank to ensure the business has the finance it needs. He deals with the customers and ensures that services such as veterinary services are available when needed. The farm employs a skilled herdsman and two other unskilled farm labourers. The farm is administered from an office run on a part-time basis by the owner's wife.

So we can define an agribusiness as either a complex business or a simple business. In fact, we can analyze and define businesses generally in this way. Let's look at another framework.

Food and agriculture is what we call a 'business sector'. We can define agribusinesses according to their economic sector too. The dairy farm belongs to the 'primary' business economic sector with things like mining, because its purpose is to produce raw materials. The secondary sector is made up of manufacturing and processing businesses. There are also tertiary and quaternary sectors, but we'll come to these later.

So we can define an agribusiness as complex or simple, as a primary or secondary sector business, or as forming part of the food and agriculture business sector. We can also define an agribusiness by its size.

Under the EU framework, a business with fewer than 10 employees is defined as a 'microbusiness'. With 10 to 50 employees it is a 'small business', and with 50 to 250 employees we call it a 'medium-sized business'. Small and medium-sized businesses or enterprises are often called 'SMEs'.

Methodology note

Up to this point, you have not mentioned how students should record information. Have a look around to see what students are doing. If some are using good methods, make a note and mention that later in the unit.

Exercise E

Point out that we often give examples of things to help clarify definitions. Give students plenty of time to look at the words in the box, then play Part 4. Feed back.

Answers

holistic	Y
farmers' market	Y
fair trade	Y
vegetable box	Y
banking	N
beauty without cruelty	Y
conventional	N
free-range	Y
factory	N
industrial	N

Note: You can ask if the concept of a farmers' market is inherently 'ethical'.

Transcript 🎧 **1.4**

Part 4

In food and agriculture in particular, we can also analyze and define a business in terms of what we call its 'philosophy'. There has been a big growth in consumer demand for organic products in recent years. Some agribusinesses make the choice to use organic raw materials to meet this demand. Similarly, some agribusinesses label their products as *fair trade*. These agribusinesses make a decision based on beliefs. We can describe them as 'ethical' businesses.

We use the word 'ethical' in general English to mean that something fits in a system of beliefs about what is right and wrong. In agribusiness, *ethical* is associated with things like the organic movement and the fair trade movement. When we use *organic* to describe a farm, we mean the farmer has made a decision to avoid use of artificial fertilizers or pesticides. When we use the term 'fair trade' about a product, we mean that the producer has a commitment to ensure the workers are treated fairly. In both cases, the producer and the farmer have made a decision that reflects their beliefs or ethics. This means caring for every part of the supply chain, including any animals, and recognizing all the people who have an interest in the business. This includes the people employed by the business. It also means the people who buy their products or who live in the local community.

Ethical businesses make decisions based on a philosophy, in contrast with mainstream companies, which tend to make decisions based on commercial interest and profit. We call these 'conventional' businesses. Conventional businesses have become aware of the importance of ethics in the marketplace and are tending to diversify into more ethical products.

Twenty years ago, supermarket food departments tended to be typical conventional agribusinesses. It was unusual, for example, to find a supermarket selling organic meat products or free-range chickens. Today, supermarkets are the biggest sellers by volume of ethical products. That doesn't mean that, for example, they sell only organic or free-range chicken or other meat products, or what we call 'green' products. It means that you will find a range of products on offer. These will include meat produced using factory methods, as well as meat produced according to strict ethical rules. They will include coffee produced on conventional commercial plantations in South America or Africa, but also coffee which has the designation 'fair trade'.

And it is not just the food section where you will find the choice of ethical products alongside conventional ones. Cosmetics are a very important part of the secondary agribusiness sector. Today, more and more people make their choices based on cost and quality on the one hand, and philosophical and moral principles on the other. We are all familiar now with the chains of shops that have sprung up in the last 30 years which specialize in organic beauty products. These, in turn, have influenced the conventional retailers so that it has become normal to find organic products in our supermarkets. In the same way, the 'beauty without cruelty' movement has been successful in persuading consumers to refuse cosmetics that use animals for testing. In both cases, we are switching from conventional to ethical choices and products.

Exercise F

Set for individual work and pairwork checking. Feed back.

Answers

Possible answer:

A sustainable agricultural or food-producing business based on philosophical principles about what is fair and right for all stakeholders in the supply chain.

Exercise G

Set for individual work and pairwork checking. You can use proper names like 'Soil Association' or words like 'ethical'. Encourage lateral thinking. When students have completed the 'Ethical' and 'Conventional' boxes, you can extend the exercises by creating a third box or list and calling it 'New approach' (you can use 'Paradigm' if you feel it is appropriate). Use the example of a modern supermarket that is aiming for the 'best of both' and get students to suggest words/expressions that reflect this thinking.

Answers

Possible answers:

Ethical	
free-range	philosophy
ethics	fair trade
beauty without cruelty	sustainable growth
Soil Association	local
organic	stakeholder

Conventional	
compliance with law	value
cost	global
profit	shareholder
growth	factory
unsustainable	intensive

↓ ↓

New approach	
global	organic options
growth	beauty without cruelty options
value	stakeholder and shareholder
ethical	productivity
philosophy	
fair trade options	

Exercise H

Refer students back to their notes from Exercise A.

Closure

1 Look at the pictures of the various aspects of the agribusiness and its activities. Ask students to brainstorm words and expressions that the pictures suggest to them from their own experience or what they've heard in the lecture.

Possible answers:

- R&D: new products; ideas, science, research
- Operations: products, cheeses, desserts, manufacturing, hygiene, quality, packaging
- Sales and Marketing: customers, liaise, consumers, meeting needs
- Finance: income, profits, budgets, payments, pay, costs, control
- Organogram: managers, departments, plans, business

2 Refer students to the *Skills bank* if you have not done so already and work through the section *Making the most of lectures*.

Introduction 5

We have talked about agribusinesses and we've looked at the very early origins of agriculture and agribusiness. Today, I'd like to look into the future. Let's imagine a time where the agriculture and food sector returns to where it began. Let's imagine a rural economy where there are local communities with high-quality, small-scale agribusinesses.

Closure

1 Test students on the pairs from Exercise B. Correct pronunciation again if necessary.

2 Refer students to the *Skills bank – Making perfect lecture notes.*

1.4 Extending skills

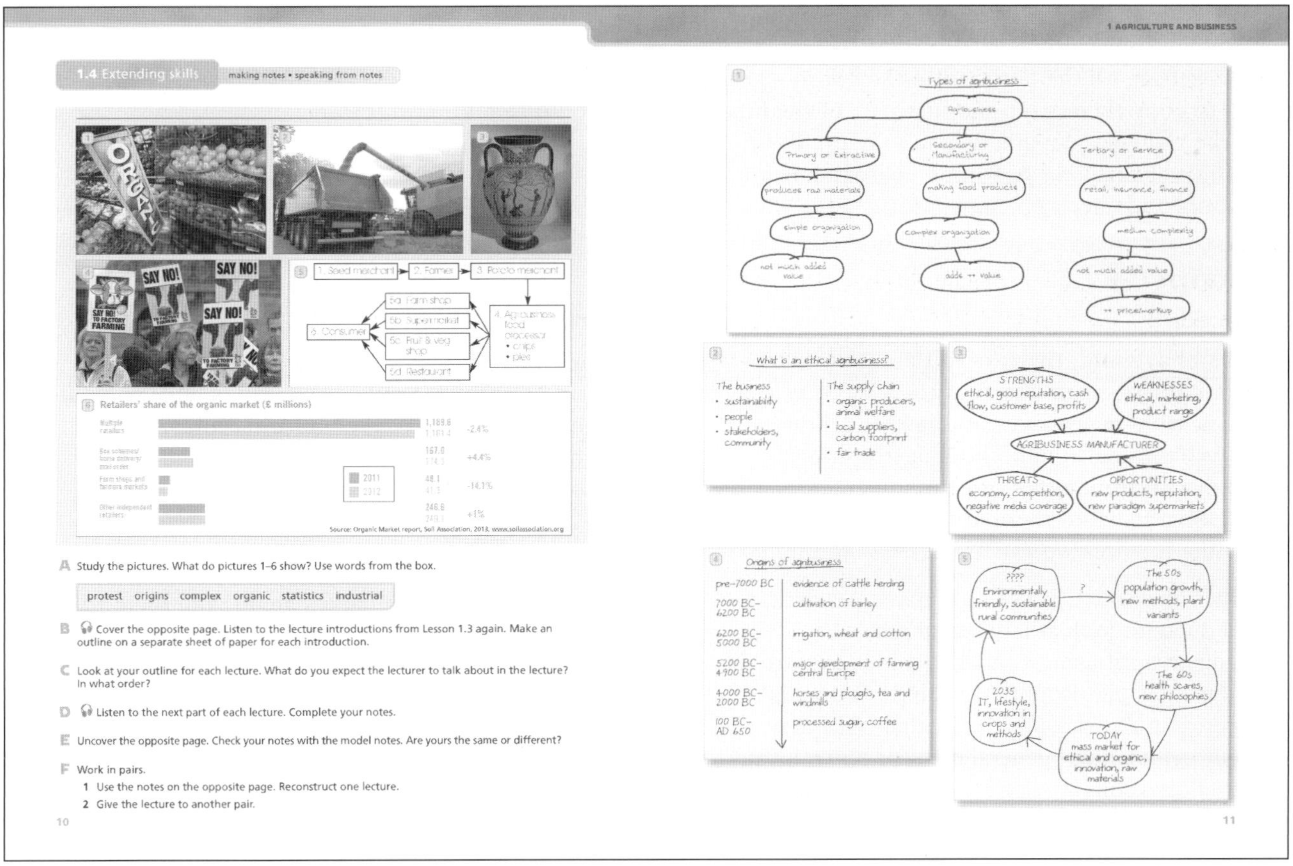

Lesson aims

- make outline notes from lecture introductions
- make notes from a variety of lecture types
- speak from notes

Further practice in:

- predicting lecture content

Introduction

Elicit as much information from the lecture in Lesson 1.2 as possible. If necessary, prompt students by reading parts of the transcript and pausing for students to complete in their own words.

Exercise A

Set for individual work and pairwork checking. Feedback orally but still do not confirm or correct. Point out that students are going to hear about all these things in today's lesson. You will return to these illustrations at the end.

For reference, the illustrations show:

1 some **organic** produce in a supermarket
2 a picture of a modern **industrial** farm
3 a vase depicting the **origins** of food and agricultural activity
4 a picture of people making a **protest** in support of an ethical principle
5 a flowchart showing the organization of a **complex** agribusiness
6 a bar graph showing some **statistics** about how organic food sales are affected by an economic down turn

Methodology note

It is best that students close the book at this stage, so they are not tempted to look at the model notes. You can give the instructions for the next few stages orally as required.

Exercise B

Make sure students understand that they are going to hear the introductions from Lesson 1.3 again. Ask them briefly if they can remember any of the content from the introductions. Spend a few moments on this if students are able to contribute. Elicit suggestions for types of notes (Lesson 1.3, Exercise E).

Explain that this time they must create an outline using an appropriate type of notes. (You can refer them again to the *Skills bank – Making perfect lecture notes.*) Make sure students understand that they don't need to write a lot at this stage – outlines may consist of just a few words, e.g., the start of a spidergram, the first part of a table or diagram. Play each introduction in turn and give students time to choose a note-type, make the outline and check it with other students.

Feed back, getting all the outlines on the board – you may wish to copy them from the first part of the model notes on the right-hand page, or you may prefer to follow your students' suggestions. Clarify the meaning of new words and check pronunciation.

Transcript 1.5

Introduction 1

Right, as I have said, there are different ways to define or categorize an agribusiness. One way is by looking at what it produces and how it produces it. Today, we're going to talk about agribusinesses, which fall into three types or economic sectors. I'd like to look at each of these in turn and try and identify the features we can associate with them.

Introduction 2

Today, I'd like to look at a different way of describing an agribusiness. And I want to explore the idea of business philosophy. We can do this by taking as our heading the idea of an ethical agribusiness, and then breaking that general concept down into the sort of questions we need to ask the business.

Introduction 3

Today, I'd like to look at a way to analyze an agribusiness further and in more depth. I'm going to use our sausage manufacturer again and put this, if you like, at the centre of our process. Then, I'm going to look at four different questions of equal importance. I want to see if we can connect the audit of an agribusiness philosophy to an examination of the business using a conventional analytic tool, the SWOT analysis. A SWOT analysis examines a business by looking at its strengths and weaknesses and its opportunities and threats.

Introduction 4

Right, we've looked in some depth at why agribusinesses are different from mainstream businesses and we've looked at different sectors of the agribusiness economy. I would like today to talk a little about the origins of farming and agribusinesses and try and pinpoint some of the key developments of the last nine centuries.

Introduction 5

We have talked about agribusinesses and we've looked at the very early origins of agriculture and agribusiness. Today, I'd like to look into the future. Let's imagine a time where the agriculture and food sector returns to where it began. Let's imagine a rural economy where there are local communities with high-quality, small-scale agribusinesses.

Methodology note

Spiral bound or stitched/stapled notebooks are not the best way to keep lecture notes. It is impossible to reorganize or add extra information at a later date, or make a clean copy of notes after a lecture. Encourage students, therefore, to use a loose-leaf file, but make sure that they organize it in a sensible way, with file dividers, and keep it tidy. Tell students to use a separate piece of paper for each outline in this lecture.

Exercise C

Set for pair or group work. Feed back, but do not confirm or correct. Students should be able to predict reasonably well the kind of information which will fit into their outline.

Exercise D

Before you play the next part of each lecture, refer students to their outline notes again. Tell them to orally reconstruct the introduction from their notes. They don't have to be able to say the exact words, but they should be able to give the gist.

Remind students that they are only going to hear the next part of each lecture once. Play each extract in turn, pausing if necessary to allow students to make notes but not replaying any section. Tell students to choose an appropriate type of notes for this part of the lecture – it could be a continuation of the type they chose for the introduction, or it could be a different type.

Transcript 1.6

Lecture 1

A primary or 'extractive' business produces raw materials. Farming is a typical primary activity. The dairy farmer 'extracts' the milk and sells it directly to consumers or to processors.

A farm is a relatively 'simple' business. We will talk more about what we mean by terms like 'simple' and 'complex'. Let's take a farm, South Farm Estates, with a turnover of, say, £1 million. We call the farm a primary or extractive business. A farm, even a large farm, has a fairly simple organizational structure. It tends to employ mainly unskilled labour. The workforce increases and decreases depending on the season. It uses the same core team of a manager and one or two skilled employees. If it supplies a supermarket with carrots, for example, the farm will probably process or sort them, but it isn't their job to add value to the raw material.

Compare that with a dairy products manufacturer, a secondary type of agribusiness. I'm going to give it a name, an acronym: 'DPM'. I want to talk about this company again.

DPM buys milk from the farm. They transform or process it into a new product. Their activities increase the value of the raw material a lot – they add value. DPM has a turnover of £3 million, but it has a complex organization. It has a number of distinct specialist departments and specialist managers.

OK. So we have primary and secondary agribusiness types. One is simple and one complex. One doesn't add significant value to the product; the other does. Now we come to the tertiary or service sector. This includes businesses such as insurance companies that specialize in agriculture. It also includes retailers, supermarkets, specialist food shops and shops selling agricultural products such as beauty products. I want to take a small retail chain selling food and other agribusiness products with a turnover somewhere between the other two. The chain will need managers with a range of skills. Let's call their organization 'medium complexity'. They employ more unskilled workers per manager than DPM, but fewer than the dairy farm. They don't add value to the product which DPM supplies. They increase the price or 'mark up' because they have high costs – the shops, staff, energy bills. They also have a bigger product range than either of the other two – they are more diverse businesses from that perspective.

1.7

Lecture 2

Philosophy and ethics are very important for the agribusiness sector. Why? Well, consumers tend to care a lot about the ingredients in their food; that is, what is in their food. And more and more today, consumers are not just concerned whether it is safe and healthy. Consumers want to know that they are buying an ethical product.

What do we mean when we say that an agribusiness is 'ethical'? It's hard to sum up in a sentence, but we can test if an agribusiness is ethical by using an audit – a formal questionnaire. Let's use our dairy product manufacture, DPM, again, as an example.

An ethical or green audit must be 'holistic'. That means we need to know every aspect of the business itself, as well as every aspect of its supply chain. The supply chain means all the businesses involved in supplying the raw materials and other materials – the 'inputs' that go to make the product.

The audit has two key, interconnected objectives. First, it has to verify that DPM meets sustainable criteria in the way it runs its business. Second, it must verify that all parts of the supply chain meet the same standards and respect the animals and people involved in making its products in the UK or other countries. And what does 'sustainable' mean? It means that the agribusiness makes its products to meet today's needs, without compromising or harming the needs of future generations – a key concept on this course!

So, starting with DPM, an auditor checks that the company has policies to look after and develop their workforce, who are key stakeholders. They check they have policies to take care of other stakeholders, such as the local community and customers. They expect the company to have policies to minimize any impact on the environment.

The audit will then check that the same policies apply in all parts of the supply chain. They'll check that suppliers meet recognized organic and animal welfare standards. Very importantly, they'll look at DPM's carbon footprint. You'll have heard of the term 'air miles' I expect. Establishing how many miles by air a product travels before you buy it is a way of focusing attention on environmental impact. It's part of checking that the company sources its ingredients and raw materials locally, where possible, to minimize the use of fossil fuels. And if a supplier is non-UK, they will check that they work to fair trade guidelines.

1.8

Lecture 3

There are many business analysis tools. Two of the best known are PEST and SWOT. PEST is an acronym standing for Political, Economic, Social and Technological. We use it to analyze the external environment of a business. SWOT stands for Strengths, Weaknesses, Opportunities and Threats, and looks at the business and how it works internally. Most business plans include a SWOT analysis.

Let's take an agribusiness called the Ethical Farming Company (EFC) and we'll assume it meets the requirements of our ethical audit. Let's now do a more conventional analysis and look at strengths, weaknesses, opportunities and threats.

First the strengths: EFC has a good reputation, it has good cash flow, a good customer base and its profits have grown consistently over the last five years. The owner is sure that profits are linked to the fact that it's known as an ethical business with good organic products. There is evidence that these products are more profitable than conventional ones.

Now move round one step to weaknesses. The first thing I should say is that a 'strength' can appear as a 'weakness' too. So the fact that this is an organic agribusiness could be a potential weakness. Why? Well, in bad economic times, consumers who usually buy an organic product, even if it costs more, may choose a cheaper alternative. The limited number of products could be a weakness. And maybe EFC aren't as good at marketing as they could be.

OK, now opportunities. The ethical/organic link is a positive, because although organic products normally cost more, it seems the market for these products is set to grow and the opportunities for selling new products will grow too. Another key point is that supermarkets and food chains are tending to sell more organic and ethical products today.

Finally, what about threats? Well, there are many linked to the things we have discussed. For example, economic problems and an increase in competition. One special threat is from bad publicity. That could be related to a case of food poisoning locally, for example. But an ethical agribusiness is committed to long-term sustainability and good business policies. The philosophy will help protect it from many of the threats that could affect conventional rivals.

1.9

Lecture 4

The first evidence of systematic wheat production that we have dates back to eleven and a half centuries ago, to around 9500 BC. For the next 2,000 years, there is evidence of cattle herding, but little else that we can call 'agribusiness'.

Around 7000 BC, we saw the beginning of a period of relative growth in the domestication of animals and crops. The earliest evidence of barley cultivation dates back to 7069 BC. Around the same period, cattle were domesticated in Turkey. Between 6200 BC and 5000 BC, irrigation was being used and inhabitants of the Indus Valley moved from growing wheat to cotton and sugar crops. People were growing rice in India.

The first major development of farming in Europe probably started between 5200 BC and 4900 BC. The communities of central Europe were probably the first to import domesticated animals and plants from central Asia into Europe.

Between 4000 BC and 2000 BC, there was a major leap forward. Horses and ploughs were used for the first time. In this period, tea was discovered and people began to use windmills. The invention of windmills seems to be evidence of the development of systematic food processing.

From 100 BC to AD 665, people acquired skills in processing sugar and the cultivation of crops in rows. This period is what is known as the Agricultural Revolution in the Arab World, and coincided with the discovery of coffee.

1.10

Lecture 5

Agribusiness and agriculture has, in the last 50 years, tended to shift its emphasis. Firstly, from an emphasis on conventional methods and mass production of foodstuffs. Secondly, to a period of emphasis on ethical and 'green' methods. Finally, to a more balanced approach today. The new approach aims to balance quantity with quality.

If we begin in the 1950s in the UK, we had a situation where the emphasis was on increasing production and 'outputs'. We can identify a number of factors or 'drivers'. There was the need to feed a rapidly growing population. And this was achieved by the development of new, high-yield crop strains and the use of efficient, industrial farming methods.

In the 1960s, there was a reaction to the idea that quantity was the most important thing. There were two main reasons for this. Firstly, people were frightened by the health scares linked to excessive use of chemicals on the land. Secondly, new, more ethical, philosophies began to become popular. These philosophies challenged the idea of industrial farming.

Today, we are seeing a new paradigm or 'model'. There are two reasons for this. Firstly, because conventional agribusinesses see a mass market for ethical products. Secondly, because new farming methods are starting to cut the cost of producing more ethical products.

What will the drivers be about 20 years from now, in 2035? Well, biotechnology will make it easier to produce more crops and better quality crops. Information technology will enable growth of online marketing. This will allow small producers to find new markets and establish sustainable commercial enterprises. People will tend to be more interested in improving the quality of their lives, not just their income. Will this encourage people back to rural areas, maybe? This could mean a regeneration of environmentally friendly, sustainable rural communities.

And would that take us back to where we started 100 years before? Perhaps!

Exercise E

Allow students to uncover the opposite page or open their books. Give them plenty of time to compare their answers with the model notes. Feed back on the final question.

Exercise F

1 Ask students to work in pairs. Assign a set of notes to each pair. They must try to reconstruct the lecture orally – including the introduction – from the notes.

2 Put the pairs together in groups of four, with different topics. Each pair should give their lecture to the other pair.

Closure

1 Work on any problems you noticed during the pairwork (Exercise F).

2 Refer back to the pictures at the top of page 10 of the Course Book. Students should now be able to name them with confidence.

Extra activities

1 Work through the *Vocabulary bank* and *Skills bank* if you have not already done so, or as a revision of previous study.

2 Use the *Activity bank* (Teacher's Book additional resources section, Resource 1A).

A Set the crossword for individual work (including homework) or pairwork.

Answers

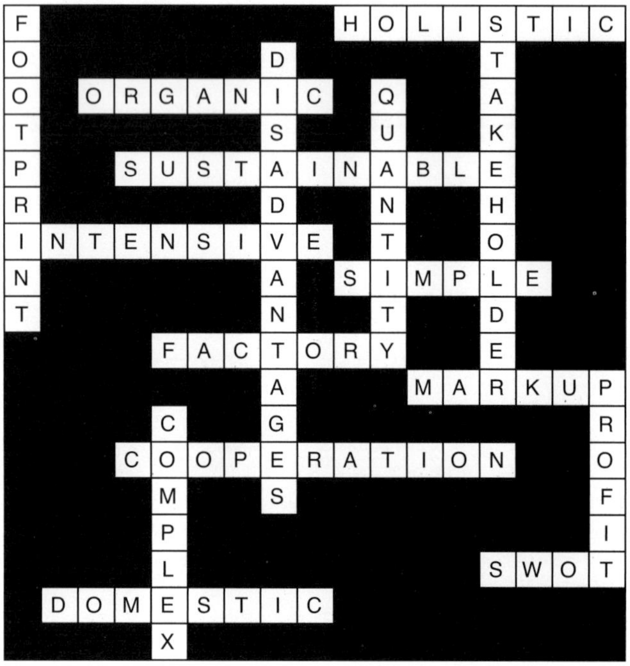

B Play noughts and crosses in pairs. There are two boards. The first contains words with affixes, the second contains agribusiness and agriculture concepts.

Teach students how to play noughts and crosses if they don't know – they take it in turns to choose a word/phrase and try to use it in context or explain what it means. If they succeed, they can put their symbol – a nought 0 or a cross X – in that box. If a person gets three of their own symbols in a line, they win.

First board: Tell students to remove the affixes to find the basic word in each case. Make sure they can tell you the meaning of the basic word (e.g., *skilled* or *unskilled*) but don't elicit the meaning of the affixed word at this stage. Put students in pairs to play the game. Monitor and adjudicate.

Second board: Put students in different pairs to play the second game. Clearly, this time they have to actually remember the facts from the lectures. Don't let them look back at notes.

3 Each of the mini-lectures from Lesson 1.4 can lead on to a great deal more work. Tell students to research one of the following, according to which group they ended in. Explain that they must come back and report to the rest of the class in the next lesson/next week.

Lecture	Research
1	Another way in which agribusinesses can be classified with examples of each type (e.g., location or size, private or public, etc.)
2	What would a green audit involve for an agribusiness? Using the Internet, check if agribusinesses have information about 'philosophy' or 'ethics'.
3	Find other examples of business analytical tools and compare and evaluate one or two against the SWOT.
4	What factors drove the development of agribusiness in early times and what factors limited the speed of development?
5	Using the Internet again, find out the origins of the organic movement. Compare your own society's agriculture and food production culture to the UK's.

4 Brainstorm note-taking techniques.
For example:

- use spacing between points
- use abbreviations
- use symbols
- underline headings
- use capital letters
- use indenting
- make ordered points
- use different colours
- use key words only

2 FOUNDATIONS OF MODERN AGRICULTURE

Unit 2 looks at how the modern agriculture and agribusiness sector was largely shaped during the period between 1750 and 1850. The first reading text looks at whether we are justified in describing the changes that happened as a 'revolution'. The second reading text looks at some of the defining events and personalities.

Note that students will need dictionaries for some exercises in this unit.

Skills focus

Reading

- using research questions to focus on relevant information in a text
- using topic sentences to get an overview of the text
- identifying paragraph structure

Writing

- writing topic sentences
- summarizing a text

Vocabulary focus

- English–English dictionaries:
  headwords
  definitions
  parts of speech
  phonemes
  stress markers
  countable/uncountable
  transitive/intransitive

Key vocabulary

agriculture	fallow	rotate
agriculturalist	fruit	rotation
barley	fumigate	seed
breed (n and v)	growth	selective
clover	import (n and v)	stock (n and v)
cultivation	network (n and v)	technologist
ditch	plough	tenant
drill (n and v)	process (n and v)	turnip
effective	proliferate	wheat
enclosure	propaganda	widespread
evolution	propagandist	yield (n and v)
expertise	revolution	

2.1 Vocabulary

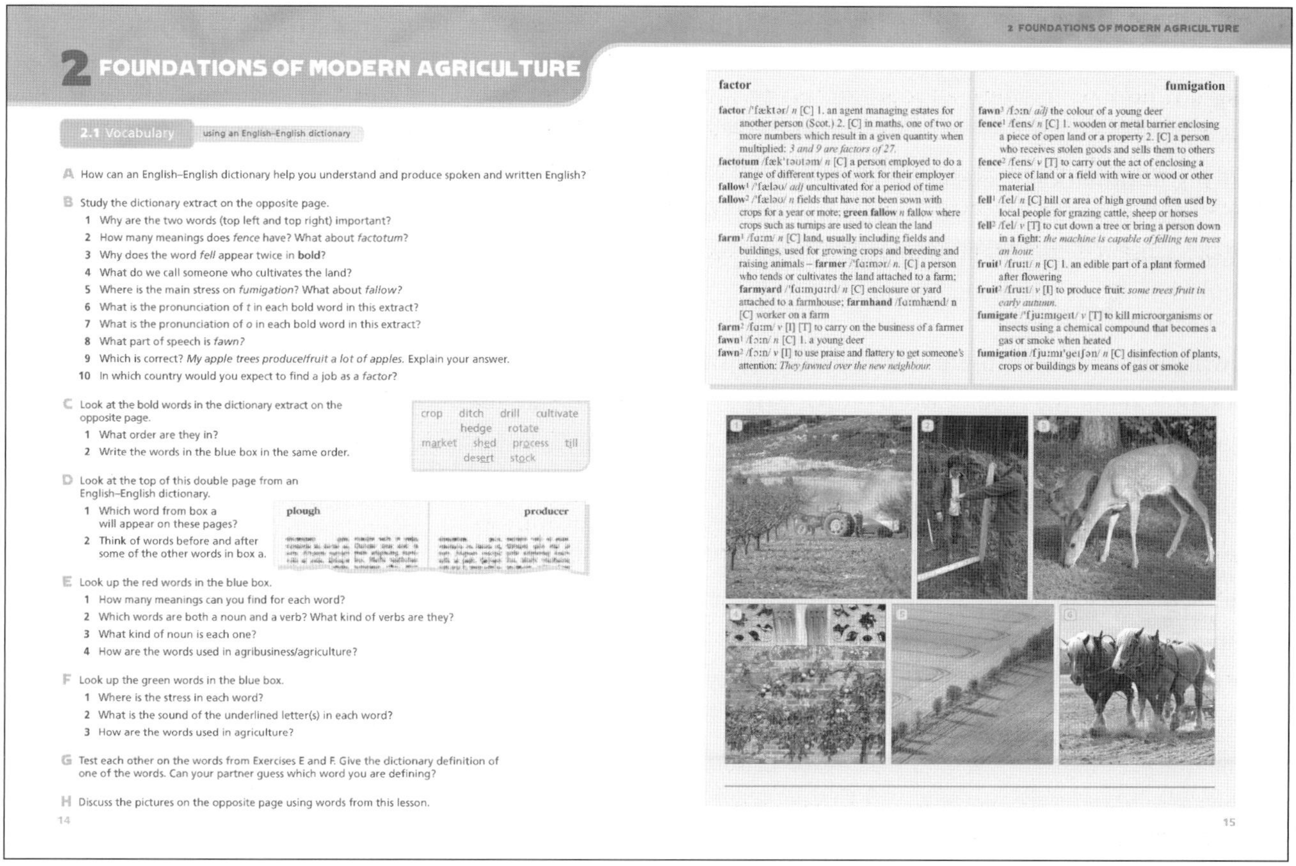

2 FOUNDATIONS OF MODERN AGRICULTURE

2.1 Vocabulary using an English–English dictionary

A How can an English–English dictionary help you understand and produce spoken and written English?

B Study the dictionary extract on the opposite page.

1 Why are the two words (top left and top right) important?
2 How many meanings does *fence* have? What about *factotum*?
3 Why does the word *fell* appear twice in **bold**?
4 What do we call someone who cultivates the land?
5 Where is the main stress on *fumigation*? What about *fallow*?
6 What is the pronunciation of *t* in each bold word in this extract?
7 What is the pronunciation of *o* in each bold word in this extract?
8 What part of speech is *fawn?*
9 Which is correct? *My apple trees produce/fruit a lot of apples.* Explain your answer.
10 In which country would you expect to find a job as a *factor*?

C Look at the bold words in the dictionary extract on the opposite page.

1 What order are they in?
2 Write the words in the blue box in the same order.

crop ditch drill cultivate hedge rotate market shed process till desert stock

D Look at the top of this double page from an English–English dictionary.

1 Which word from box a will appear on these pages?
2 Think of words before and after some of the other words in box a.

plough producer

E Look up the red words in the blue box.

1 How many meanings can you find for each word?
2 Which words are both a noun and a verb? What kind of verbs are they?
3 What kind of noun is each one?
4 How are the words used in agribusiness/agriculture?

F Look up the green words in the blue box.

1 Where is the stress in each word?
2 What is the sound of the underlined letter(s) in each word?
3 How are the words used in agriculture?

G Test each other on the words from Exercises E and F. Give the dictionary definition of one of the words. Can your partner guess which word you are defining?

H Discuss the pictures on the opposite page using words from this lesson.

14

2 FOUNDATIONS OF MODERN AGRICULTURE

factor **fumigation**

factor /ˈfæktər/ *n* [C] 1. an agent managing estates for another person (Scot.) 2. [C] in maths, one of two or more numbers which result in a given quantity when multiplied: *3 and 9 are factors of 27.*
factotum /fækˈtəʊtəm/ *n* [C] a person employed to do a range of different types of work for their employer
fallow[1] /ˈfæləʊ/ *adj* uncultivated for a period of time
fallow[2] /ˈfæləʊ/ *n* fields that have not been sown with crops for a year or more; **green fallow** *n* fallow where crops such as turnips are used to clean the land
farm[1] /fɑːm/ *n* [C] land, usually including fields and buildings, used for growing crops and breeding and raising animals – **farmer** /ˈfɑːmər/ *n*. [C] a person who tends or cultivates the land attached to a farm; **farmyard** /ˈfɑːmjɑːrd/ *n* [C] enclosure or yard attached to a farmhouse; **farmhand** /fɑːmhænd/ n [C] worker on a farm
farm[2] /fɑːm/ *v* [I] [T] to carry on the business of a farmer
fawn[1] /fɔːn/ *n* [C] 1. a young deer
fawn[2] /fɔːn/ *v* [I] to use praise and flattery to get someone's attention: *They fawned over the new neighbour.*
fawn[3] /fɔːn/ *adj* the colour of a young deer
fence[1] /fens/ *n* [C] 1. wooden or metal barrier enclosing a piece of open land or a property 2. [C] a person who receives stolen goods and sells them to others
fence[2] /fens/ *v* [T] to carry out the act of enclosing a piece of land or a field with wire or wood or other material
fell[1] /fel/ *n* [C] hill or area of high ground often used by local people for grazing cattle, sheep or horses
fell[2] /fel/ *v* [T] to cut down a tree or bring a person down in a fight: *the machine is capable of felling ten trees an hour.*
fruit[1] /fruːt/ *n* [C] 1. an edible part of a plant formed after flowering
fruit[2] /fruːt/ *v* [I] to produce fruit: *some trees fruit in early autumn.*
fumigate /ˈfjuːmɪgeɪt/ *v* [T] to kill microorganisms or insects using a chemical compound that becomes a gas or smoke when heated
fumigation /fjuːmɪˈgeɪʃən/ *n* [C] disinfection of plants, crops or buildings by means of gas or smoke

15

General note

Take in a set of English–English dictionaries.

Read the *Vocabulary bank* at the end of the Course Book unit. Decide when, if at all, to refer students to it. The best time is probably at the very end of the lesson or the beginning of the next lesson, as a summary/revision.

Lesson aims

- learn how to make full use of an English–English dictionary
- gain fluency in the target vocabulary

Introduction

1 Revise the vocabulary from the last unit. Check:
   - meaning
   - pronunciation
   - spelling

2 Ask students whether they use a translation (bilingual) dictionary or an English–English (monolingual) dictionary. Elicit the advantages and disadvantages of a translation dictionary.

Answers

Possible answers:

Advantages	Disadvantages
good when you know the word in your own language and need a translation into English	not good when there is more than one possible translation of a word – which is the correct one?
when you look up an English word, the translation into your language is easy to understand	English–English dictionaries often have more examples and precise definitions of each word

Methodology note

Recent research has shown that, despite the insistence of generations of language teachers on the use of English–English dictionaries in class, nearly 90 per cent of students use a translation dictionary when studying on their own.

Exercise A

Ask the question as a general discussion. Confirm but do not explain anything. Point out that the next exercise will make the value of this kind of dictionary clear.

Answers

Model answers:

The following information is useful for spoken English:

- stress
- pronunciation of individual phonemes – particularly when a phoneme has multiple pronunciations

The following information is useful for written English:

- information about the type of word – C/U; T/I
- the spelling – although students might make the point that if you don't know the spelling, you can't find the word in the first place; point out that you can often guess the possible spelling – for example, *management* could be *managment* but if you don't find it there, you can try it with an *e* in the middle …
- examples of the word in use to memorize
- some synonyms for lexical cohesion – this is a very important point, although you may not want to elaborate on this now

Exercise B

Set for individual work and pairwork checking. Feed back, ideally using an OHT or other visual display of the dictionary extract to highlight points. You might suggest that students annotate the dictionary extract in their books, highlighting symbols, etc., and writing notes on the meaning and value.

Answers

Model answers:

1 They tell you the first and last words on the pages to help you locate the word you want.

2 *fence* – three meanings; *factotum* – one.

3 Because the same word can be a noun or a verb.

4 A *farmer.*

5 *fumigation* – on the third syllable; *fallow* – on the first syllable.

6 Sometimes /t/, sometimes /ʃ/ – can students spot the pattern: *factor, factotum, fumigate* = /t/, while *fumigation* = /t/ + /sh/?

7 Three pronunciations – /ɔː/, /əʊ/ and /ə/

8 A noun or a verb.

9 The correct verb is *produce,* since the dictionary tells us that *fruit,* when used as a verb, does not take an object, i.e., it is intransitive.

10 Scotland.

Exercise C

Note: If students are from a Roman alphabet background, you may want to omit this exercise.

1 Students should quickly be able to identify alphabetical order.

2 Set for individual work and pairwork checking. Feed back, getting the words on the board in the correct order. Don't worry about stress and individual phonemes at this point – students will check this later with their dictionaries.

Language note

It may seem self-evident that words in a dictionary are in alphabetical order. But students from certain languages may not automatically recognize this. In the famous Hans Wehr dictionary of written Arabic, for example, you must first convert a given word to its root and look that up, then find the derived form. So *aflaaj* (the plural of *falaj* = irrigation channel) will not be found under A but under F since the root is *f-l-j*.

Exercise D

1 Set for pairwork. Feed back orally, explaining the principle if necessary.

2 Set for pairwork. Ask each pair to choose five words to work on. Ask them to find words connected with agriculture if they can. Feed back orally.

Answers

1 *Process* will appear on the double page spread.

2 Answers depend on which words students choose.

Exercise E

Give out the dictionaries, if you have not already done so.

Remind students that dictionaries number multiple meanings of the same part of speech and multiple parts of speech. Remind them also of the countable/uncountable and transitive/intransitive markers.

(Note that different dictionaries may use different methods for indicating these things. The *Oxford Advanced Learner's Dictionary*, for example, uses [V] for intransitive verbs and [Vn] for transitive verbs.)

Write the headings of the table in the Answers section on the board, and work through the first word as an example.

Set for pairwork. Feed back, building up the table in the Answers section on the board. (Students' answers will vary – accept any appropriate meanings and definitions.)

Answers

Model answers:

Word	Part of speech	Type	Main meaning in agriculture	Main meaning(s) in general English
crop	n	C	total yield of a cereal, fruit, etc.: *the maize crop*	a stick used by horse riders
	v	T		to cut off the top or ends of something: *She cropped the heads off the roses.*
ditch	n	C	a trench dug to carry water away	
	v	T		to abandon: *The thieves ditched the car.*
drill	n	C	a shallow furrow in which seeds are planted; a piece of agricultural machinery	a machine for making holes in hard material
	v	T		to make a hole in something by boring
cultivate	v	T	to prepare land or fields for crops	to develop: *She cultivated his support.*
hedge	n	C	a row of low bushes or trees	to offset risk: *They hedged their bets.*
	v	I	to grow closely planted bushes or low trees to form a barrier	
rotate	v	T	to turn something on its axis	planting crops in order

Exercise F

Remind students how stress and the pronunciation of individual phonemes are shown in a dictionary. Refer them to the key to symbols in the dictionary if necessary. Write the headings of the table in the Answers section on the board, and work through the first word as an example.

Set for pairwork. Feed back, building up the table in the Answers section on the board.

Answers

Model answers:

Stress	Sound	Part of speech	Type	Main meaning in agriculture
'market	/ɑː/	n	C	place where buying or selling of livestock happens
		v	T	to advertise or sell a product
'shed	/e/	n	C	a single storey farm building: *lambing shed*
		v	T	to lose: *The animal shed its fur.*
'process	/əʊ/	v	T	to make raw ingredients into food
		n	C	a way of doing something: *The pasteurization process.*
'till	/ɪ/	v	T	to turn or cultivate land
'desert*	/ɜː/	n	C	areas of land with very little rain or vegetation
		v	T	to leave when you mean 'abandon': *They abandoned the tractor when it broke down and walked back to the farm.*
'stock	/ɒ/	n	U/C	animals for breeding: *blood stock, breeding stock*
		v	T	to keep a supply for future use

*Stress changes to *de'sert* in verb form

Exercise G

Demonstrate how to do the exercise by giving a few definitions and getting students to tell you the word (without reading from the board or their books, if possible). Stick to agriculture rather than general English definitions and encourage students to do the same. Point out that dictionaries often use a small set of words that help to define, e.g., *place, set, method, kind, type, principle, person.* Give definitions using these words and ask students to identify what you are defining, e.g., *It's a place where cattle are bought and sold (market); It's the person who treats sick livestock (vet)*; etc.

Exercise H

Work through the pictures. Check that students understand what they show and their relationship to agribusiness and agriculture.

For reference, the pictures show:

1 two men **fumigating** trees in an orchard
2 labourers at work building **fences**
3 a herd of deer, including **fawn**
4 a picture of a pear tree with **fruit**
5 an arial picture of an arable **farm** showing crop **rotation**
6 a plough **tilling** the earth

Closure

1 Remind students that you can identify the part of speech of an unknown word by looking at the words before or after the word, i.e.,

- nouns often come before and after verbs, so if you know that X is a verb, the next content word before or after is probably a noun
- nouns often come immediately after articles
- verbs often come after names and pronouns
- adjectives come before nouns or after the verb *be*

Come back to this point when you are feeding back on the reading texts in this unit.

2 Ask students if they have encountered any of the items in the pictures. Can they think of examples of any of the other agricultural items or processes described in these exercises?

2.2 Reading

2 FOUNDATIONS OF MODERN AGRICULTURE

2.2 Reading using research questions

A What are the greatest agricultural advances of all time?

B Study the results of the Hadford University student survey on the right.

1 Define each agricultural achievement.
2 How did each achievement change human life?
3 Which is the greatest achievement?

C You are going to read a text. What should you do before you read a text in detail? See *Skills bank*

D This text is about changes in agriculture in recent centuries.

1 Think of some research questions before you read.
2 Compare your questions with those in the Hadford University assignment on this page.

E Study these topic sentences from the text and answer the questions below.

In the 1750s, a number of factors began to change agriculture in Britain.

Firstly, from 1750, there were growing population levels sustained by self-sufficiency in food.

Secondly, productivity rose after 1750, even though the proportion of the population employed in agriculture dropped consistently.

This period of rapid change is called the Agrarian or Agricultural Revolution and it came about for different reasons.

The Enclosures were important in encouraging the growth of larger, more efficient farms.

Some, but not all, of the revolutionary ideas originated in England.

There is general agreement that the century from 1750 was one of *revolution*, not *evolution*.

1 What period in history is the text about?
2 Where do you expect to find the answer to each question in the Hadford University assignment? Write 1, 2 or 3 next to the topic sentence.
3 What do you expect to find in the other paragraphs?

F Read the text on the opposite page and check your ideas. See *Skills bank*

Top 10 agricultural advances

- harvesting of seeds
- four-course crop rotation
- invention of the plough
- automated milking parlour
- modern fertilizers
- development of selective breeding
- introduction of irrigation
- invention of the combine harvester
- GM foods
- CAP (Common Agricultural Policy)

Student survey, Hadford University Department of Agribusiness and Agriculture, 2012

HADFORD *University*

Faculty: Agribusiness and Agriculture

Assignment

Do some research into key events in agriculture. Make notes to answer these questions:

1 Why is the period from 1750 to 1850 described as a 'revolution'?
2 What was the main benefit of the CAP?
3 How do you think the changes affected the lives of people in rural communities in England?

16

1750 to 1850: Agricultural revolution or evolution?

In the 1750s, a number of factors began to change agriculture in England. One of these factors was the urgent need to feed a growing population. People disagree about the changes that took place. Some believe that they were completely new. Others argue that they were a continuation of a process that had started much earlier. However, there are two reasons for saying that events in this period comprised a revolution in agriculture.

Firstly, from 1750, there were growing population levels sustained by self-sufficiency in food. The population at this point was approximately 6 million. This was a level probably achieved twice before in England's history. However, in previous periods, growth in population to this level had been followed by a corresponding decline. This was largely because agriculture at the time could not sustain the extra numbers. But after 1750, the population continued to grow, reaching nearly 17 million by 1850. This time, food production grew at a sufficient rate to maintain this level of population.

Secondly, productivity rose after 1750, even though the proportion of the population employed in agriculture dropped consistently. There were fewer people working on the land, but they were producing more food. This was due to increases in the amount of land under cultivation and the intensification of its use. In 1850, with the population at record levels, agriculture employed only 22 per cent of the workforce. This is the lowest percentage in English history, and a lower percentage than any of England's European competitors.

This period of rapid change is called the Agrarian or Agricultural Revolution and it came about for different reasons. Social changes were one factor in starting and sustaining the Revolution. For example, the Enclosure Laws led to many ordinary people losing their right to use common land for grazing their animals. Landowners used the laws to increase their access to the land for their own use. The Enclosures, as they are known, were bitterly resented by many and are still contentious today.

The Enclosures were important in encouraging the growth of larger, more efficient farms. Of course, many ordinary people were forced to leave rural areas as a result. But there is no doubt that life for ordinary people was very hard anyway during this period. And the cities must have offered many poor farm labourers and their families an attractive alternative, with new types of jobs and better wages.

Some, but not all, of the revolutionary ideas originated in England. However, some were imported from Holland and Denmark and developed in East Anglia. They fall roughly into four categories. The first relates to the science of agriculture and includes innovations such as crop rotation and improvements to soil quality. The second comprises the technological developments and new machines that began to spread quickly. Thirdly, word of mouth was important to assure the rapid spread of new practices from East Anglia to the rest of the country. Finally, new stock breeding techniques were significant. These involved the use of only the best animals. They led to the availability of better blood lines, higher productivity and survival rates, and greater profits. This, in turn, led to more investment in farming.

There is general agreement that the century from 1750 to 1850 was one of *revolution*, not *evolution*. There is substantial evidence that the period was a turning point. Agriculture in England moved from a low-intensity system to a high-intensity system, which was based on arable crops. The template for our modern agricultural industry was established.

17

General note

Take in an English–English dictionary.

Read the *Skills bank* section on doing reading research at the end of the Course Book unit. Decide when, if at all, to refer students to it. The best time is probably at the very end of the lesson or the beginning of the next lesson, as a summary/revision. Alternatively, you could refer students to the *Skills bank* after Exercise C.

Lesson aims

- prepare for reading research
- use research questions to structure reading research

Introduction

1 Hold up an English–English dictionary and say a word from Lesson 2.1. Ask students where approximately they will find it in the dictionary – i.e., beginning, middle, two-thirds of the way through, etc. Follow their advice and read the word at the top left. Ask students if the target word will be before or after. Continue until you get to the right page. Repeat with several more words from Lesson 2.1.

2 Give definitions of some of the words from Lesson 2.1 for students to identify.

Exercise A

Set the question for pairwork. Feed back ideas to the whole class.

Refer students to the list of Top 10 agricultural advances. Point out that this is a survey of students' opinions; it is not *scientific*. Try to get them to think laterally and identify examples from their own and different cultures and perspectives – green, economic, etc.

Exercise B

Encourage discussion of the survey results in pairs. Give an example of each, e.g.:

1 *The development of four-field rotation was the basis on which farmers were able to make a quantum improvement in productivity, creating the template for modern farming.*

2 *This development led to a significant rise in productivity, so cut the cost of food and reduced the likelihood of famines.*

3 *I'd say that the harvesting of seeds was the greatest because it provided the basis of modern farming.*

Subject note

It is not unusual to find Norfolk rotation 'four-field', 'four-course' or 'four-year' crop rotation used interchangeably.

Answers

Model answers:

1/2

Achievement	Definition	Effect
harvesting of seeds	systematic collection and storing of plant seeds	development of the concept of the farm
four-course crop rotation	arable system which *rotates*, e.g., root crops (turnip), then cereal (barley), then ley (clover) and cereal (wheat)	more and cheaper food, better use of available land, better soil, more employment
invention of the plough	tool for turning soil to create a bed for seed planting	more and cheaper food and reduction in manual labour
the automated milking parlour	milking sheds which have eliminated need for manual labour	
modern fertilizers	chemical compounds designed to boost production of arable crops	improved soil quality, reduction in food shortages/destruction of habitat and damage to health
development of selective breeding	systematic process of identifying best farm animals and breeding only from them	better-quality food in shops at a lower price
introduction of irrigation	process of artificially applying water to the land	allows crops to grow in times of low rainfall
invention of the combine harvester	invented by Hiran Moore as a harvesting machine in 1834; combines three processes in one	reduced the cost of harvesting; essential for the growth of industrial farming
GM foods	foods derived from genetically modified crops	offered crop strains that are more disease resistant and high yielding/possible threat to health
CAP (Common Agricultural Policy)	established in 1957 to stabilize prices and supplies of food in the European Union	debatable; established the prototype for future policy frameworks and established minimum prices, set quotas and import tariffs

Exercise C

Students may or may not be able to articulate preparation for reading. Elicit ideas. One thing they must identify – reading for a purpose. Point out that they should always be clear about the purpose of their reading. A series of questions to answer, or **research questions**, is one of the best purposes.

Refer students to the *Skills bank* at this stage if you wish.

Exercise D

1 Set for pairwork. Elicit some ideas, but do not confirm or correct.

2 Refer students to the Hadford University research questions on the right of the page. Check comprehension. If students have come up with better research questions, write them on the board for consideration during the actual reading.

Exercise E

Remind students about topic sentences if they haven't mentioned them already in Exercise C. Give them time to read the topic sentences in this exercise. Point out that the topic sentences are in order, so they give a rough overview of the whole text. Some topic sentences clearly announce what the paragraph will be about. Others may only give a hint of how it will develop.

1 Set for group discussion.

2 Remind students of the research questions. Look at the first research question as an example, then set for pairwork. Point out that they may match a research question to more than one topic sentence, and that some topic sentences may not relate to the research questions (i.e., they don't have to write a number for each topic sentence).

3 Explain that here students look at the topic sentences they *didn't* number in question 2, and try to work out the likely content of each paragraph. Do the first as an example, then set for pairwork.

Feed back, eliciting and checking that they are reasonable possibilities, based on the topic sentence. You can accept multiple ideas for the same paragraph provided they are all possible.

Answers

Possible answers:

1 1750 to 1850 or the Agricultural Revolution.

2 Different interpretations are possible:

In the 1750s, a number of factors began to change agriculture in England.	1/3
Firstly, from 1750, there were growing population levels sustained by self-sufficiency in food.	1/3
Secondly, productivity rose after 1750, even though the proportion of the population employed in agriculture dropped consistently.	1/3
This period of rapid change is called the Agrarian or Agricultural Revolution and it came about for different reasons.	
The Enclosures were important in encouraging the growth of larger, more efficient farms.	
Some, but not all, of the revolutionary ideas originated in England.	
There is general agreement that the century from 1750 to 1850 was one of *revolution*, not *evolution*.	1

3 Answers depend on the students. Discuss.

Exercise F

Point out, if students have not already said this, that the topic sentences are normally the first sentences of each paragraph. Tell students to compare the contents of each paragraph with their predictions. Encourage them to take notes as they read.

If necessary, the reading can be set for homework.

Closure

1 Unless you have set the reading for homework, do some extra work on oral summarizing as a comprehension check after reading (see *Skills bank – Using topic sentences to summarize*). Students work in pairs. One student says a topic sentence and the other student summarizes the paragraph from memory in his/her own words, or if necessary reads the paragraph again and then summarizes it without looking.

2 You may also want to redo the text as a jigsaw – the text is reproduced in the additional resources section at the back of this Teacher's Book (Resource 2B) to facilitate this.

3 As a further activity after reading, remind the students of the note-taking skills practised in Unit 1. Discuss appropriate note-taking forms for this text. They can then write notes on the text. Tell them to keep their notes, as they will be useful for the summary exercise in Lesson 2.3.

2.3 Extending skills

2.3 Extending skills — paragraph structure • topic sentences • summarizing

A Study the words in box a. They are all used in some form in the text in Lesson 2.2.

1 Give two common meanings of each word.
2 Check with your dictionary.

a: rate land hard type rotation spread template

B Study the words in box b. They are all from the text in Lesson 2.2.

1 What is the base word in each case? What part of speech is the base word?
2 Does the prefix/suffix change the part of speech?
3 How does the prefix/suffix change the meaning of the base word?

b: enclosure encourage imported technological assure availability

C Look back at the topic sentences from the text in Lesson 2.2 (Exercise E, page 16). Don't look at the text on page 17. What information comes after each topic sentence? Suggest possible content.

Example:
There is general agreement that the century from 1750 to 1850 was one of *revolution*, not *evolution*.
Evidence for this statement will follow the topic sentence.

D Write a summary of the text on page 17. Paraphrase the topic sentences. Add extra information and examples. See *Skills bank*

2.4 Extending skills — using research questions • writing topic sentences • summarizing

A What are the most important advancements in agriculture from Lesson 2.2?

B The lecturer has asked you to do some research into *landmarks in agriculture*.

1 What do you understand by the term?
2 Think of good research questions before you read the text on the opposite page.
3 Look quickly at the text on the opposite page. What is the best way to record information while you are reading?

C Study the text on the opposite page.

1 Highlight the topic sentences.
2 Read each topic sentence. What will you find in the rest of the paragraph?
3 Which paragraph(s) will probably answer each research question? Read those paragraphs and make notes.
4 Have you got all the information you need? If not, read other paragraphs.

D Use the Internet to research one of the most important advancements in agriculture from the list in Lesson 2.2.

1 Make notes.
2 Write a series of topic sentences which summarize your findings.
3 Report back to the other students. Read out each topic sentence, then add extra details.

18

General note

Take in a set of English–English dictionaries.

Lesson aims

- identify paragraph structure
- produce good topic sentences
- summarize a text

Further practice in:

- vocabulary from Lesson 2.2

Introduction

Test students on the factual information in the text from the previous lesson, e.g. *What do we mean when we talk about the Enclosures? How long is it generally assumed the Agricultural Revolution lasted?* If a student says, accurately, *I didn't read about that. It wasn't relevant to my research*, accept it and praise the student.

Exercise A

Set for pairwork. Feed back orally.

Methodology note

Don't help students to find words in a text. It's a key reading skill to be able to pattern match, i.e., get a word in your mind's eye and then find it on the page.

Answers

Possible answers:

Word		Meaning 1		Meaning 2
rate	n	level: *The survival rate of young animals in winter is low.*	v	to assess: *The judges rated her performance highly.*
land	n	farm production area: *farm land*	v	an aeroplane 'lands' at the end of a journey
hard	adj	difficult: *a hard life*	adj	dense: *Oak is a hard wood.*
type	n	strain of: *Jerseys are a type of cow.*	v	to write using a machine: *She types her essays.*
rotation	n	turning around on an axis	n	varying crops in a particular order to maximize soil quality and yield
spread	v	to disseminate: *They spread the idea.*	v	to distribute something evenly over a surface: *We spread butter on toast.*
template	n	a model: *The revolution was a template for …*	v	create a model to work from: *Before you manufacture a part, you need to template it.*

Exercise B

Set for individual work and pairwork checking. Students can check these points in a dictionary. Feed back, taking apart the words and showing how the affixes can change the meaning.

Answers

Model answers:

Word	Base word	Affix and meaning
enclosure	close (v)	*en~* = in/within: *to put something within*
encourage	courage (n)	*en~* = in/into: *to put courage into someone;* noun → verb
imported	port (n)	*im~* = in: *We produce no oil – we import it.*
technological	technology (n)	*~ical* = pertaining to; noun → adjective
assure	sure (adj)	*a~* = to make (sure to/of): *I assure you of my support;* adjective → verb
availability	avail (v)	*~ity* = condition or state of being; adjective → noun – here the noun is formed from the adjective 'available'.

Exercise C

Do the first paragraph first, then set the rest for individual work followed by pairwork checking.

Feed back with the whole class. Ask the students to locate the discourse markers used which help to identify the way in which the paragraph is continued. Build the table in the Answers section below on the board (or on an OHT or PowerPoint slide).

An alternative procedure would be to look at each topic sentence and feed back before moving on to the next topic sentence.

Answers

Possible answers: see table below.

Discourse note

In academic writing, topic sentences often consist of a general point. The sentences that follow then support the general statement in various ways, such as:

- giving a definition and/or a description
- giving examples
- giving lists of points (e.g., arguments or reasons)
- restating the topic sentence in a different way to help clarify it
- giving more information and detail on the topic sentence to clarify it

Often – but not always – the type of sentence is shown by a 'discourse marker' – e.g., *for example*, *first of all*, etc. This helps to signal to the reader how the writer sees the link between the sentences and is therefore a good clue as to the purpose of the sentences following the topic sentence.

Topic sentence	Possible paragraph content	Discourse markers
In the 1750s, a number of factors began to change agriculture in England.	list of factors	One of these … Some … Others … However …
Firstly, from 1750, there were growing population levels sustained by self-sufficiency in food.	support for factor 1	Firstly, … However, But …
Secondly, productivity rose after 1750, even though the proportion of the population employed in agriculture dropped consistently.	support for factor 2	Secondly, …
This period of rapid change is called the Agrarian or Agricultural Revolution and it came about for different reasons.	some background to the innovations	For example, …
The Enclosures were important in encouraging the growth of larger, more efficient farms.	background on the Enclosures	Of course, … But there is no doubt that …
Some, but not all, of the revolutionary ideas originated in England.	background on the types of innovation and where they came from	Some, but not all, … However, … The first … The second … Thirdly, … Finally, …
There is general agreement that the century from 1750 to 1850 was one of *revolution*, not *evolution*.	evidence for this statement and summary remarks	There is general agreement that … substantial evidence that …

Exercise D

Students can work individually (for homework) or in pairs (in class). Ask them to write a summary in about 150 words. They should use their own words as far as possible, but they should also try to incorporate the vocabulary they have practised so far. Refer students to the *Skills bank – Using topic sentences to summarize.*

Methodology note

There are two reasons for students to use their own words in written work (except when quoting and acknowledging sources):

1. The work involved in rewording information and ideas helps us to mentally process them and to retain them in memory.
2. Copying whole sentences from the work of other writers is plagiarism (unless the quotation is acknowledged). Universities disapprove of plagiarism and may mark down students who plagiarize. In the commercial world, an accusation of plagiarism can cause legal problems, and in the academic world, it can severely damage a teacher's reputation and career.

Closure

Tell students to define some of the agriculture words from the text on page 17. Alternatively, give definitions of some of the words and tell students to identify the words.

2.4 Extending skills

2.3 Extending skills paragraph structure • topic sentences • summarizing

A Study the words in box a. They are all used in some form in the text in Lesson 2.2.

1. Give two common meanings of each word.
2. Check with your dictionary.

a
rate land hard type rotation spread template

B Study the words in box b. They are all from the text in Lesson 2.2.

1. What is the base word in each case? What part of speech is the base word?
2. Does the prefix/suffix change the part of speech?
3. How does the prefix/suffix change the meaning of the base word?

b
enclosure encourage imported technological assure availability

C Look back at the topic sentences from the text in Lesson 2.2 (Exercise E, page 16). Don't look at the text on page 17. What information comes after each topic sentence? Suggest possible content.

Example:
There is general agreement that the century from 1750 to 1850 was one of *revolution*, not *evolution*.
Evidence for this statement will follow the topic sentence.

D Write a summary of the text on page 17. Paraphrase the topic sentences. Add extra information and examples. See *Skills bank*

2.4 Extending skills using research questions • writing topic sentences • summarizing

A What are the most important advancements in agriculture from Lesson 2.2?

B The lecturer has asked you to do some research into *landmarks in agriculture*.

1. What do you understand by the term?
2. Think of good research questions before you read the text on the opposite page.
3. Look quickly at the text on the opposite page. What is the best way to record information while you are reading?

C Study the text on the opposite page.

1. Highlight the topic sentences.
2. Read each topic sentence. What will you find in the rest of the paragraph?
3. Which paragraph(s) will probably answer each research question? Read those paragraphs and make notes.
4. Have you got all the information you need? If not, read other paragraphs.

D Use the Internet to research one of the most important advancements in agriculture from the list in Lesson 2.2.

1. Make notes.
2. Write a series of topic sentences which summarize your findings.
3. Report back to the other students. Read out each topic sentence, then add extra details.

Landmarks in agriculture

From the 1750s onwards, farming productivity was improved in a number of ways. These improvements were numerous developments in the techniques which farmers employed to increase productivity. They had an impact on the lives of the rural workforce and on the wider economy and social life of England.

A small number of key individuals, who became known as 'agriculturalists', started many of the changes. There is some argument over which of them contributed most. The group brought new expertise, and shared an interest in the development of agriculture and a commitment to change. They tend to fall into three categories. First, there were men like Robert Blakewell who developed the science of farming. Jethro Tull was typical of a second group who improved the technology. Finally, Arthur Young pioneered the use of publicity.

Lord 'Turnip' Townshend introduced the Norfolk four-course crop rotation system to England. Under his system, farmers rotated wheat, turnips, barley and clover annually. The turnips were used as a cleansing crop. Clover replaced the nutrients in the soil. Turnips and clover were used for animal fodder, enabling farmers to feed animals over the winter. This meant fresh meat all year. Townshend also introduced a method called 'marling' to increase productivity where topsoil is mixed with sandy soil to give better cropping.

Jethro Tull is well known today, though many of the innovations he proposed were not implemented in his lifetime. His contributions include a horse-powered seed drill. This replaced a system where workers distributed seeds by hand, with a machine that used a rotating cylinder. The seed drill distributed the seeds into ploughed furrows, and closed the furrows.

Robert Bakewell pioneered selective breeding, using simple and effective methods. He bred from the best animal. His methods created the Leicester Longwool sheep and the Dishley Longhorn cow. He used record keeping to ensure good health and this became the template for breeders. The Collings brothers used his ideas as a basis to breed Durham Shorthorn cattle, which combined high milk yields and lean meat.

Innovation was crucial to the Agricultural Revolution, but people were also needed to spread the word. Arthur Young and Thomas Coke were examples of early agricultural propagandists. They both recognized the need to establish good channels of communication but they used contrasting methods. Young edited a journal and was Secretary of the Board of Agriculture. He spread the message. Coke was a Norfolk landowner who used his own farm to demonstrate the benefits of change. His tenant farmers were offered leases of 20 to 40 years and were encouraged to try new methods. He organized regular events on his estate to demonstrate these methods.

It is difficult to identify the most influential innovation. Periods of rapid and dramatic change tend to be characterized by a range of events and influences. The period from 1750 to 1850 is a good example.

Landmarks in agriculture

Lesson aims

- use research questions to structure reading research
- write topic sentences for a short research report/summary

Introduction

Give a word from the text in Lesson 2.2 which is part of a phrase. Ask students to try to complete the phrase. It's probably better if you give the first word in the phrase, but you might also try giving the second word at times or at the end of the exercise.

Possible two-word phrases:

Enclosure Laws
common land
crop rotation
rapid spread
higher productivity
survival rates
substantial evidence
low-intensity
arable crops

Exercise A

Group discussion. Build up a list on the board, students' books closed.

Answers

Answers depend on the students.

Exercise B

1. This question relates to the reading text on page 19 and acts as a pre-question. Set for pairwork. Feed back orally.
2. Remind students of the importance of research questions – reading for a purpose. Set for pairwork. Feed back, writing up suitable questions on the board.
3. Elicit the different kinds of notes you can use – see Unit 1 *Skills bank*. Remind students to think about the best kind of notes before and while they are reading.

Methodology note

It is good for students to get into the habit of thinking about the form of their notes before they read a text in detail. If they don't do this, they will tend to be drawn into narrative notes rather than notes which are specifically designed to help them answer their research questions.

Answers

Possible answers:

1 Important achievements and advancements in agriculture.
2 The three questions from Lesson 2.2 would be fine, although students may come up with better ones.
3 See Unit 1 *Skills bank*. A timeline would be an appropriate form of notes.

Exercise C

1/2 Remind students of the importance of topic sentences. Set for individual work and pairwork checking. Encourage students not to read ahead. Perhaps you should ask students to cover the text and only reveal each topic sentence in turn, then discuss possible contents of the paragraph. Remind them that this is a technique for previewing a text and at this point they do not need to read every part of the text. This will come later. If you have an OHP or other visual display, you can tell students to close their books and just display the topic sentences (additional resources section, Resource 2C), or you can give them out on a handout.

3 Set the choice of paragraphs for pairwork. Students then read individually, make notes and compare them. Monitor and assist.

4 Give students time to read other paragraphs if they need to.

Answers

Possible answers:

2

Topic sentence	Possible paragraph content
From the 1750s onwards, farming productivity was improved in a number of ways.	indications about the type of development
A small number of key individuals, who became known as 'agriculturalists', started many of the changes.	information about who were the prime movers
Lord 'Turnip' Townshend introduced the Norfolk four-course crop rotation system to England.	biographical details about Townshend
Jethro Tull is well known today, though many of the innovations he proposed were not implemented in his lifetime.	explanation as to why Tull is famous
Robert Bakewell pioneered selective breeding, using simple and effective methods.	information about selective breeding
Innovation was crucial to the Agricultural Revolution, but people were also needed to spread the word.	some background on how the information about changes was broadcast
It is difficult to identify the most influential innovation.	a summary statement

3 The appropriate paragraphs to read depend on the research questions from Exercise B.

Discourse note

It is as well to be aware (though you may not feel it is appropriate to discuss with students at this point) that in real academic texts, the topic sentence may not be as obvious as in the texts in this unit. Sometimes there is not an explicit topic sentence, so that the overall topic of the paragraph must be inferred. Or the actual topic sentence for the paragraph can be *near* rather than *at* the beginning of the paragraph. Sometimes, also, the first sentence of a paragraph acts as a topic statement for a succession of paragraphs.

Exercise D

1 If it is possible to research on the Internet during the lesson, send students to the computers now. They can work in groups. If not, set the task for homework and feed back next lesson.
2 Set for individual work and group work checking.
3 The idea is that students, on the basis of the topic sentences, present their information to fellow students.
Make sure students realize that they only have to write the topic sentences. They can add the details orally.

Closure

1 Focus on some of the vocabulary from the text, including:

agriculturalist
cleansing
cylinder
dramatic
horse-powered
innovation
journal
lean meat
marling
means
nutrients
prioritize
propagandist
record keeping
selective
tenant
topsoil

2 You may also want to redo the text as a jigsaw, as before – the text is reproduced in the additional resources section (Resource 2D) to facilitate this.

Extra activities

1 Work through the *Vocabulary bank* and *Skills bank* if you have not already done so, or as a revision of previous study.

2 Use the *Activity bank* (Teacher's Book additional resources section, Resource 2A).

A Set the wordsearch for individual work (including homework) or pairwork.

Answers

R	O	T	A	T	I	N	G	Q	K	N	L	U	R	N	H	L
Z	E	I	R	R	I	G	A	T	I	O	N	E	A	E	J	N
C	P	R	P	I	O	N	E	E	R	R	Z	W	G	F	W	S
M	U	T	O	S	Q	L	D	L	K	I	S	S	R	F	V	R
Z	U	L	M	T	A	P	J	F	L	S	E	E	I	I	T	V
R	Y	N	T	I	A	D	B	I	E	T	F	L	C	C	E	J
D	X	A	E	I	L	T	T	D	A	C	C	E	U	I	C	E
N	N	T	E	E	V	R	I	L	W	H	O	C	L	E	H	N
B	F	R	I	Q	E	A	P	O	C	B	M	T	T	N	N	C
J	Z	Y	S	F	W	M	T	T	N	O	B	I	U	T	O	L
P	I	A	K	W	E	J	I	I	H	M	I	V	R	E	L	O
R	P	F	O	T	U	D	F	Z	O	L	N	E	E	R	O	S
T	J	S	O	E	N	T	K	F	K	N	E	K	J	K	G	U
P	R	O	P	A	G	A	N	D	I	S	T	Z	D	P	I	R
D	R	I	L	L	I	N	N	O	V	A	T	I	O	N	C	E
O	A	R	E	V	O	L	U	T	I	O	N	U	S	V	A	M
A	L	G	H	L	Z	P	L	O	U	G	H	K	A	R	L	O

B Do the quiz as a whole class, or in teams, or set for homework – students can reread the texts to get the answers if necessary.

Answers

1. Hiran Moore
2. 1834
3. Thomas Coke
4. Jethro Tull
5. Turnip; turnips were one of the crops used under his four-course crop rotation system
6. selective breeding
7. Dishley Longhorn
8. Collings brothers
9. turnips
10. clover

3 Ask students to work in small groups to research and feed back to the group on agricultural advancements from the 20th century.

If students are going to do research on the Internet, suggest that they type in *History*, then their topic, to get some potential texts. Alternatively, you can do this research before the lesson and print off some pages for students to work from.

Remind students that they can't possibly read everything they find, so they must use the topic sentences to decide if a paragraph is worth reading.

4 You can get students to practise their reading aloud – a skill which is not vital but is sometimes useful – by following this approach.

Photocopy and cut up one of the jigsaw texts in the additional resources section (Resources 2B and 2D). Give topic sentences to Student A and the corresponding paragraphs to Student B.

Student A reads out a topic sentence.

Student B finds the corresponding paragraph and reads it out.

An alternative is to give Student A the topic sentences and Student B a set of sentences chosen from each paragraph (one sentence per paragraph). Student A reads out the topic sentences one by one. Student B decides which of his/her sentences is likely to appear in the same paragraph as the topic sentence. Both students have to agree that the paragraph sentence matches the topic sentence.

5 Have a competition to practise finding words in a monolingual dictionary. Requirements:

- an English–English dictionary for each student (or pair of students if necessary)
- Unit 2 key vocabulary list

Put students in teams with their dictionaries closed. Select a word from the vocabulary list and instruct students to open their dictionaries and find the word. The first student to find the word is awarded a point for their team. Additional points can be awarded if the student can give the correct pronunciation and meaning.

3 SOIL, LAND AND THE ENVIRONMENT

Unit 3 looks at soil, land and the environment and how they relate to the challenge of producing healthy food to meet our needs. The first lecture looks at the key topics of food security and self-sufficiency and what they mean in developed and developing countries. The second lecture explores the contribution organic methods can make in reducing global warming.

Skills focus

Listening

- preparing for a lecture
- predicting lecture content
- making lecture notes
- using different information sources

Speaking

- reporting research findings
- asking for information
- formulating questions

Vocabulary focus

- stress patterns in multi-syllable words
- prefixes

Key vocabulary

access (n and v)	ecomarketing	nutritious
adequate	enable	optimum
affordability	ensure	potato blight
aquarial	famine	principle
availability	fluctuation	promote
biodiversity	food chain	public relations
beetroot	food security	ratio
benefit (n and v)	genetic modification	self-sufficiency
carbon footprint	global	shortage
coherent	greenwash	soil quality
commodity	indicators	staple
conservation	impact	starvation
cost	losses	trend
disease	malnutrition	
distribution network	necessity	

3.1 Vocabulary

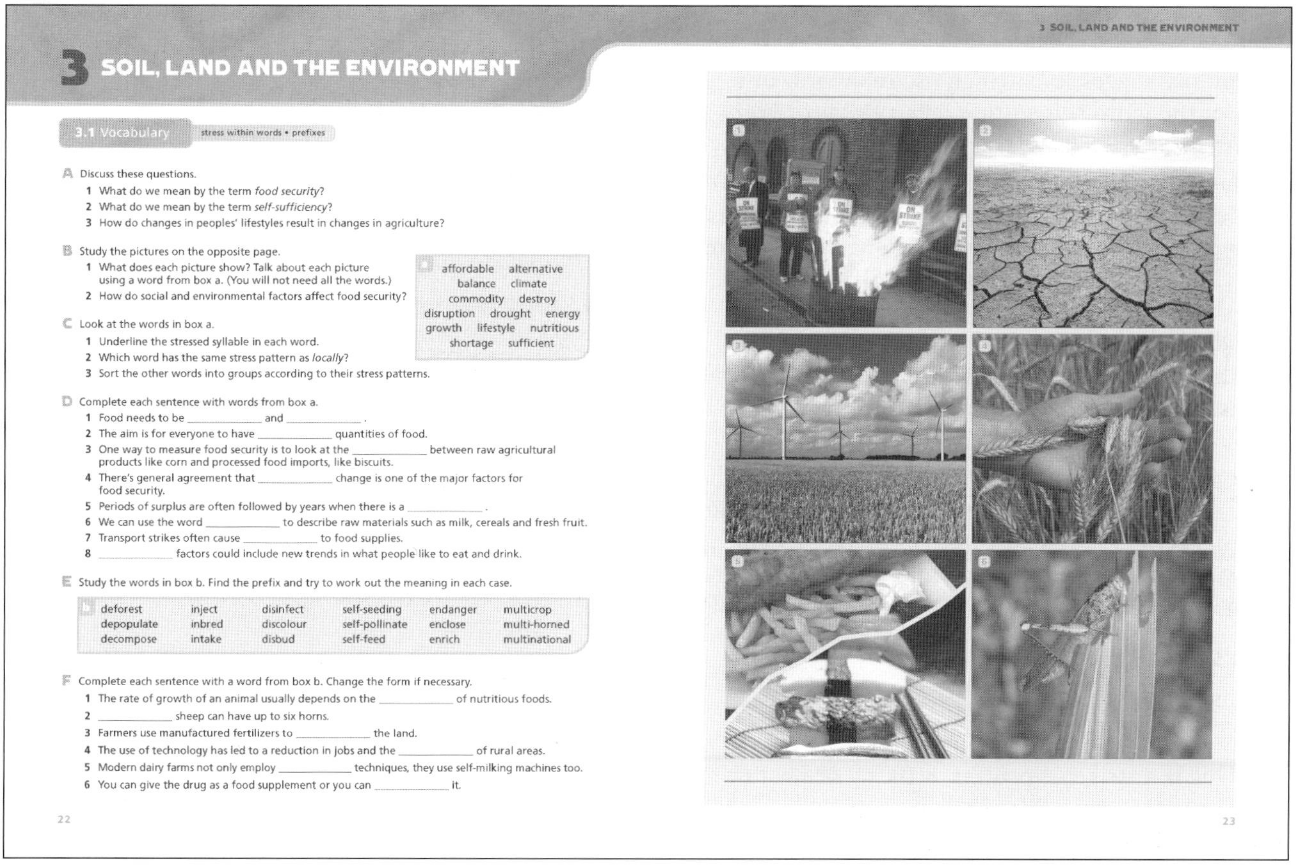

3 SOIL, LAND AND THE ENVIRONMENT

3.1 Vocabulary stress within words • prefixes

A Discuss these questions.

1 What do we mean by the term *food security*?
2 What do we mean by the term *self-sufficiency*?
3 How do changes in peoples' lifestyles result in changes in agriculture?

B Study the pictures on the opposite page.

1 What does each picture show? Talk about each picture using a word from box a. (You will not need all the words.)
2 How do social and environmental factors affect food security?

a
affordable alternative
balance climate
commodity destroy
disruption drought energy
growth lifestyle nutritious
shortage sufficient

C Look at the words in box a.

1 Underline the stressed syllable in each word.
2 Which word has the same stress pattern as *locally*?
3 Sort the other words into groups according to their stress patterns.

D Complete each sentence with words from box a.

1 Food needs to be ______ and ______ .
2 The aim is for everyone to have ______ quantities of food.
3 One way to measure food security is to look at the ______ between raw agricultural products like corn and processed food imports, like biscuits.
4 There's general agreement that ______ change is one of the major factors for food security.
5 Periods of surplus are often followed by years when there is a ______ .
6 We can use the word ______ to describe raw materials such as milk, cereals and fresh fruit.
7 Transport strikes often cause ______ to food supplies.
8 ______ factors could include new trends in what people like to eat and drink.

E Study the words in box b. Find the prefix and try to work out the meaning in each case.

b					
deforest	inject	disinfect	self-seeding	endanger	multicrop
depopulate	inbred	discolour	self-pollinate	enclose	multi-horned
decompose	intake	disbud	self-feed	enrich	multinational

F Complete each sentence with a word from box b. Change the form if necessary.

1 The rate of growth of an animal usually depends on the ______ of nutritious foods.
2 ______ sheep can have up to six horns.
3 Farmers use manufactured fertilizers to ______ the land.
4 The use of technology has led to a reduction in jobs and the ______ of rural areas.
5 Modern dairy farms not only employ ______ techniques, they use self-milking machines too.
6 You can give the drug as a food supplement or you can ______ it.

22

23

General note

Read the *Vocabulary bank* at the end of the Course Book unit. Decide when, if at all, to refer your students to it. The best time is probably at the very end of the lesson or the beginning of the next lesson, as a summary/revision.

Dictionaries will be useful in this lesson.

Lesson aims

- gain a greater understanding of the importance of stress within words and some of the common patterns
- extend knowledge of words which contain prefixes
- gain fluency in the target vocabulary

Introduction

1 Revise the vocabulary from the first two units. Check:
   - meaning
   - pronunciation
   - spelling

2 Revise the concepts from Unit 1. Ask students to give a definition of the difference between an ethical and a conventional agribusiness. Write the clearest definition on the board. Then write two columns headed 'ethical' and 'conventional' and ask the students to suggest words or expressions that they associate with one or the other. Give some examples if you need to, e.g., *quality, quantity, fair trade, profit, cost.*

Exercise A

Put students in pairs to discuss the questions. Feed back orally.

Answers

Possible answers:

1 *Food security* means protecting the food supply to make sure that people have enough to eat.
2 *Self-sufficiency* means being able to support yourself without outside help.
3 Agribusinesses must adapt to changes in the demands of the market brought about by the changing lifestyles of their customers.

Exercise B

1 Refer students to the pictures on the opposite page. Ask what students can see in the first picture. Elicit *It's a protest.* Ask which word from box a could be used to say something *more* about the picture. Accept any suggestions which use the word, or a form of the word, *disruption*, e.g., *The protesters are disrupting the business.*

Set the remaining pictures for pairwork. Students should make two statements: First, what they can see and then a further comment about each picture using at least one word from box a. If necessary, they should check the meanings of the words in their dictionaries. Not all the words are relevant. Feed back with the whole class. Accept any reasonable suggestions. Check/correct pronunciation, especially the stress patterns.

2 Set for pairwork. Students should consider how the words they have used in question 1 can affect food security. (For example, if there is a strike, it can affect the food security of a country.) Feed back orally, but do not confirm or correct at this stage.

Answers

Possible answers:

1

1. This is a strike or protest. Striking or protesting employees may **dis'rupt** deliveries.
2. This is a damaged crop. **'Drought** has destroyed the crops in this field.
3. This is a windfarm. Wind power is an example of **al'ternative 'energy**.
4. These are crops in a field. The **'growth** rate of a crop depends on factors such as the weather and soil quality.
5. This shows fish and chips and sushi. The decline of the fish and chip shop, and the rising popularity of sushi is an example of the way changing **'lifestyles** can affect agribusinesses.
6. This is a locust. Swarms of locusts can **destroy** fields of crops in minutes.

2 Accept all reasonable suggestions.

Methodology note

From now on, whenever you present a group of words in a box, as here, ask students for the part of speech of each word. This is good practice and also good preparation for changing the form of the word if a different part of speech is required in the associated exercise(s).

Exercise C

Write *locally* on the board. Ask students to say how many syllables there are in the word (there are three). Draw vertical lines to divide the syllables. Then ask students to say where the main stress is and draw a line under the syllable:

| <u>lo</u> | cal | ly |

Point out the importance of stressed syllables in words – see *Language note* below.

1 Set for pairwork. Tell students to divide the words into syllables first, then to underline the strongest stress. Feed back.

2 Ask students to find the word which has the same stress pattern as *locally.* Write it on the board like this:

| <u>lo</u> | cal | ly |

| <u>en</u> | er | gy |

3 Set for pairwork. Students should match words with the same number of syllables and with main stresses in the same place.

Language note

In English, speakers emphasize the stressed syllable in a multi-syllable word. Sometimes listeners may not even hear the unstressed syllables. Vowels, in any case, often change to schwa or a reduced form in unstressed syllables. Therefore it is essential that students can recognize key words from the stressed syllable alone when they hear them in context. Multi-syllable words may seem to have more than one stressed syllable. This is a secondary stress, e.g. *su,staina'bility*. For the present purposes, students should identify only the primary, or strongest, stress in the word.

Stress sometimes moves to fit common patterns when you add a suffix, *e.g., 'capable, capa'bility.* Other suffixes, such as *~ment* or *~al*, don't affect the stress of the root word, e.g., *em'ploy, em'ployment*; *'person, 'personal.*

Sometimes it is difficult to be sure exactly how a word should be divided into syllables. Use vowel sounds as a guide to the number of syllables. If in doubt, consult a dictionary.

Answers

1 affordable
alternative
balance
climate
commodity
destroy
disruption
drought
energy
growth
lifestyle
nutritious
shortage
sufficient

2 energy

3 affordable, alternative, commodity
balance, climate, lifestyle, shortage
destroy
drought
growth
energy
disruption, nutritious, sufficient

Exercise D

Set for individual work and pairwork checking. Not all the words are needed. Feed back orally.

Answers

1. Food needs to be affordable and nutritious.
2. The aim is for everyone to have sufficient quantities of food.
3. One way to measure food security is to look at the balance between raw agricultural products like corn and processed food imports, like biscuits.
4. There's general agreement that climate change is one of the major factors for food security.
5. Periods of surplus are often followed by years when there is a shortage.
6. We can use the word commodity to describe raw materials such as milk, cereals and fresh fruit.
7. Transport strikes often cause disruption to food supplies.
8. Lifestyle factors could include new trends in what people like to eat and drink.

Exercise E

Set for pairwork. Students should look at all three words in each group, to find and then deduce the meaning of the prefix. Encourage them to use a phrase as a definition rather than a single-word translation. They need to develop a sense of the broader meaning of the prefix. Feed back, getting the meanings on the board.

Answers

Model answers:

de~ = to remove from, to decrease, to change in the opposite direction

in~ = inside*

dis~ = to show the opposite or negative

self~ = through or by means of itself or oneself

en~ = to make or cause to become

multi~ = many

*this prefix also means *not* in some cases, e.g., *invalid, inaccurate, indecisive*

Methodology note

While not the case here, with some of these words, it is difficult to work out the base word, e.g., *trinsic* or *intrinsic*. However, you can point out that you can sometimes understand roughly what a word means if you understand the prefix, e.g., *intrinsic* must be something to do with being *in* something, so context will help you to guess the rough meaning.

Exercise F

This is further practice in using words with prefixes. Remind students that they must make sure the form of the word fits into the sentence. Feed back, checking pronunciation and stress patterns.

Answers

1. The rate of growth of an animal usually depends on the intake of nutritious foods.
2. Multi-horned sheep can have up to six horns.
3. Farmers use manufactured fertilizers to enrich the land.
4. The use of technology has led to a reduction in jobs and the depopulation of rural areas.
5. Modern dairy farms not only employ self-feeding techniques, they use self-milking machines too.
6. You can give the drug as a food supplement or you can inject it.

Closure

1. Ask students to decide whether the sentences in Exercise F are facts or opinions. Discuss ideas with the whole class.
2. If you have not already done so, refer students to the *Vocabulary bank* at the end of Unit 3. Work through some or all of the stress patterns.

Language note

The patterns shown in the *Vocabulary bank* in Unit 3 are productive, i.e., they enable you to make more words or apply the rules accurately to other words. The words with unusual patterns tend to be the more common ones, so if students come across a new multi-syllable word at this level, it is likely to conform to the patterns shown. Native speakers recognize the patterns and will naturally apply them to unusual words, e.g., proper nouns. How, for example, would you pronounce these nonsense words?

felacom
bornessity
shimafy
emtonology
scolobility
nemponary
cagoral
andimakinise
ortepanimation

3.2 Listening

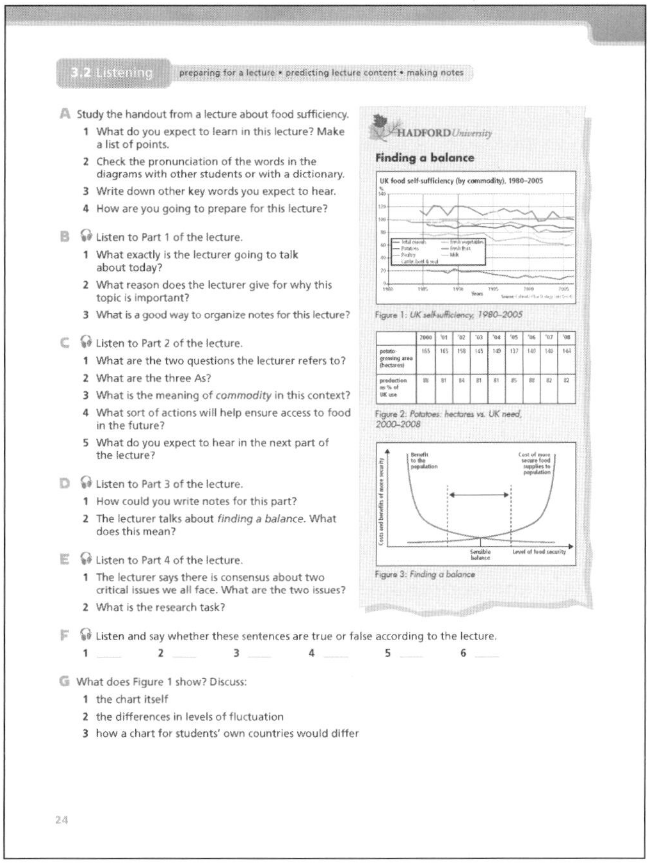

3.2 Listening preparing for a lecture • predicting lecture content • making notes

A Study the handout from a lecture about food sufficiency.
1 What do you expect to learn in this lecture? Make a list of points.
2 Check the pronunciation of the words in the diagrams with other students or with a dictionary.
3 Write down other key words you expect to hear.
4 How are you going to prepare for this lecture?

B Listen to Part 1 of the lecture.
1 What exactly is the lecturer going to talk about today?
2 What reason does the lecturer give for why this topic is important?
3 What is a good way to organize notes for this lecture?

C Listen to Part 2 of the lecture.
1 What are the two questions the lecturer refers to?
2 What are the three As?
3 What is the meaning of *commodity* in this context?
4 What sort of actions will help ensure access to food in the future?
5 What do you expect to hear in the next part of the lecture?

D Listen to Part 3 of the lecture.
1 How could you write notes for this part?
2 The lecturer talks about *finding a balance*. What does this mean?

E Listen to Part 4 of the lecture.
1 The lecturer says there is consensus about two critical issues we all face. What are the two issues?
2 What is the research task?

F Listen and say whether these sentences are true or false according to the lecture.
1 ___ 2 ___ 3 ___ 4 ___ 5 ___ 6 ___

G What does Figure 1 show? Discuss:
1 the chart itself
2 the differences in levels of fluctuation
3 how a chart for students' own countries would differ

HADFORD University

Finding a balance

Figure 1: *UK self-sufficiency, 1980–2005*

	2000	'01	'02	'03	'04	'05	'06	'07	'08
potato-growing area (hectares)	165	165	158	145	149	137	140	140	144
production as % of UK use	88	81	84	81	81	85	88	82	82

Figure 2: *Potatoes: hectares vs. UK need, 2000–2008*

Figure 3: *Finding a balance*

24

Lesson aims

Further practice in:

- planning and preparing for a lecture
- predicting lecture content
- choosing the best form of notes
- making notes

Introduction

Review key vocabulary by:

- using flashcards
- playing the alphabet game in the extra activities section at the end of this unit

Exercise A

Refer students to the handout with Figures 1, 2 and 3. Write the title *Food security* or *Self-sufficiency* on the board.

1 Set for individual work and pairwork checking. Feed back, eliciting some ideas.

2 Set for pairwork.

3 Brainstorm to elicit key words. Allow the class to decide whether a word should be included.

4 Elicit some points – the four Ps (Plan, Prepare, Predict, Produce). If necessary, refer students to Unit 1 *Skills bank* to review the preparation for a lecture. One way to help the students to make provisional notes is to:
   - brainstorm what they would include
   - organize their topics into a logical sequence

Answers

Answers depend on the students.

Exercise B

1 Tell students they are only going to hear the introduction to the lecture. Ask what information they expect to get from the introduction (i.e., the outline of the lecture).

Give students time to read the choices of topics. Check that they understand the meaning and relevance. Remind them they will only hear the introduction once, as in a lecture. Play Part 1. Allow them to compare answers.

Feed back. Ask them to justify their choice by saying what they heard related to it. Confirm the correct answer.

2 Elicit ideas. Confirm or correct.

3 Elicit ideas.

Answers

Model answers:

1 Food security.

2 Food security is important because it involves all aspects of the agribusiness sector and the food chain.

3 Since the lecture breaks down into introduction, key question 1, key question 2, additional information and conclusion, headings and bullet points could be used. For example:

Intro

- bullet 1
- bullet 2

Key question 1

- bullet 1
- bullet 2, etc.

Transcript 1.11

Part 1

Good morning, everybody. Today, I'm going to look at some related issues. Firstly, I'm going to talk about the key question and main points relating to food security. Then, I'm going to talk about the key question of self-sufficiency and I'll then be giving additional relevant information and trying to draw conclusions. OK. The first thing to say is, food security is not about being safe from stealing or attack. Food security is about having access to safe and nutritious food. The UK government has recently focused on food security and land use because of the global food price rises that took place between 2006 and 2008. Self-sufficiency is one aspect of food security. Self-sufficiency means meeting your own country's food needs without help from other countries. Self-sufficiency and food security are connected to land use. And land use – the way we use land to produce the food we need – is closely linked to issues such as organic farming and soil quality, which I'll be talking about in the next lecture. Food security is important because it involves all aspects of the agribusiness sector and the food chain. There are two key questions a country like the UK must ask about food security. First, how much food does the UK actually need to produce? Second, what are the factors the UK needs to consider to ensure food security in the future?

Exercise C

Before playing Part 2, refer students to Figure 1. Ask students what they expect to hear. Give them time to read questions 1–4. Tell them to write only brief notes. The main task is to absorb the meaning.

Play Part 2. Give them time to answer questions 1–4. Allow them to compare their answers. Feed back.

Ask them what they expect in the next part of the lecture (question 5). Elicit ideas but do not confirm or correct.

Answers

Model answers:

1 How much food does the UK actually need to produce? What are the factors the UK needs to consider to ensure food security in the future?

2 Availability, access, affordability.

3 A commodity is a substance such as wheat or potatoes sold in very large quantities.

4 Improving industrial relations to avoid the threat of strikes; encouraging people to source their food locally to reduce environmental impact; helping developing countries improve productivity to ensure alternative supplies.

5 Answers depend on the students.

Transcript 1.12

Part 2

OK. The first question is this, how much food does a country like the UK need to produce to be 'secure'? Let's use a *commodity* as an example – that is, a substance sold in very large quantities, such as potatoes. Potatoes were first introduced into the UK in the late 1400s. They've been a staple part of the UK diet ever since. Let's take an example of their importance from history. Ireland in the 1800s was dependent on potatoes for its food security. Most of the population in Ireland used at least some of their land to grow potatoes. But potatoes are susceptible to diseases. Potato blight, one of the worst of the diseases, caused terrible famine in Ireland in the mid-1800s. As a result of the potato blight, a million people died of starvation, and the same number left Ireland. What do we learn from this? Ireland was not *food secure*. Let's compare this with the UK today. Potatoes are still a staple food. If we take the period 2001–2008, the UK produced roughly 80 per cent of the potatoes it needed. Interestingly, in the same period, the amount of land used to grow potatoes dropped by more than 10 per cent. But the UK had no significant problem with the shortfall. New techniques in disease control and improved strains – types of disease-resistant seed potatoes – increased productivity. At the same time, people had money to spend on alternatives to potatoes, such as pasta and rice. As a result, the UK probably has a higher level of general food security than in the past, although the UK is not self-sufficient in potatoes.

OK, let's look at question two. What are the factors the UK needs to consider to ensure food security in the future? There are three aspects to this question which we need to consider. First there is *availability*, next there is *access* and finally *affordability*. OK, let's start with availability. Availability is about whether there is *enough* nutritious, healthy food, and whether the supply is secure. Access is about getting the food we have to the people who need it. And affordability is about whether the people who need it have enough money to buy the food.

Let's look at availability in more detail. Let's think about potatoes again. If we want to ensure availability without increasing land use, we need to increase productivity. That means, as I've said, using strains that resist disease and deliver higher yields per hectare. And that means, in turn, investing in research and development, the science of farming. It also means investing in training the people involved in farming to ensure we have a skilled workforce. And we need, in the long term, to promote sustainable farming methods that help ensure healthy crops and food today without compromising our ability to meet our needs tomorrow.

So, what about access to supplies of potatoes? Well, the UK, for example, has a good distribution and retail network. The important thing is to think about how to make distribution more sustainable and secure. That includes improving industrial relations to avoid the threat of strikes. It means encouraging people to source their food locally to reduce the carbon footprint. It also means helping developing countries improve productivity to ensure alternative supplies. The more countries that can supply the products we all need, the more food secure all nations and communities become. And these measures will tend to impact on the price or affordability of potatoes.

There is growing consensus that food security isn't about one country's self-sufficiency. It's a *global* issue. Food security in the UK is linked to food security in every other country in the developed, and also in the developing, world. It's all about balance.

Exercise D

Play the first two sentences of Part 3. Ask the first question. Set the second question for individual work and pairwork checking. Play the rest of the recording. Tell students to take notes. Allow students to compare their definitions. Don't, at this stage, confirm the answers.

Answers

Model answers:

1 A table is good for comparing, e.g., the approach to food security in developed and developing countries.

2 A balance between the amount of food we produce ourselves and the amount we import; between the land we use for agriculture and the land we use for building; between the *benefits* to society of food security and the *cost* to society of food security.

Transcript 1.13

Part 3

Developing countries have a very different approach compared with the approach in *developed* countries. In developing countries, food security is used to describe the challenges of providing people with sufficient food to eat. And with more than 850 million people in the world classified as living in hunger, food security in developing countries is a critical issue.

As we've seen, in the developed world, food security is about choices and finding a balance. And there are several elements which must balance. Firstly, the amount of food we produce ourselves and the amount we import. Secondly, the land we use for agriculture and the land we use for building. We must also think about the balance between the benefits and the cost to society of food security; for example, the risk of climate change and the cost of action to prevent it.

So, we have 'the three As' – availability, access and affordability – and in developed countries we assume that we can choose how we balance these things against the cost to our population. But for the developing world, food security is not about balancing costs and benefits. There are no choices. It is most often about simply finding ways to ensure an adequate supply of nutritious, healthy food, by any possible means. There are no spare resources to help build strategies to deal with emergencies that arise from, for example, the effects of climate change on raw material production. Dealing with climate change is not an option or choice; it's a matter of survival. Choice is, in reality, limited or non-existent. And one reason for this tends to be that developing countries do not receive a fair return or income, in return for the commodities and products they produce, in the current global economy. This means they are not able to invest in modernizing their agriculture and agribusiness sector – in developing a sustainable food economy. In developing nations, indicators of food security therefore tend to be about very simple factors such as the proportion of their populations' income spent on food and on levels of malnutrition and hunger. These are the things that decide if the country's food supply is secure, and if its people will live or die when their economies are hit by any new or unforeseen problem caused by natural cycles or global warming.

Exercise E

Part 4 summarizes key issues in the discussion about food security and defines two critical issues facing all countries. Tell students that this is the last part of the lecture. What do they expect to hear? What are the two critical issues? Play Part 4.

1 Students should check what they expected to hear as they listen. After the summary has finished, they should correct and complete their notes. Guide them to the correct answer.

2 Elicit ideas. (If you wish, you could ask your students to do some research on this topic themselves. However, it is not essential at this stage. Students are asked to do a separate research task at the end of Lesson 3.3 and feed back on that in Lesson 3.4.)

Answers

Model answers:

1 How to reduce the impact of climate change and its impact on food production, and how to make the global economy fairer so developing countries can invest in modernizing their agricultural systems and agribusinesses.

2 The research task is to look at why organic principles are important for food security – in particular, soil quality.

Transcript 1.14

Part 4

OK, so there is growing consensus that food security means different things depending on where you live, but food security *is* a global challenge. There is also a growing consensus that there are two issues we all face which are critical for a secure food future. The first concerns the impact of climate change on global food production, and the question is: What can we do to reduce this? The second issue is about the way our global community currently favours the interests of developed countries at the expense of developing ones. And the issue here is how we move towards a fairer global economy so that countries in the developing world receive a fair price for their agricultural products and labour so they can invest in modernizing their agriculture and agribusiness sector.

Right. That's it for today. Next time I want to look at why *organic* principles are important for food security. And, in particular, I want to look at the issue of soil quality and how it relates to the topics we have discussed today. So don't forget to do some research on this topic before you come.

Exercise F

These are sentences about the ideas in the lecture.

Set for pairwork. Say or play the sentences. Give time for students to discuss and then respond. Students must justify their answers.

Answers

1	true	Food prices rose sharply between 2006 and 2008.
2	false	The Irish potato famine took place in the 1800s.
3	true	The UK uses less land to grow potatoes now than in 2001.
4	false	Developing countries see food security very differently to developed countries.
5	false	More than 850 million people in the world are classified as living in hunger.
6	false	Self-sufficiency means meeting your own country's food needs without other countries' help.

Transcript 1.15

1 Food prices rose sharply between 2006 and 2008.

2 The Irish potato famine took place in the 1700s.

3 The UK uses less land to grow potatoes now than in 2001.

4 Developing countries see food security in the same way as developed ones.

5 Fewer than 300 million people are classified as living in hunger.

6 Self-sufficiency means meeting others' needs as well as your own.

Exercise G

This exercise practises making questions and describing information in a table.

Ask students to think of three questions they could ask about the chart using Wh~ question words such as *What …? Where …? When …? Who …? How many …? What proportion/percentage …?* etc. Elicit some examples and write these on the board. For example:

1 *What information does Figure 1 show?*

2 *Which commodity does the UK produce least of?*

3 *What time span does Figure 1 represent?*

4 *Which commodity shows the least fluctuation?*

5 *What proportion of our poultry requirements did the UK produce in 1995?*

Put students in pairs to ask each other their questions. Check.

Answers

Possible answers to example questions above:

1 the UK's levels of self-sufficiency in food by commodity

2 fresh fruit

3 1980–2005 (25 years)

4 milk

5 roughly 92%

Closure

Ask students to:

- give a definition of principle. A possibility is: *A fundamental or truth; a general truth.*
- discuss whether they think principles are important as part of an agribusiness. See if students think that principles will help us solve problems like global warming or food shortages. If they do, you could ask students to write, in one sentence, the sort of principles they would like to propose that they think would achieve a global consensus.

Note: Students will need their lecture notes from Lesson 3.2 in the next lesson.

3.3 Extending skills

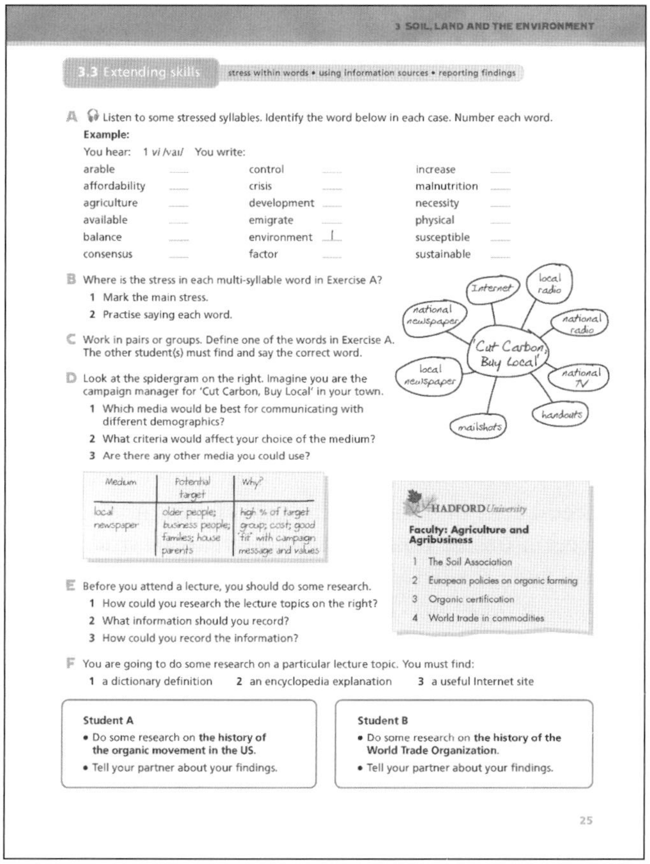
3 SOIL, LAND AND THE ENVIRONMENT

3.3 Extending skills — stress within words • using information sources • reporting findings

A Listen to some stressed syllables. Identify the word below in each case. Number each word.

Example:

You hear: 1 *vi* /vaɪ/ You write:

arable ___	control ___	increase ___
affordability ___	crisis ___	malnutrition ___
agriculture ___	development ___	necessity ___
available ___	emigrate ___	physical ___
balance ___	environment 1	susceptible ___
consensus ___	factor ___	sustainable ___

B Where is the stress in each multi-syllable word in Exercise A?

1 Mark the main stress.

2 Practise saying each word.

C Work in pairs or groups. Define one of the words in Exercise A. The other student(s) must find and say the correct word.

D Look at the spidergram on the right. Imagine you are the campaign manager for 'Cut Carbon, Buy Local' in your town.

1 Which media would be best for communicating with different demographics?

2 What criteria would affect your choice of the medium?

3 Are there any other media you could use?

Medium	Potential target	Why?
local newspaper	older people; business people; families; house parents	high % of target group; cost; good 'fit' with campaign message and values

HADFORD University

Faculty: Agriculture and Agribusiness

1 The Soil Association

2 European policies on organic farming

3 Organic certification

4 World trade in commodities

E Before you attend a lecture, you should do some research.

1 How could you research the lecture topics on the right?

2 What information should you record?

3 How could you record the information?

F You are going to do some research on a particular lecture topic. You must find:

1 a dictionary definition 2 an encyclopedia explanation 3 a useful Internet site

Student A

- Do some research on **the history of the organic movement in the US.**
- Tell your partner about your findings.

Student B

- Do some research on **the history of the World Trade Organization.**
- Tell your partner about your findings.

25

General note

Read the *Skills bank* at the end of the Course Book unit. Decide when, if at all, to refer students to it. The best time is probably at the beginning of this lesson or the end of the next lesson, as a summary/revision.

Lesson aim

This lesson is the first in a series about writing an assignment or giving a presentation based on research. The principal aim of this lesson is to introduce students to sources of information.

Introduction

1 Tell students to ask you questions about the information in the lecture in Lesson 3.2 as if you were the lecturer. Refer them to the *Skills bank* for typical language if you wish.

2 Put students in pairs. Student A must ask Student B about the information in the lecture in Lesson 3.2 to help him/her complete the notes from the lecture. Then they reverse roles. Go round, helping students to identify gaps in their notes and to think of good questions to get the missing information. Refer them to the *Skills bank* if you wish for language they can use in the pairwork.

Pairs then compare notes and decide what other information would be useful and where they could get it from. For example, information from the Internet on the level and sources of imports of commodities might be useful. In the feedback, write a list of research sources on the board, including dictionaries, encyclopedias, specialist reference books (about agriculture, politics, business management) and the Internet.

Point out that dictionaries are good for definitions, although you may need to go to a specialist dictionary for a technical word. Otherwise, try an encyclopedia, because technical words are often defined in entries when they are first used. You could also try Google's 'define' feature, i.e., type *define: principle*. But remember you will get definitions from other disciplines, not just your own, so you need to scan to check the relevant one. (*Principle* has a special legal meaning as well as a more general philosophical or political one.)

When doing an Internet search, it is also important to check spellings – princip*al* has a different meaning to princip*le*, and UK English often uses an 's' where US English uses a 'z', e.g., specialise/specialize.

Exercise A

Point out the importance of stressed syllables in words – see *Language note* on the next page.

In this exercise, students will hear each word with the stressed syllable emphasized, and the rest of the syllables underspoken.

Play the recording, pausing after the first few to check that students understand the task. Feed back, perhaps playing the recording again for each word before checking. Ideally, mark up an OHT of the words.

Language note

In English, speakers emphasize the stressed syllable in a multi-syllable word. Sometimes listeners may not even hear the unstressed syllables. Vowels, in any case, often change to schwa or a reduced form in unstressed syllables. Therefore it is essential that students can recognize key words from the stressed syllable alone when they hear them in context.

Answers

necessity	3
development	7
environment	1
sustainable	8
susceptible	12
control	10
factor	17
arable	2
emigrate	14

increase	6
agriculture	18
balance	9
consensus	5
physical	11
affordability	16
malnutrition	4
available	13
crisis	15

Transcript 1.16

1 en'vironment
2 'arable
3 ne'cessity
4 malnu'trition
5 con'sensus
6 'increase
7 de'velopment
8 su'stainable
9 'balance
10 con'trol
11 'physical
12 su'sceptible
13 a'vailable
14 'emigrate
15 'crisis
16 afforda'bility
17 'factor
18 'agriculture

Exercise B

Erase the words or turn off the OHP. Ask students to guess or remember where the stressed syllable is on each word. Tell them to mark their idea with a light vertical stroke in pencil. Elicit and drill. Refer students to the *Vocabulary bank* at this stage if you wish.

Answers

See transcript for Exercise A.

Exercise C

Set for pair or group work. Go round and assist/correct.

Exercise D

1/2 Refer students to the spidergram and the table. Elicit question forms for this discussion such as:

Which medium is most …?

Why do you think … would choose …?

Why do businesses tend to use …?

Where would you go to find prices for …?

Put students in small groups or pairs to discuss the questions.

Feed back, building up the table in the Answers section on the board. The more reasons students can give, the better.

3 Discuss with the whole class. Accept any reasonable suggestions.

Answers

Model answers:

1/2

Medium	Potential target	Why?
local newspaper	older people; business people; families; houseparents	high % of target group; cost; good 'fit' with campaign message and values
national newspaper	families; business people	coverage of all, including target groups, campaign profile; very high cost
Internet	young people; employed; self-employed	high % of target group; high cost; good 'fit' with campaign message and values
local radio	older people; non-management workers; self-employed; families	high % of target group; cost; good 'fit' with campaign message and values; cost
national radio	families; business people	raise profile for campaign; target and related interest groups; high cost
national TV	younger people; families; business people	high % of target group; high impact; very high cost
handouts	families; houseparents	impact from, e.g., volunteers at supermarket; feedback; cost
mailshots	families	very good coverage and proven success; cost

Exercise E

Remind students again about the four Ps. Refer students to the lecture topics and the questions. Make sure they understand that all three questions relate to before, rather than during, the lecture. Work through as a whole class if you wish.

Answers

Model answers:

1 Look up key words in a dictionary/encyclopedia/on the Internet. Check pronunciation so you will recognize the words in the lecture.

2 Lecture 1: meanings of these key words; information about the history of the Soil Association

Lecture 2: notes on key policies affecting organic agribusiness

Lecture 3: main certificate-awarding organizations; notes on aspects of an agribusiness the certificates cover

Lecture 4: notes of the major commodities; key data on volume of trade

3 Perhaps do a spidergram so that it is easier to brainstorm with fellow students and cover all the possible areas that the lecturer might focus on.

Exercise F

Set for pairwork, giving each member of the pair a different research task. If students have access in class to reference material, allow them to at least start the activity in class. Otherwise, set for homework. Before the feed back to partner stage, refer students to the *Skills bank – Reporting information to other people.*

Closure

Dictate sentences with words from Exercise A in context for students to identify the words again.

3.4 Extending skills

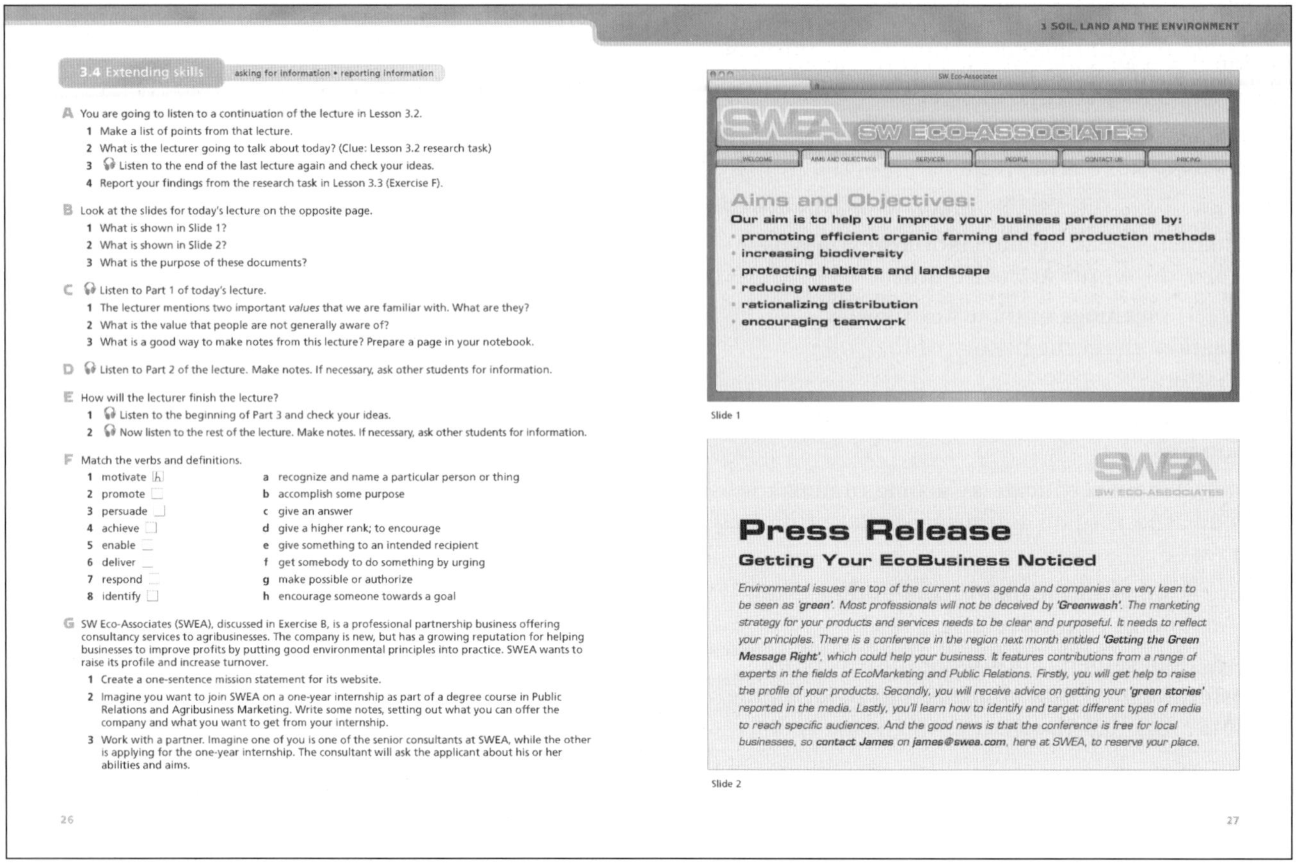

3.4 Extending skills — asking for information • reporting information

A You are going to listen to a continuation of the lecture in Lesson 3.2.

1 Make a list of points from that lecture.
2 What is the lecturer going to talk about today? (Clue: Lesson 3.2 research task)
3 Listen to the end of the last lecture again and check your ideas.
4 Report your findings from the research task in Lesson 3.3 (Exercise F).

B Look at the slides for today's lecture on the opposite page.

1 What is shown in Slide 1?
2 What is shown in Slide 2?
3 What is the purpose of these documents?

C Listen to Part 1 of today's lecture.

1 The lecturer mentions two important *values* that we are familiar with. What are they?
2 What is the value that people are not generally aware of?
3 What is a good way to make notes from this lecture? Prepare a page in your notebook.

D Listen to Part 2 of the lecture. Make notes. If necessary, ask other students for information.

E How will the lecturer finish the lecture?

1 Listen to the beginning of Part 3 and check your ideas.
2 Now listen to the rest of the lecture. Make notes. If necessary, ask other students for information.

F Match the verbs and definitions.

1 motivate h	a recognize and name a particular person or thing
2 promote	b accomplish some purpose
3 persuade	c give an answer
4 achieve	d give a higher rank; to encourage
5 enable	e give something to an intended recipient
6 deliver	f get somebody to do something by urging
7 respond	g make possible or authorize
8 identify	h encourage someone towards a goal

G SW Eco-Associates (SWEA), discussed in Exercise B, is a professional partnership business offering consultancy services to agribusinesses. The company is new, but has a growing reputation for helping businesses to improve profits by putting good environmental principles into practice. SWEA wants to raise its profile and increase turnover.

1 Create a one-sentence mission statement for its website.
2 Imagine you want to join SWEA on a one-year internship as part of a degree course in Public Relations and Agribusiness Marketing. Write some notes, setting out what you can offer the company and what you want to get from your internship.
3 Work with a partner. Imagine one of you is one of the senior consultants at SWEA, while the other is applying for the one-year internship. The consultant will ask the applicant about his or her abilities and aims.

SWEA SW ECO-ASSOCIATES

WELCOME | AIMS AND OBJECTIVES | SERVICES | PEOPLE | CONTACT US | PRICING

Aims and Objectives:

Our aim is to help you improve your business performance by:

- promoting efficient organic farming and food production methods
- increasing biodiversity
- protecting habitats and landscape
- reducing waste
- rationalizing distribution
- encouraging teamwork

Slide 1

SWEA SW ECO-ASSOCIATES

Press Release

Getting Your EcoBusiness Noticed

Environmental issues are top of the current news agenda and companies are very keen to be seen as ***'green'****. Most professionals will not be deceived by* ***'Greenwash'****. The marketing strategy for your products and services needs to be clear and purposeful. It needs to reflect your principles. There is a conference in the region next month entitled* ***'Getting the Green Message Right'****, which could help your business. It features contributions from a range of experts in the fields of EcoMarketing and Public Relations. Firstly, you will get help to raise the profile of your products. Secondly, you will receive advice on getting your* ***'green stories'*** *reported in the media. Lastly, you'll learn how to identify and target different types of media to reach specific audiences. And the good news is that the conference is free for local businesses, so* ***contact James*** *on* ***james@swea.com****, here at SWEA, to reserve your place.*

Slide 2

Lesson aims

- ask other people for information

Further practice in:

- choosing the best form of notes
- making notes
- reporting information

Introduction

1 Elicit as much information from the lecture in Lesson 3.2 as possible. If necessary, prompt students by reading parts of the transcript and pausing for students to complete in their own words.

2 Remind students of the language involved in asking for information from other people – see *Skills bank*. Drill some of the sentences if you wish.

Exercise A

1/2 Set for pairwork. Encourage students to ask each other for information.

3 Play Part 4 of the lecture from Lesson 3.2 to enable students to check their answers. Feed back.

4 Elicit information from the students' research task (Lesson 3.3). Do not confirm or correct at this stage except pronunciation mistakes on key words.

Answers

Model answers:

1 Food security means something different in developed and developing countries. Self-sufficiency and food security aren't the same; there needs to be a balance.

Three factors to consider to ensure food security are access, availability and affordability. Food security is a global issue.

There are two major issues: the impact of climate change and the impact of trading and distribution systems.

2 Organic principles and soil quality.

Transcript 1.14

Part 4

OK, so there is growing consensus that food security means different things depending on where you live, but food security *is* a global challenge. There is also a growing consensus that there are two issues we all face which are critical for a secure food future. The first concerns the impact of climate change on global food production, and the question is: What can we do to reduce this? The second issue is about the way our global community currently favours the interests of developed countries at the expense of developing ones. And the issue here is how we move towards a fairer global economy so that countries in the developing world receive a fair price for their agricultural products and labour so they can invest in modernizing their agriculture and agribusiness sector.

Right. That's it for today. Next time I want to look at why *organic* principles are important for food security. And, in particular, I want to look at the issue of soil quality and how it relates to the topics we have discussed today. So don't forget to do some research on this topic before you come.

Exercise B

Refer students to the lecture slides. Set for pairwork discussion. Feed back.

Answers

Model answers:

1 SWEA aims and objectives page from the company website.

2 A press release giving information about a regional green marketing conference being organized by SWEA.

3 Answer depends on the students. Suggestions could be 1) the website gives information to prospective clients about the company to encourage them to contact and use SWEA services, and 2) to provide different media with information about an upcoming event and through them, engage potential attendees in the business community.

Exercise C

Set for individual work, then pair or group discussion. Play Part 1 of the lecture.

Methodology note

Don't tell students words they can't remember. It would be quite normal for students not to write down every keyword in a lecture. If they don't remember them all this time, they should at least put the key words they remember in order. They can then listen for the other key words as the text develops.

Answers

Model answers:

1 The value of soil quality in assuring the long-term sustainability of an agricultural system, and in ensuring a supply of healthy food.

2 The benefits of soil quality in reducing carbon emissions and combating global warming.

3 The values could be headings, with bullet points below each heading.

The organizations and their aims/services, etc. could be a spidergram.

The statistics could be a diagram showing the increase in organic land use against reduction in carbon emissions.

(See answers to Exercises D and E in this lesson.)

Transcript 1.17

Part 1

In the last lecture, we talked about food security. Today, I'd like to look at what we mean by 'organic principles' and 'soil quality' and how these relate to food security. Most people understand the value of soil quality in assuring the long-term sustainability of an agricultural system. Most people also understand its value in ensuring a healthy supply of food for the population. Perhaps a lot of people are *not* aware of the value of soil quality in reducing carbon emissions and combating global warming.

Exercise D

Play Part 2 of the lecture. Students should recognize the rhetorical structure – see Answers section below – and complete in effect a spidergram and a table. When students have done their best individually, put them in pairs or small groups to complete their notes by asking for information from the other student(s).

In the feedback, allow the correct meaning, not just these words.

Answers

Model answers:

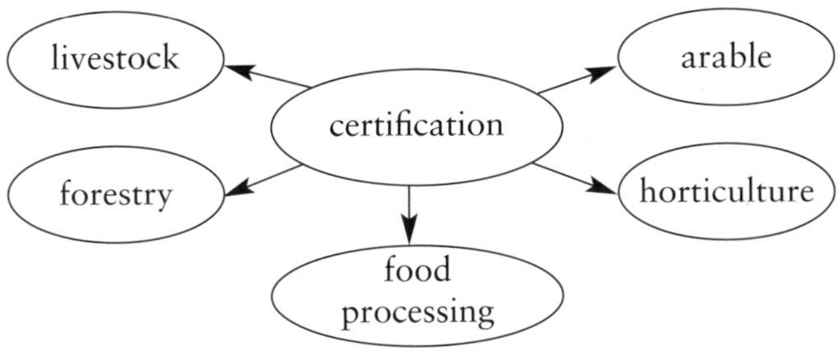

Heading	Categories	Principles
livestock	beef cattle dairy cattle sheep pigs poultry (eggs) poultry (meat)	• suppliers to the agribusiness must be certified • you must display labels showing where your organic products were farmed • you must ensure that at least 95% of the produce ingredients are organic to label a produce as 'organic'

Transcript 1.18

Part 2

Soil quality is closely linked to food production. It is fundamental to a successful agribusiness sector. UN research indicates that more than 3 billion acres of agricultural land – an area larger than China and India put together – has been damaged by human actions in the last 60 years. These include deforestation, creation of industrial-scale farms involving the removal of hedgerows and the overuse of agrochemicals, with the attendant loss of the insects and other organisms that help maintain soil quality. And the average soil erosion rates – that is, the amount of soil lost annually – in Europe are estimated to exceed the average rate of soil formation, the amount of soil 'put back'. In other words, we are losing more soil than we are able to create.

The UK government wants to stop this process, and encourage agribusinesses to work to protect and improve soil quality. One of their strategies is to get agribusinesses to sell organic products and to get consumers to buy them. And to avoid people cheating, that means we need products that are *certified* as organic. What does certification mean in practice, then? A certified product has to be made from raw materials or ingredients grown on organic soil using only organic farming methods. I'll say that again, a certified product must be made from materials or ingredients grown on organic soil and using organic farming methods only. To ensure the process of certification, the government works with a number of organic certification bodies, or 'CBs'. Certification covers five agribusiness sectors. These sectors are livestock, arable, horticulture, food processing and forestry. If we take livestock as an example, then certification includes primary sector agribusinesses in these categories: beef cattle, dairy cattle, pigs, sheep, and poultry for meat and eggs. So, if you are an agribusiness that makes *organic* products that use eggs, you must make sure your egg supplier is certified. That means that the chickens that lay your eggs are reared on organic land and eat only organic food. They can't eat chicken food made from non-organic cereal, for example. They must have labels showing where they came from if they're organic. And you must ensure that at least 95 per cent of your finished product ingredients are organic. Then you can sell your product as 'organic'.

OK, that's one part of a strategy for sustainability. Now, before I finish, I mentioned that soil has a very important role to play in two other ways.

Exercise E

Ask the initial question and elicit ideas, but do not confirm or correct at this stage.

1 Play the first two sentences of Part 3. Feed back.

2 Play the rest of Part 3. Give students time to do their own work, then set for pair or group completion.

Answers

Model answers:

1 Talking about the importance of soil quality and giving some statistics.

2 Describing recent research showing the potential benefits of an organic approach to farming in improving food quality and reducing greenhouse gas emissions – the amount of carbon released into the atmosphere.

Transcript 1.19

Part 3

Yes, soil really has a very important part to play, not only in providing the medium for growing good quality, healthy agriproducts, but also in helping to reduce global warming. The key point is that good organic soil *captures* carbon much more efficiently than soil that has been farmed non-organically. And I want to give you a couple of statistics which I think make the point.

Recent research indicates that if all UK farmland was farmed used organic methods, then something like 3 million more tonnes of carbon would be absorbed by the soil each year. That is the equivalent to taking 1 million cars off the road in the UK. The other statistic I want to quote is that globally, if we switch to organic farming methods, we'll not only improve the quality of the food we produce, but we'll cut more than 10 per cent of our global greenhouse gas emissions.

Exercise F

Set for pairwork. Monitor and assist. Feed back, writing the words on the board as the students correctly identify them. Check pronunciation and stress patterns.

Answers

1	motivate	h encourage someone towards a goal
2	promote	d give a higher rank; to encourage
3	persuade	f get somebody to do something by urging them
4	achieve	b accomplish some purpose
5	enable	g make possible or authorize
6	deliver	e give something to an intended recipient
7	respond	c give an answer
8	identify	a recognize and name a particular person or thing

Exercise G

Refer students to Slide 1, which provides a model mission statement.

1 Set for individual work or pairwork. Students should write two or three sentences beginning *To ...* . They should use their own ideas. Feed back. Accept any reasonable suggestions. Write a few on the board.

2 Set for pairwork. Get one student from each pair to think about what they have to *offer* and the other to think about what they want to *get* from the internship. 'Offer' might include new thinking, enthusiasm, creativity, energy. 'Get' might include experience dealing with real situations, an improved CV, contacts, a reference.

3 Put students in different pairs to do the role-play. Remind students of the ways of asking politely for information (refer to *Skills bank* if necessary).

Answers

Answers depend on the students.

Methodology note

End all listening lessons by referring students to the transcript at the back of the Course Book, so they can read the text while the aural memory is still clear. You could set this as standard homework after a listening lesson. You can also get students to highlight key sections and underline key sentences.

Closure

Ask students to list ways that they can use their power as consumers of food and agricultural products to:

a improve the balance between the economies of developed and developing countries;

and

b reduce greenhouse gas emissions and agriculture's potential for causing global warming.

Extra activities

1 Work through the *Vocabulary bank* and *Skills bank* if you have not already done so, or as a revision of previous study.

2 Use the *Activity bank* (Teacher's Book additional resources section, Resource 3A).

A Set the crossword for individual work (including homework) or pairwork.

Answers

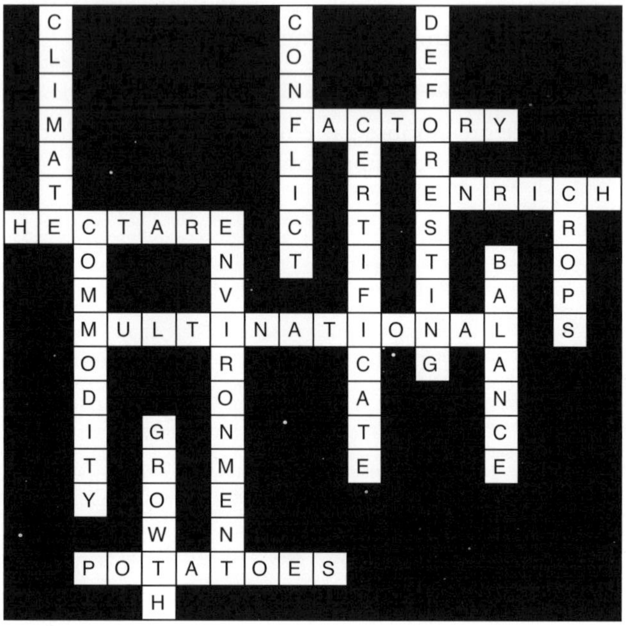

B This game practises pronunciation and meaning recognition. It can only be played in groups in class.

Students must think of one word for each of the categories on the bingo card. Allow them to use any of the vocabulary from this unit. They should write their words on card 1, or copy the bingo grid into their notebooks.

Each student says one of their own words at random once only, concentrating on the pronunciation. The others must identify the category and cross it out on card 2.

The winner is the first student to identify the correct category for all the words. If the teacher keeps a record of which words have been said, he/she can say when a successful card could have been completed.

3 Students can play this alphabet game by themselves or as a group/class. The aim is to think of a word related to agribusiness for each letter of the alphabet. For example:

Student A: agriculture

Student B: agriculture, biodiversity

Student C: agriculture, biodiversity, conservation

Each student adds something for the next letter of the alphabet. They should try to use words from the unit if possible. A student misses a turn if he/she can't remember the items, or add another letter.

4 Tell students to do some Internet research on newspaper coverage of the following topics in the UK and in their own country (how much coverage, positive or negative):

- fair-trade products
- animal rights
- genetic modification

Useful websites (at the time of writing):

www.guardian.co.uk;

www.telegraph.co.uk;

www.foe.co.uk

Note that a lot of the information will be in very complex English, but students should be able to record the basic details and report back in the next lesson.

4 TECHNOLOGY AND AGRIBUSINESS

In this unit, one type of external influence on agribusiness is covered: technology. Lesson 4.2 looks at how technology impacts on agribusiness, from the technological advancements of the Industrial Revolution to the current revolution in information technology. The use of computers in education is also covered: Lessons 4.1, 4.3 and 4.4 guide students to a more efficient use of the Internet and computers in research.

Note that students will need access to a computer with an Internet connection for some exercises in this unit.

Skills focus

Reading

- identifying topic development within a paragraph
- using the Internet effectively
- evaluating Internet search results

Writing

- reporting research findings

Vocabulary focus

- computer jargon
- abbreviations and acronyms
- discourse and stance markers
- verb and noun suffixes

Key vocabulary

access (n and v)	index	register
browse	intranet	search (n and v)
data	keyword	search engine
database	log in/log on	software
default	log off	technology
document	login (n)	username/ID
electronic	media	web page
exit (v)	menu	
hyperlink	password	

Abbreviations and acronyms

The Computer Jargon Buster on page 31 of the Course Book gives the meanings of most of these.

CAL	PDF	USB
HTML	PIN	WAN
HTTP	PPT	WWW
ISP	ROM	
LCD	URL	

4.1 Vocabulary

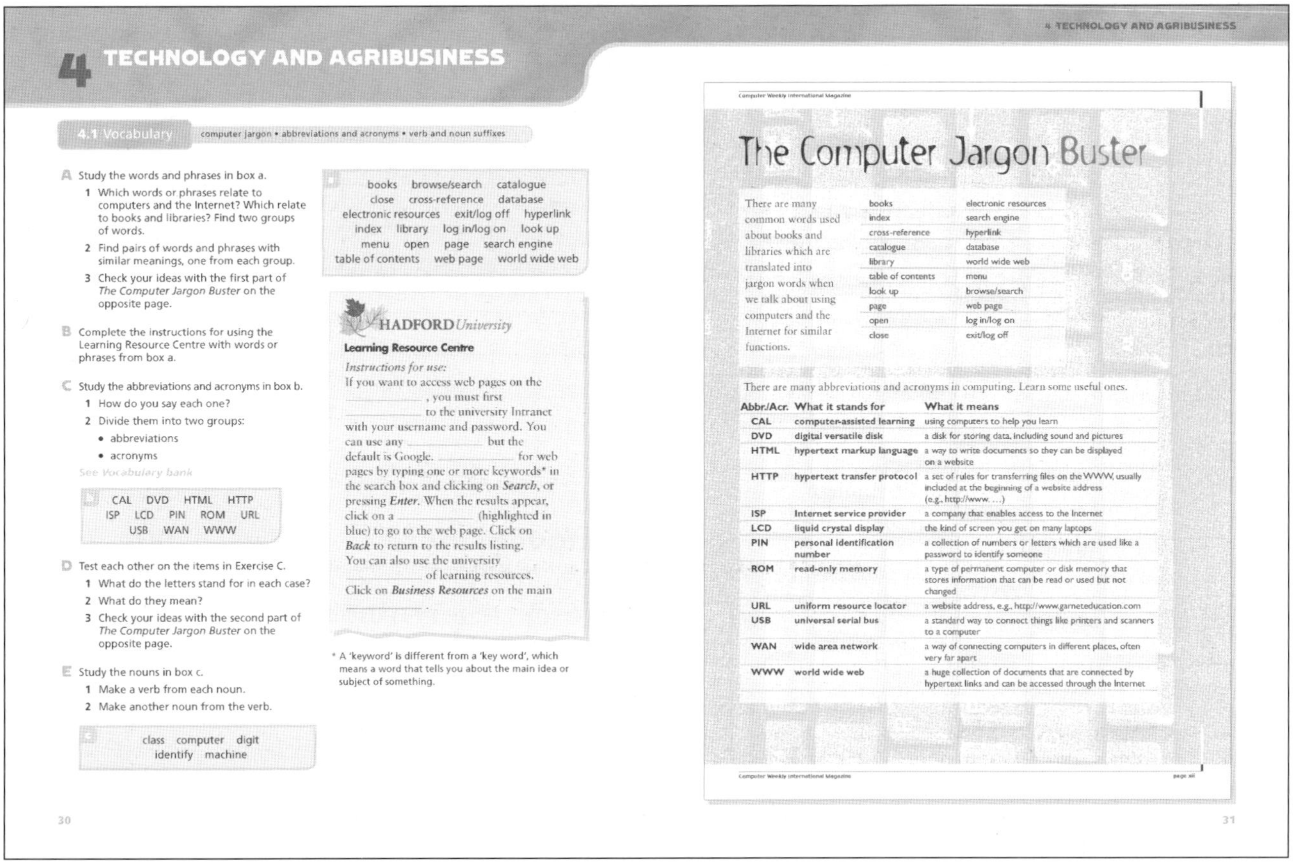

4 TECHNOLOGY AND AGRIBUSINESS

4.1 Vocabulary — computer jargon • abbreviations and acronyms • verb and noun suffixes

A Study the words and phrases in box a.

1 Which words or phrases relate to computers and the Internet? Which relate to books and libraries? Find two groups of words.
2 Find pairs of words and phrases with similar meanings, one from each group.
3 Check your ideas with the first part of *The Computer Jargon Buster* on the opposite page.

B Complete the instructions for using the Learning Resource Centre with words or phrases from box a.

C Study the abbreviations and acronyms in box b.

1 How do you say each one?
2 Divide them into two groups:
   - abbreviations
   - acronyms

See *Vocabulary bank*

b: CAL DVD HTML HTTP ISP LCD PIN ROM URL USB WAN WWW

D Test each other on the items in Exercise C.

1 What do the letters stand for in each case?
2 What do they mean?
3 Check your ideas with the second part of *The Computer Jargon Buster* on the opposite page.

E Study the nouns in box c.

1 Make a verb from each noun.
2 Make another noun from the verb.

c: class computer digit identify machine

a: books browse/search catalogue close cross-reference database electronic resources exit/log off hyperlink index library log in/log on look up menu open page search engine table of contents web page world wide web

HADFORD University
Learning Resource Centre
Instructions for use:
If you want to access web pages on the ______, you must first ______ to the university Intranet with your username and password. You can use any ______ but the default is Google. ______ for web pages by typing one or more keywords* in the search box and clicking on *Search*, or pressing *Enter*. When the results appear, click on a ______ (highlighted in blue) to go to the web page. Click on *Back* to return to the results listing. You can also use the university ______ of learning resources. Click on *Business Resources* on the main ______.

* A 'keyword' is different from a 'key word', which means a word that tells you about the main idea or subject of something.

30

4 TECHNOLOGY AND AGRIBUSINESS

Computer Weekly International Magazine

The Computer Jargon Buster

There are many common words used about books and libraries which are translated into jargon words when we talk about using computers and the Internet for similar functions.

books	electronic resources
index	search engine
cross-reference	hyperlink
catalogue	database
library	world wide web
table of contents	menu
look up	browse/search
page	web page
open	log in/log on
close	exit/log off

There are many abbreviations and acronyms in computing. Learn some useful ones.

Abbr./Acr.	What it stands for	What it means
CAL	computer-assisted learning	using computers to help you learn
DVD	digital versatile disk	a disk for storing data, including sound and pictures
HTML	hypertext markup language	a way to write documents so they can be displayed on a website
HTTP	hypertext transfer protocol	a set of rules for transferring files on the WWW, usually included at the beginning of a website address (e.g., http://www. …)
ISP	Internet service provider	a company that enables access to the Internet
LCD	liquid crystal display	the kind of screen you get on many laptops
PIN	personal identification number	a collection of numbers or letters which are used like a password to identify someone
ROM	read-only memory	a type of permanent computer or disk memory that stores information that can be read or used but not changed
URL	uniform resource locator	a website address, e.g., http://www.garneteducation.com
USB	universal serial bus	a standard way to connect things like printers and scanners to a computer
WAN	wide area network	a way of connecting computers in different places, often very far apart
WWW	world wide web	a huge collection of documents that are connected by hypertext links and can be accessed through the Internet

Computer Weekly International Magazine — page xii

31

General note

If possible, hold this lesson in a room where there is a computer, or bring in a computer.

Read the *Vocabulary bank* at the end of the Course Book unit. Decide when, if at all, to refer your students to it. The best point is probably Exercise C, or at the very end of this lesson or the beginning of the next lesson, as a summary/revision.

Lesson aims

- gain fluency in the meaning, pronunciation and spelling of key computing terms, abbreviations and acronyms
- understand how verbs can be formed from nouns, and nouns from verbs, through the addition of suffixes

Introduction

Familiarize students with computer terminology using some or all of the following activities.

1 Using a computer or a picture of a computer as a starting point, elicit some or all of the following:

PC	*CPU (central processing unit)*	*USB port*
laptop	*hard disk*	*accessory*
monitor	*program*	*printer*
screen	*database*	*scanner*
desktop	*slot*	*CD burner*
icon	*CD*	*Internet*
keyboard	*DVD*	*e-mail*
keys		*the web*
mouse		

2 Ask students to suggest verbs used in computing. Elicit some or all of the following. A good way to do this is to open a program such as Word (in English) and look at the words and symbols on the toolbars.

switch on	*press*	*paste*
start up	*open*	*enter*
shut down	*close*	*delete*
log on/log off	*exit*	*insert*
click	*save*	*highlight*
double-click	*select*	*undo*
hold	*copy*	

3 Ask students whether they normally use the library or the Internet to find information. Elicit the advantages and disadvantages of both. (There is so much emphasis on using computers nowadays, students often forget that there is a lot of information readily to hand in the library.)

Answers

Possible answers:

Library

Advantages	Disadvantages
easy to look things up in a dictionary or an encyclopedia	books can be out of date
you can find information in your own language	the book may not be in the library when you want it
information is usually correct	most books can't be accessed from home (though this is now starting to change)

Internet

Advantages	Disadvantages
a lot of information from different sources	difficult to find the right keywords
information is usually more up-to-date than books	information is often not correct
can be accessed from home	difficult to know which results are the best
you can quickly and easily get copies of books or journal articles not in your library	you may have to pay for the books/articles/information

Exercise A

Ask students to study the words in box a and elicit that they all relate to research.

Set for pairwork. Tell students to decide *and justify* the pairs they choose. If necessary, give an example: *index, search engines*.

To help students understand what a database is, refer to ones they are familiar with in your college, e.g., student records, exam results, library catalogues, etc.

Students may argue that some terms are not exact equivalents, e.g., *catalogue/database*. Discuss any objections as they arise.

Answers

Model answers:

Common word or phrase for books and libraries	Word or phrase for Internet and electronic information
books	electronic resources
index	search engine
cross-reference	hyperlink
catalogue	database
library	world wide web
table of contents	menu
look up	browse/search
page	web page
open	log in/log on
close	exit/log off

Language note

Log in and *log on*: these two verbs are used a little differently. *Log in* is used when accessing a closed system such as a college Intranet. *Log on* is used for open systems such as the Internet in general, as in *You can log on to the Internet with a hand-held computer*. Note also that the related noun has now become one word (*login*). The opposite of *log in* is *log out*, while the opposite of *log on* is *log off*.

Exercise B

Set for individual work and pairwork checking. Ensure that students read *all* the text and have a general understanding of it before they insert the missing words.

Feed back by reading the paragraph or by using an OHT or other visual display of the text. Discuss alternative ideas and decide whether they are acceptable. Verify whether errors are due to using new words or to misunderstanding the text.

Answers

If you want to access web pages on the world wide web, you must first log in to the university Intranet with your username and password. You can use any search engine but the default is Google. Browse/Search for web pages by typing one or more keywords in the search box and clicking on ***Search***, or pressing ***Enter***. When the results appear, click on a hyperlink (highlighted in blue) to go to the web page. Click on ***Back*** to return to the results listing.

You can also use the university database of learning resources. Click on ***Business Resources*** on the main menu.

Exercise C

Set for pairwork. Feed back, eliciting ideas on pronunciation and confirming or correcting. Build up the two lists on the board. Establish that one group are acronyms, i.e., they can be pronounced as words: PIN = /pɪn/. The other group are abbreviations, i.e., they are pronounced as letters: HTTP = H-T-T-P. Drill all the abbreviations and acronyms. Make sure students can say letter names and vowel sounds correctly.

Elicit that words with normal consonant/vowel patterns are *normally* pronounced as a word and those with unusual patterns are *normally* pronounced with single letters. Refer to the *Vocabulary bank* at this stage if you wish.

Methodology note

Don't discuss the meaning at this point. This is covered in the next activity.

Answers

Acronyms: CAL /kæl/, PIN /pɪn/, ROM /rɒm/, WAN /wæn/.

Abbreviations: DVD, HTML, HTTP, ISP, LCD, URL (not pronounced /ɜːl/), USB, WWW.

Exercise D

1 Introduce the verb *stand for*. Elicit examples of common abbreviations and ask what they stand for. Set for pairwork. Tell students to pick out the ones they already know first. Next, they pick out the ones they are familiar with but don't know what they stand for – and guess.

2 Elicit the meanings without reference to *The Computer Jargon Buster* if possible.

3 Refer students to the *The Computer Jargon Buster* to verify their answers. As a follow-up, elicit other common abbreviations from IT or agribusiness.

Language note

If students don't use acronyms or initial abbreviations in their language, a discussion about the reasons for using them is useful. They will then know how to find the meaning of new ones when they meet them. You might point out that abbreviations can sometimes be longer than the thing they abbreviate! For example, world wide web is three syllables, whereas WWW is nine. It evolved because it is quicker to write, but it is longer, and harder, to say. Note that WWW is usually written in lower case letters (www), presumably because it is lower case in URLs.

It is also possible to mix acronyms with abbreviations: for example, JPEG – J /peg/. Point out the field of ICT is developing at an incredible speed and new acronyms and abbreviations are constantly being created.

Exercise E

Set for individual work and pairwork checking. Feed back, highlighting the changes from noun form to verb in the case of *identity/identify* and *machine/mechanize*.

Answers

Model answers:

Noun 1	Verb	Noun 2
class	classify	classification
computer	computerize	computerization
digit	digitize	digitization
identity	identify	identification
machine	mechanize	mechanization

Language note

Both *~ise/~ize* (*~isation/~ization*) forms are acceptable in British English. American English usage is *~ize* (*~ization*).

Closure

Ask students whether they agree with the following statements.

1 Every college student must have a computer.

2 The college library uses a computer to help students find information.

3 College departments use computers to store research data.

4 Students can't do research without a computer.

5 College computers can access research data from other colleges and universities.

6 College computers can access research data from businesses and the media.

7 A personal computer can store information students think is important.

8 Computers have generally improved the quality of education.

9 Computers have made learning more interesting and enjoyable.

4.2 Reading

4.2 Reading preparation for reading research • topic development

A Discuss these questions.

1. What has been the most important technological development for agribusiness in the last 50 years?
2. How can computers help to reduce agribusiness costs?

B Look at the title of the text on the opposite page.

1. What will the text be about?
2. In what ways can new technology cause problems at work? Make a list.
3. Write some questions that you would like the text to answer.

C Work in pairs. Look at pictures 1, 2 and 3.

1. Choose a picture. Describe it. Can your partner guess which one it is?
2. What do the pictures show?

D One student wrote some ideas about new technology before reading the text on the opposite page. Write **A** (I agree), **D** (I disagree) or **?** (I'm not sure) next to the ideas about opportunities and threats on the right.

E Look carefully at the topic sentences in the text on the opposite page.

1. Identify the topic and the comment about the topic. See *Skills bank*
2. What do you think each paragraph will be about?

F Read the text carefully. Were your questions from Exercise B answered?

G Does the writer of the text agree or disagree with the ideas in Exercise D? Which ideas are not mentioned?

H Study the notes a student made in the margin of the text on the opposite page.

1. What ideas are in the other paragraphs? Write some key words.
2. Which words introduce new ideas in each paragraph? See *Skills bank*

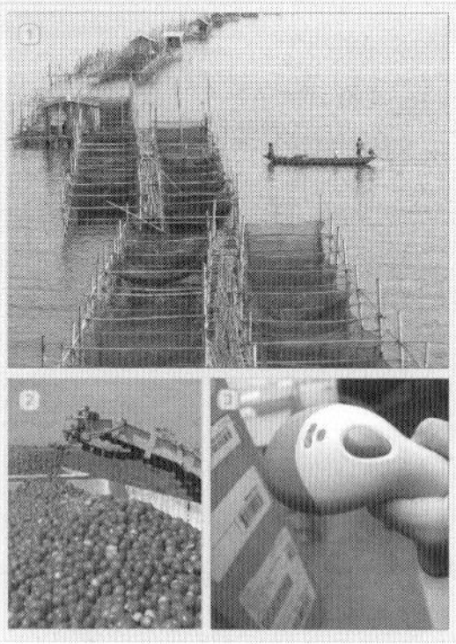

Opportunities/threats

New technology changes society. ____

New technology means people lose their jobs. ____

New technology makes life better for everyone. ____

Everyone likes new technology. ____

Information technology means more, not better, information. ____

Developments in information technology are taking place too fast. ____

Agribusinesses which don't keep up with the latest developments will fail. ____

In agribusinesses, people are more important than technology. ____

The new revolution?

example / *result of tech.*

Although we live in a time of rapid technological change, new technologies have been having an impact on agribusinesses and agriculture for centuries. Technology has been a key aspect of the industry from the time of the Industrial Revolution onwards. For example, in the 18th century, the United Kingdom was responsible for inventions that contributed significantly to the industrialization of farming in Europe. New technologies gave rise to farm machines that increased productivity. This in turn changed agriculture and society, enabling countries to meet the needs of growing populations.

New technologies have also always brought with them both opportunities and threats. During the Industrial Revolution in England, farmers and land owners who recognized these new opportunities often became rich. The lives of some of their workers also improved. Unfortunately, change can sometimes bring negative consequences as well. The increased industrialization of farming has tended to lead to small farms being forced out of business. Those that remain often struggle to earn enough to live on.

change / *problems + benefits*

Another example of technology impacting on agribusiness is industrial salmon farming. Stocks of wild fish have been significantly reduced by overfishing. Some entrepreneurs have seen the potential of farming salmon. They have used new technologies to keep and feed salmon and control pests and diseases in fish farms. But this has had an impact on traditional salmon fisheries and other stakeholders. Clearly, those opposing the development of alternative methods have good reasons to fear the loss of traditional jobs. But there is a positive side to the debate. The new technologies and methods of farming have evolved. They have also reduced the cost of buying salmon for millions of consumers. This in turn has increased the demand for salmon, which has created new job opportunities in the fish farming sector.

The revolution of the 20th and 21st centuries has been in information technology. Wine producers have recently devoted a lot of attention to organic and biodynamic viticulture, and have been using information technology to help them grow crops in a more environmentally friendly way. This involves using technology to reduce the need for pesticides and weed killers. Weather stations in the vineyard feed data back to a computer that predicts when the risk of pests or diseases is highest. Chemicals are used only when required. In addition, growers complement these interventions by introducing wasps and ladybirds to eliminate pests.

The use of computers is widespread among agribusinesses. Food producers commonly use bar codes to provide information about where the product originated. Agribusinesses use the Internet to market their produce and a growing number of people buy food products online. Online sales are having a major impact on the retail sector. One benefit is that small specialist producers can build market share. This, in turn, is encouraging diversity and traditional crafts. New communication technologies aren't all good news, of course. Companies have to invest to maintain competitive IT systems and in people who have the skills to use them. You can, of course, have the cleverest computers in the world, but if you don't have someone with skills adapted to new ways of using them, the business will struggle to benefit. Charles Darwin said that species that survive are usually not the smartest or the strongest, but the ones most responsive to change. He would probably agree that this applies to agribusinesses as well.

General note

Read the *Skills bank – Developing ideas in a paragraph* at the end of the Course Book unit. Decide when, if at all, to refer students to it. The best time is probably Exercise E, or at the very end of the lesson or the beginning of the next lesson, as a summary/revision.

Lesson aims

- prepare to read a text by looking at title, topic sentences
- understand the purpose of discourse markers and stance markers in the development of a topic

Introduction

Ask students how, where and why they use computers. They should answer in some detail with examples. Encourage them to use the vocabulary, abbreviations and acronyms from Lesson 4.1.

Exercise A

Set for general discussion. Allow students to debate differences of opinion. Encourage them to give examples if they can.

Answers

Possible answers:

1. Probably information technology, i.e., the use of computers in all aspects of agribusinesses and services. For example: keeping records, controlling production and distribution processes, assisting in design and planning, monitoring quality and cash flow, storing data, communicating with customers and the workforce.
2. Business costs include labour, time and materials. A computer can reduce the number of workers needed. It can make the production process quicker and more efficient; it can reduce wastage of materials.

Exercise B

1 Write the title of the reading text on the board. Discuss with the whole class, eliciting suggestions.

2 Set for pairwork and whole class feedback. Tell students to think of four or five possible problem areas. Do not confirm or correct at this point.

3 Set for pairwork. Tell students to think of four or five questions with different *Wh~* question words:

What …?

Where …?

When …?

Why …?

How …?

If you want, you could choose some questions to write on one side of the board.

Answers

Possible answers:

1 If an agribusiness does not take advantage of new technology, it will lose competitive edge and it will not survive.

2 Possible problems associated with the introduction of new technology:

- cost: it may be expensive to introduce
- labour relations: people may resist new technology if it threatens their jobs
- skills: people may need to learn new skills
- reliability: the new technology may not work well
- motivation: staff may not like the new machines
- managing the introduction of the technology: it may be difficult to decide when to introduce it because of the speed of new developments
- unemployment: this may be the result of increased automation

3 Possible questions:

Why does new technology mean change? etc.

What are the current challenges of new technology?

What were the results of new technology in the past?

Where have there been the most changes caused by new technology?

When has there been the most change resulting from new technology?

Exercise C

1 Set for pairwork.

2 Set for pairwork and whole class feedback. Write the answers on the board.

Answers

Model answers:

1 A fish farm: for the intensive rearing of fish.

2 A tomato wagon for harvesting tomatoes on an industrial scale.

3 A bar code reader for use in, e.g., product traceability.

Exercise D

Set for individual work and pairwork checking. Feed back, trying to get consensus on each point, but do not actually confirm or correct. Preface your remarks with phrases like: *So most of you think … You all agree/disagree that …*

Point out that the statements are very strong. Draw students' attention to words like *everyone*. Point out also that the use of plurals with countable nouns when making a generalizing statement implies *all*, e.g., *agribusinesses* means *all agribusinesses*. The truth may actually be better expressed with a limiting word, e.g., *most/some/many*, or with words which express possibility such as *may* or *seem*, or adverbs such as *sometimes, usually, often*.

Remind students to look back at these predictions while they are reading the text (Exercise F).

Exercise E

Review paragraph structure – i.e., paragraphs usually begin with a topic sentence which makes a statement that is then expanded in the following sentences. Thus, topic sentences give an indication of the contents of the paragraph. You may wish to refer students to the *Skills bank* at this point.

1 Write the topic sentences from the text on an OHT, or use Resource 4B from the additional resources section. Take the first sentence and identify the subject of the main clause with a box. This is the **topic**. For example:

Although we live in a time of rapid technological change, new technologies have been having an impact on agribusinesses and agriculture for centuries.

What is the sentence saying about *new technologies*? The answer is: they *have been having an impact on agribusinesses and agriculture for centuries*. Underline these words. These words constitute the comment which the sentence is making about the topic. Note that the subordinate clause simply provides more information about the topic: it is not the focus of the sentence. The focus of the sentence is on the **comment** being made. Thus a topic sentence consists of both a topic and a comment about the topic which is explained and expanded on in the rest of the paragraph.

Set the remaining sentences for individual work and pairwork checking.

2 Set for pairwork. Tell students that their analysis of the topic sentences may help them. Feed back with the whole class. Point out any language features which led them to draw their conclusions.

Answers

Possible answers:

1

	Topic	Comment
Para 1	… new technologies have been having an impact on agribusiness and agriculture	for centuries.
Para 2	New technologies	have also always brought with them both opportunities and threats.
Para 3	Another example of technology impacting on agribusiness	is industrial salmon farming.
Para 4	The revolution of the 20th and 21st centuries	has been in information technology.
Para 5	The use of computers	is widespread among the agribusinesses.

2

	Predicted content	Notes
Para 1	a historical view of technology and agribusiness	the words *for centuries* point to the idea of a historical perspective
Para 2	new technologies aren't always good news	the use of the present perfect and *always* point to the historical view
Para 3	another example of new technology which brought benefits and problems	the word *another* links this paragraph with the previous one in which opportunities and threats are discussed
Para 4	information technology in the 20th and 21st centuries	this revolution is contrasted with the Industrial Revolution mentioned in the first paragraph
Para 5	the influence of computers	the focus of the sentence is the level of uptake in the agribusiness sector

Exercise F

Set the reading. Tell students to read for good understanding. When everyone has read the text, discuss any vocabulary items that may have caused difficulties.

If you have previously written questions on the board, ask students to say which were answered. Then tell the same pairs as in Exercise B to discuss which of their questions were answered. Feed back with the whole class, asking a few pairs to say which of their questions were answered and which were not.

Exercise G

Set for individual work and pairwork checking. Feed back with the whole class, asking students to say which parts of the text discuss the ideas. Ask students to say how the sentences in Exercise D could be better worded to reflect the ideas in the text.

Answers

Possible answers:

See table on next page.

Exercise H

The purpose of this exercise is for students to try to identify the information structure of each paragraph and to see how a new step in the progression of ideas may be signalled by a rhetorical marker or phrase. Direct students' attention to the handwritten notes in the left margin of the text. Explain that the notes are key words which summarize (in the reader's own words) the ideas in the text. The notes are written next to the relevant parts of the text. A good idea is to make an OHT of the text and use a highlighter to indicate which are the relevant parts of the text. Point out that often (but not always) a new step in the development of ideas is shown by a rhetorical marker or a phrase. The development may be more information: In paragraph 1 we have *for example*. Or the development may be a contrast: In paragraph 2 we have *unfortunately*.

1 Set for pairwork. Tell students to decide on some similar key words for the ideas in the remaining three paragraphs and to write these words next to the appropriate part of the text. Feed back with the whole class, eliciting suitable key words for the left margin and using a highlighter or OHT pen to indicate the phrases or sentences in the text which correspond to the key words.

2 Set for individual work and pairwork checking. Feed back with the whole class, identifying the discourse and stance markers.

Answers

Possible answers for Exercise G:

	Agree/disagree/not mentioned	Possible rewording
New technology changes society.	agree – para 1, last sentence	New technology ***can*** change society.
New technology means people lose their jobs.	agree – para 3, sentence 5 and 6 disagree – para 3, last sentence	New technology ***can*** mean that ***some*** people lose their jobs.
New technology makes life better for everyone.	agree – para 1, last sentence	New technology ***tends to*** make life better for ***more people.***
Everyone likes new technology.	disagree – para 2, last sentence	There are ***usually/often some people*** who dislike new technology.
Information technology means more, not better, information.	not mentioned	
Developments in information technology are taking place too fast.	not mentioned/disagree – the text says only that 'use of computers is widespread'	
Agribusinesses which don't keep up with the latest developments will fail.	agree – para 5, penultimate and last sentences	The message is that agribusinesses that don't respond to change are unlikely to survive.
In agribusinesses, people are more important than technology.	agree – para 5, sentence 9	Technology is important, but only when it is used effectively.

Answers

Possible answers for Exercise H:

1 Para 2: benefits of new tech./problems
  Para 4: tech more accurate/useful data
  Para 5: implications for business/people = still more important
2 Para 1: although, for example
  Para 2: unfortunately
  Para 3: another example, but
  Para 4: in addition

Language note

The relation of one sentence to the previous sentence is not always made explicit by rhetorical markers or phrases. In fact, overuse of markers is to be avoided.

If there is no marker or phrase, the relationship can be deduced usually by the position of the sentence in the paragraph, or by the meaning. For example, in the first and last paragraphs, the second sentences expand on the first sentences.

Closure

1 Divide the class into groups. Write the five topic sentences on strips, or photocopy them from the additional resources section (Resource 4B). Make a copy for each group. Students must put them into the correct order.

Alternatively, divide the class into two teams. One team chooses a topic sentence and reads it aloud. The other team must give the information triggered by that topic sentence. Accept a prediction or the actual paragraph content. However, ask students which it is – prediction or actual.

Language note

There is no universal logic to the structuring of information in a text. The order of information is language-specific. For example, oriental languages tend to have a topic sentence or paragraph summary at the end, not the beginning, of the paragraph. Equally, students whose first language is Arabic might structure a particular type of discourse in a different way from native English speakers. So it is important for students to see what a native speaker writer would consider to be a 'logical' ordering.

2 Refer students back to the sentences in Exercise D. Students should find it easier to comment on these now that they have read the text.

3 Focus on some of the vocabulary from the text, including:

bar codes	*market share*
biodynamic	*interventions*
diversity	*ongoing*
eliminate	*online marketing*
fisheries	*revolution*
generate	*stakeholders*
impact (n)	*viticulture*

4.3 Extending skills

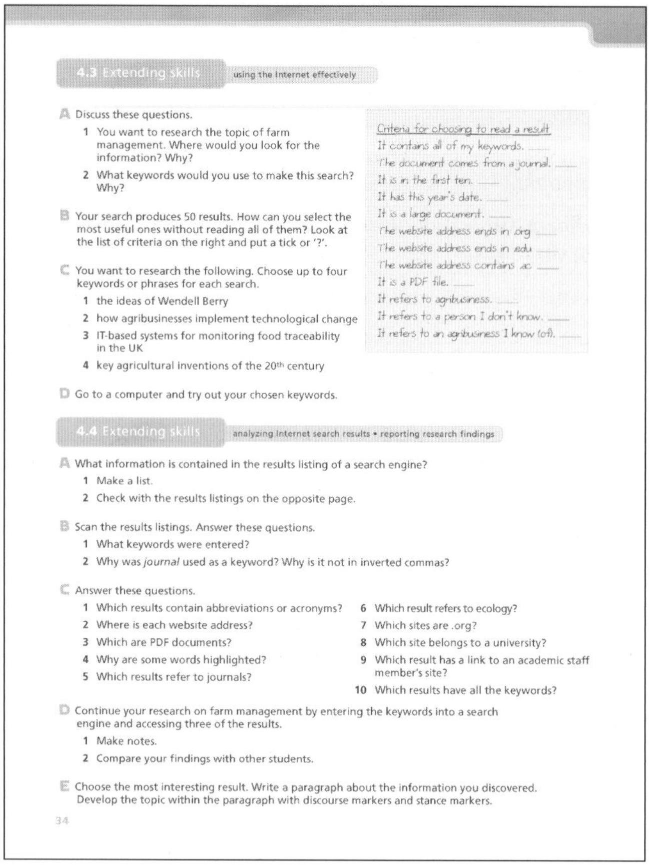

4.3 Extending skills using the Internet effectively

A Discuss these questions.

1 You want to research the topic of farm management. Where would you look for the information? Why?

2 What keywords would you use to make this search? Why?

B Your search produces 50 results. How can you select the most useful ones without reading all of them? Look at the list of criteria on the right and put a tick or '?'.

Criteria for choosing to read a result

It contains all of my keywords. ___
The document comes from a journal. ___
It is in the first ten. ___
It has this year's date. ___
It is a large document. ___
The website address ends in .org ___
The website address ends in .edu ___
The website address contains .ac ___
It is a PDF file. ___
It refers to agribusiness. ___
It refers to a person I don't know. ___
It refers to an agribusiness I know (of). ___

C You want to research the following. Choose up to four keywords or phrases for each search.

1 the ideas of Wendell Berry

2 how agribusinesses implement technological change

3 IT-based systems for monitoring food traceability in the UK

4 key agricultural inventions of the 20th century

D Go to a computer and try out your chosen keywords.

4.4 Extending skills analyzing Internet search results • reporting research findings

A What information is contained in the results listing of a search engine?

1 Make a list.

2 Check with the results listings on the opposite page.

B Scan the results listings. Answer these questions.

1 What keywords were entered?

2 Why was *journal* used as a keyword? Why is it not in inverted commas?

C Answer these questions.

1 Which results contain abbreviations or acronyms?

2 Where is each website address?

3 Which are PDF documents?

4 Why are some words highlighted?

5 Which results refer to journals?

6 Which result refers to ecology?

7 Which sites are .org?

8 Which site belongs to a university?

9 Which result has a link to an academic staff member's site?

10 Which results have all the keywords?

D Continue your research on farm management by entering the keywords into a search engine and accessing three of the results.

1 Make notes.

2 Compare your findings with other students.

E Choose the most interesting result. Write a paragraph about the information you discovered. Develop the topic within the paragraph with discourse markers and stance markers.

34

General note

Students will need access to a computer with an Internet connection. If computers are not available during the lesson, part of the lesson can be set for private study.

Lesson aim

- learn or practise how to use the Internet efficiently for research

Introduction

Brainstorm the uses of the Internet. Then brainstorm what the important factors are when using the Internet. These should include:

- the search engines students use and why. Note that there are now a large number of search engines to suit different purposes. It is not necessarily a good idea to use Google exclusively.
- how to choose *and write* keywords in their preferred search engine
- how they extract the information they want from the results

Put students in groups and ask them to compare how they normally use a computer to find information. Ask each group to produce a set of tips for using the Internet. Then, as a class, produce an accepted set of tips.

Key words to elicit: *search engine, keyword, website, web page, website address, search result, subject directory*

Note: Where the subject is a new one or a fairly general topic, it is a good idea to start first with a **subject directory** which evaluates sites related to the topic and collects them in one place. Some examples are: Academic Info; BUBL LINK; INFOMINE; The WWW Virtual Library.

Exercise A

Write *farm management* on the board.

1 Set for class discussion. Make sure students give reasons for their answers. Accept their answers at this stage.

2 Remind students that words in English often have more than one meaning, so care must be taken to get the desired result.

Answers

Possible answers:

1 In a current journal – very useful, as recent articles give the latest information, and many journals have a specialist/academic focus.

On the Internet – good if the correct keywords are used and a careful selection of results is made. Since it is a general topic, it would benefit from a search with a subject directory such as The WWW Virtual Library on http://vlib.org/.

In a textbook – useful if there is an up-to-date one, but books take time to publish, so even the latest may be out of date in these technologically fast-moving times.

2 In this list of possible keywords, the first is an obvious starting point; others are also possible.

farm management	very specific to the task
journal	specialist/academic; can provide up-to-date information
this year (the year's date) current state nowadays	the time factor

Exercise B

Set for pairwork. Remind students of the research topic.

Feed back, encouraging students to give reasons for their decisions. Emphasize that we only know what *might* be useful at this stage.

Establish that company sites often end in *.com*.

Answers

Model answers:

- ✓ It contains all of my keywords. *(but check that the meaning is the same)*
- ✓ The document comes from a journal. *(current information)*
- ? It is in the first ten. *(a web page can have codes attached to put it high on the list)*
- ✓ It has this year's date. *(current information)*
- ? It is a large document. *(size is no indication of quality)*
- ✓ The website address ends in .org *(because it is a non-profit organization)*
- ✓ The website address ends in .edu *(because it is an educational establishment)*
- ✓ The website address contains .ac *(because it is an educational establishment)*
- ? It is a PDF file. *(file type is no indication of quality)*
- ✓ It refers to agribusiness. *(relevant)*
- ? It refers to a person I don't know. *(may not be reliable)*
- ✓ It refers to an agribusiness I know (of). *(reliable)*

Language note

PDF stands for *portable document format*. PDF documents look exactly like the original documents, and can be viewed and printed on most computers, without the need for each computer to have the same software, fonts, etc. They are created with Adobe Acrobat software.

Exercise C

Set for individual work and pairwork checking. Ask students to compare their choice of keywords with their partner, and justify their choice.

Answers

Possible answers:

The following combinations will produce results provided the words in bold are included.

1. **Wendell Berry** farmer / Wendell Berry writer / Wendell Berry sustainability
2. **technological change agribusiness / managing** technological change in agribusiness / agribusiness **new technology**
3. monitoring **traceability** using **IT systems** in **UK**
4. **20th century agricultural inventors** / most important 20th century agricultural **inventions** / 20th century agricultural **pioneers**

Exercise D

Students should try out different combinations to discover for themselves which gives the best results.

Closure

Tell students to think of their own question for research, as in Exercise C, and find the best web page for the data by entering appropriate keywords.

Ask students to write their question on a piece of paper and sign it. Put all the questions in a box. Students pick out one of the questions at random and go online to find the best page of search results. From those results they can find the most useful web page. They should ask the questioner for verification.

4.4 Extending skills

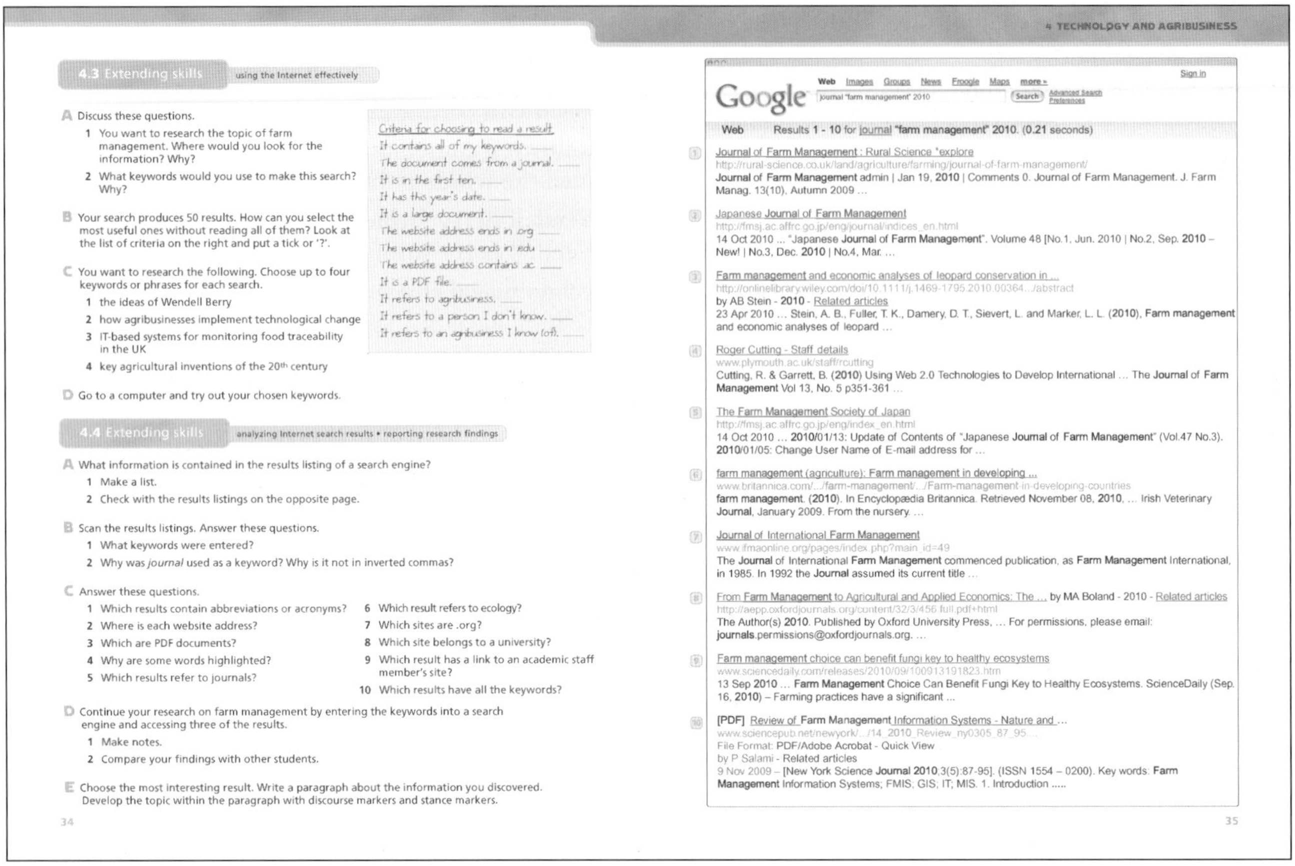
4 TECHNOLOGY AND AGRIBUSINESS

4.3 Extending skills using the Internet effectively

A Discuss these questions.

1 You want to research the topic of farm management. Where would you look for the information? Why?

2 What keywords would you use to make this search? Why?

B Your search produces 50 results. How can you select the most useful ones without reading all of them? Look at the list of criteria on the right and put a tick or '?'.

Criteria for choosing to read a result
It contains all of my keywords. ___
The document comes from a journal. ___
It is in the first ten. ___
It has this year's date. ___
It is a large document. ___
The website address ends in .org ___
The website address ends in .edu ___
The website address contains .ac ___
It is a PDF file. ___
It refers to agribusiness. ___
It refers to a person I don't know. ___
It refers to an agribusiness I know (of). ___

C You want to research the following. Choose up to four keywords or phrases for each search.

1 the ideas of Wendell Berry
2 how agribusinesses implement technological change
3 IT-based systems for monitoring food traceability in the UK
4 key agricultural inventions of the 20th century

D Go to a computer and try out your chosen keywords.

4.4 Extending skills analyzing Internet search results • reporting research findings

A What information is contained in the results listing of a search engine?

1 Make a list.
2 Check with the results listings on the opposite page.

B Scan the results listings. Answer these questions.

1 What keywords were entered?
2 Why was *journal* used as a keyword? Why is it not in inverted commas?

C Answer these questions.

1 Which results contain abbreviations or acronyms?
2 Where is each website address?
3 Which are PDF documents?
4 Why are some words highlighted?
5 Which results refer to journals?
6 Which result refers to ecology?
7 Which sites are .org?
8 Which site belongs to a university?
9 Which result has a link to an academic staff member's site?
10 Which results have all the keywords?

D Continue your research on farm management by entering the keywords into a search engine and accessing three of the results.

1 Make notes.
2 Compare your findings with other students.

E Choose the most interesting result. Write a paragraph about the information you discovered. Develop the topic within the paragraph with discourse markers and stance markers.

34

35

General note

Students will need access to a computer with an Internet connection. If computers are not available during the lesson, part of the lesson can be set for private study.

Lesson aims

- examine a page of Internet search results critically
- report Internet search findings in a short written summary

Introduction

Ask students what problems they had, what lessons they have learnt and what advice they can give from their Internet search experience in Lesson 4.3. Brainstorm the important factors when searching for information on the Internet and put them in order of importance.

Exercise A

Set for pairwork. Students should first make a list of information they expect to find in search engine results. (They should do this before they look at the search engine results on the right-hand page.) They should then look at the page of results and identify any other information that is there.

Answers

Possible answers:

number of results
keywords used
time taken
title of document
type of document
quotations from the text with keywords highlighted
date
web address/URL

Exercise B

This is further reinforcement on key words. Set for pairwork or whole class discussion.

Answers

Model answers:

1 journal, "farm management", 2010
2 Inverted commas are put round a phrase to indicate that it is all a meaningful lexical item. In Google, *journal* does not need them, as it is one word.

Exercise C

This detailed examination of the results page should make students aware of the content, so that they can make an educated selection of a web page with useful information. Set for pairwork.

Make sure in feedback that students are aware of what the following abbreviations stand for: PDF (portable document format), HTML (HyperText Markup Language), ppt (PowerPoint), php (hypertext preprocessor)

Answers

Possible answers:

1 Acronyms/abbreviations:
   Result 1 admin, Manag.
   Result 2 Jun., Sep., Dec., Mar., html, No.
   Result 4 Vol, No.
   Result 5 Vol, No.
   Result 8 org.
   Result 9 Sep.
   Result 10 PDF

Note: Students may identify further abbreviations, e.g., in the website addresses.

2 Immediately below the result heading.
3 10 (portable document format = viewed as a real text page).
4 Blue = titles and viewing information; green = website address; bold black = key words.
5 1, 2, 4, 5, 6, 7, 8 and 10
6 9
7 7 and 8
8 4
9 4
10 1, 2, 4, 5, 6, 8 and 10

Exercise D

1 Set for individual work. Students should input the keywords again. They will not get exactly the same results page as here, but the results should be comparable. Tell them to take notes.
2 Set for pairwork. Feed back, getting students to tell the rest of the class about their most interesting findings. Encourage other students to ask questions.

Exercise E

Set for individual work. Students can complete it in class or for homework.

Closure

1 Focus on some of the vocabulary connected with using the Internet, including:
   website
   web address/URL
   search engine
   search results
   input
   keyword
   key in
   log in/log on
   username
   password
   access
2 The importance of the care needed when selecting keywords can be demonstrated by a simple classroom activity. Tell the class you are thinking of a particular student who you want to stand up. Say (for example):
   It's a man. (all the men stand up and remain standing)
   He has dark hair. (only those with dark hair remain standing)
   He has a beard.
   He has glasses.
   He's tall.
   His name begins with A.
   And so on.

   When only one student remains, ask the class to list the minimum number of keywords necessary to identify only that student. Make sure they discard unnecessary ones. For example, if all students have dark hair, that is unnecessary.
3 Finding the keywords for familiar topics is another activity, done in groups. For example, they could:
   find their own college record (name, ID number or date of entry)
   find their last exam results (name, class, subject, date)
   find a book in the library about methods used in viticulture (viticulture, viticulture methods, etc.)

Extra activities

1 Work through the *Vocabulary bank* and *Skills bank* if you have not already done so, or as a revision of previous study.

2 Use the *Activity bank* (Teacher's Book additional resources section, Resource 4A).

A Set the wordsearch for individual work (including homework) or pairwork.

Answers

A	V	V	E	N	P	R	O	T	E	C	T	Z	P	X	N
H	L	A	C	I	N	F	E	C	T	C	W	Z	C	T	C
P	R	O	M	O	T	E	G	B	W	O	D	A	B	K	O
R	B	I	N	N	O	V	A	T	E	N	E	W	I	C	O
E	S	Q	W	J	W	E	M	I	T	T	F	R	W	C	R
D	E	G	R	A	D	E	Z	M	A	A	O	O	X	O	D
C	P	F	M	A	G	G	H	V	S	M	L	T	T	O	I
O	A	J	I	O	M	R	I	N	E	I	I	A	R	P	N
N	R	I	O	I	T	T	O	C	L	N	A	T	A	E	A
S	A	V	Z	J	L	I	P	W	E	A	T	E	N	R	T
E	T	N	J	U	U	Z	V	X	C	T	E	C	S	A	E
R	E	I	C	C	Z	I	S	A	T	E	K	U	F	T	K
V	Q	G	K	O	P	E	R	A	T	E	K	L	O	E	U
E	F	B	C	K	D	I	S	G	B	E	Z	T	R	I	V
E	Z	T	E	J	B	G	K	P	H	T	O	A	M	Q	K
I	R	R	I	G	A	T	E	V	M	B	A	N	T	Y	Y

Verb	Noun
conserve	conservation
contaminate	contamination
cooperate	cooperation
coordinate	coordination
cultivate	cultivation
defoliate	defoliation
degrade	degradation
emit	emission
grow	growth
infect	infection
innovate	innovation
irrigate	irrigation
motivate	motivation
operate	operation
promote	promotion
protect	protection
rotate	rotation
select	selection
separate	separation
transform	transformation

B Set for pairwork. Teach students how to play noughts and crosses if they don't know – they take it in turns to say the abbreviation or acronym, and what it stands for. If they succeed, they can put their symbol – a nought 0 or a cross **X** – in that box. If a person gets three of their own symbols in a line, they win.

3 Write the acronyms and abbreviations from the unit on cards, or photocopy them from the additional resources section (Resource 4C). Divide the class into two teams. A student selects a card and reads it correctly. (Speed is of the essence.) Alternatively, one team picks a card with an acronym or abbreviation; the other team gives the actual words.

You can follow this up by eliciting other acronyms and abbreviations from the students – in particular, common/useful ones from the field of agribusiness.

4 Have a class debate: 'If an agribusiness does not have the latest technology, it cannot possibly compete successfully.' Ask two students to prepare an opening argument for and against.

Some points: remind students of the points in Lesson 4.2 Exercises A and B2. If new technology seems so great, why doesn't every company use it? Although new businesses may be able to set up with the latest technology, established companies already have facilities in place which may not be compatible. For them, the introduction of new technology can only occur when the budget allows. There is also the question of workers losing their jobs, or lack of people with appropriate skills. It is possible to make gradual changes which may be highly effective without spending large sums of money on the latest equipment.

5 Ask students to work in small groups to research and feed back to the group on some other commercially successful innovations in agriculture and agribusiness. Three research questions could be:

1 *What problem did people face in this area?*
2 *What was the turning point in solving the problem?*
3 *How did the solution change business or life?*

If students are going to do research on the Internet, suggest that they type in *history*, then their topic, to get some potential texts. Alternatively, you can do this research before the lesson and print off some pages for students to work from.
Remind students that they can't possibly read everything they find, so they must use the topic sentences to decide if a paragraph is worth reading.

5 AGRIBUSINESS MARKETING

Marketing is covered in this unit and in Unit 6. In Unit 5, the focus is on the nature of marketing, types of markets, market segmentation and market research. The first listening extract, from a lecture, looks at what marketing is and why it is important, and gives an overview of some key marketing concepts. The second listening extract is from a seminar about market mapping. This leads into work on the four types of market research: primary, secondary, qualitative and quantitative.

Skills focus

Listening

- understanding 'signpost language' in lectures
- using symbols and abbreviations in note-taking

Speaking

- making effective contributions to a seminar

Vocabulary focus

- word sets: synonyms, antonyms, etc.
- the language of trends
- common lecture language

Key vocabulary

Marketing	Market segmentation and target markets	Statistical/market trends	Research methods
advertising	category	decline (n and v)	analyze
brand (n)	gender	decrease (n and v)	data analysis
consumer	manual	drop (n and v)	focus group
mailing	mass market	gradual	interview
make (n)	niche market	growth	market research
market leader	professional	increase (n and v)	primary research
market share	segment (n and v)	outperform	qualitative research
marketing mix (n)	skilled	rise (n and v)	quantitative research
product	socio-economic status	steady (adj)	questionnaire
promotion	unskilled		reliable
satisfy (demand)			response rate
special offer			sample (n and v)
sponsorship			secondary research
strategy			statistical trends
target (n and v)			statistics
			survey (n and v)

5.1 Vocabulary

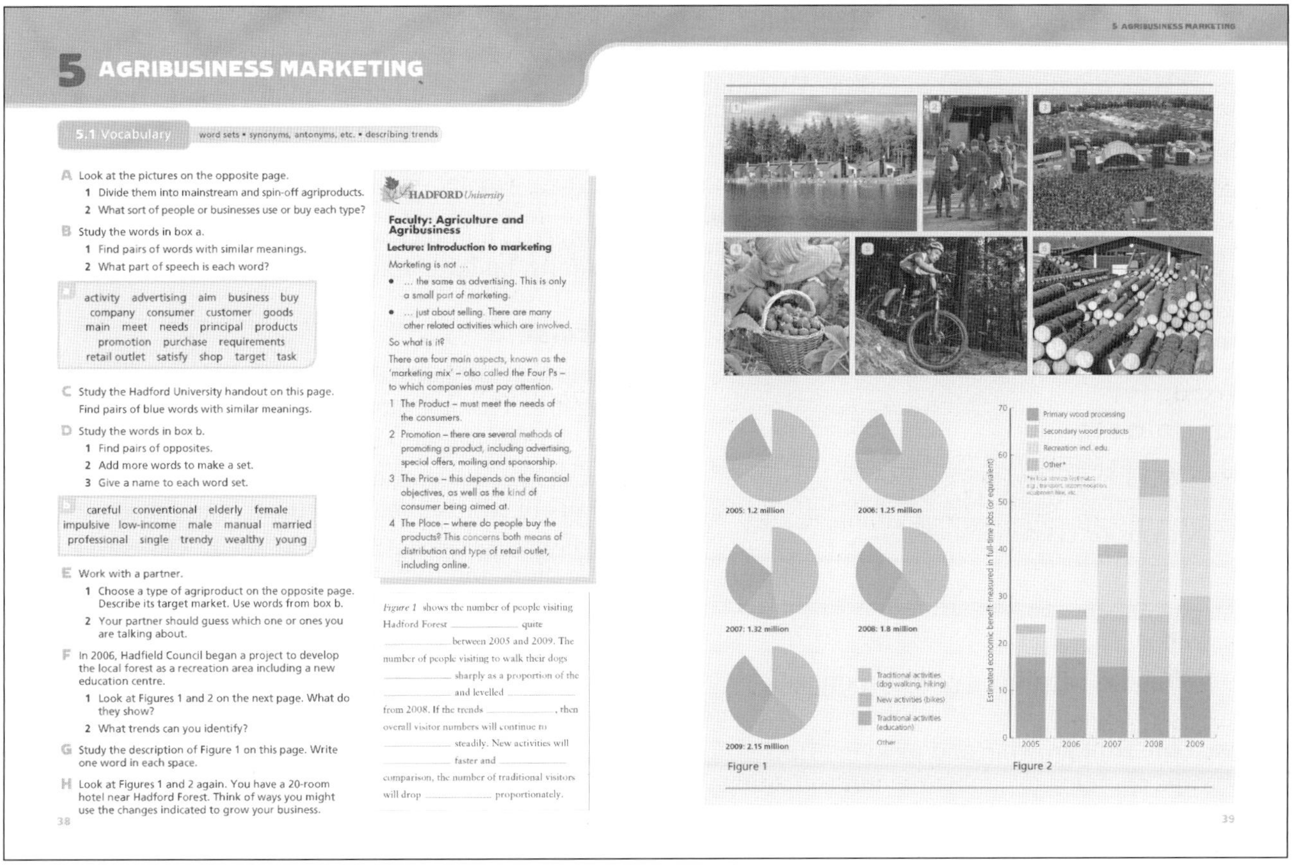

5 AGRIBUSINESS MARKETING

5.1 Vocabulary word sets • synonyms, antonyms, etc. • describing trends

A Look at the pictures on the opposite page.
1 Divide them into mainstream and spin-off agriproducts.
2 What sort of people or businesses use or buy each type?

B Study the words in box a.
1 Find pairs of words with similar meanings.
2 What part of speech is each word?

activity advertising aim business buy company consumer customer goods main meet needs principal products promotion purchase requirements retail outlet satisfy shop target task

C Study the Hadford University handout on this page. Find pairs of blue words with similar meanings.

D Study the words in box b.
1 Find pairs of opposites.
2 Add more words to make a set.
3 Give a name to each word set.

careful conventional elderly female impulsive low-income male manual married professional single trendy wealthy young

E Work with a partner.
1 Choose a type of agriproduct on the opposite page. Describe its target market. Use words from box b.
2 Your partner should guess which one or ones you are talking about.

F In 2006, Hadfield Council began a project to develop the local forest as a recreation area including a new education centre.
1 Look at Figures 1 and 2 on the next page. What do they show?
2 What trends can you identify?

G Study the description of Figure 1 on this page. Write one word in each space.

H Look at Figures 1 and 2 again. You have a 20-room hotel near Hadford Forest. Think of ways you might use the changes indicated to grow your business.

38

HADFORD University

Faculty: Agriculture and Agribusiness

Lecture: Introduction to marketing

Marketing is not ...

- ... the same as advertising. This is only a small part of marketing.
- ... just about selling. There are many other related activities which are involved.

So what is it?

There are four main aspects, known as the 'marketing mix' – also called the Four Ps – to which companies must pay attention.

1 The Product – must meet the needs of the consumers.
2 Promotion – there are several methods of promoting a product, including advertising, special offers, mailing and sponsorship.
3 The Price – this depends on the financial objectives, as well as the kind of consumer being aimed at.
4 The Place – where do people buy the products? This concerns both means of distribution and type of retail outlet, including online.

Figure 1 shows the number of people visiting Hadford Forest ________ quite ________ between 2005 and 2009. The number of people visiting to walk their dogs ________ sharply as a proportion of the ________ and levelled ________ from 2008. If the trends ________, then overall visitor numbers will continue to ________ steadily. New activities will ________ faster and ________ comparison, the number of traditional visitors will drop ________ proportionately.

Figure 1

Figure 2

39

General note

Read the *Vocabulary bank – Vocabulary sets* and *Describing trends* at the end of the Course Book unit. Decide when, if at all, to refer your students to it. The best time is probably at the very end of the lesson or the beginning of the next lesson, as a summary/revision.

Lesson aims

- gain an understanding of lexical cohesion in texts through building word sets, synonyms and opposites/converses
- use appropriate language for describing trends

Introduction

Do some vocabulary revision from the previous units. For example:

1 Choose some words with different meanings in general and agribusiness English (see Units 1 and 2). Ask students to say or write two sentences using each word with a different meaning. Some examples are: *battery, intensive, market, drill, hedge, cultivate, desert, stock, head,* etc. If necessary, students can work with their dictionaries.

2 Choose some prefixes and suffixes (see Units 1 and 4). Write them on the board. Ask students to give the meaning of the affix and an example of a word.

3 Dictate some of the key vocabulary from Unit 3. Ask students to check their spellings (with a dictionary) and group the words according to their stress patterns.

Exercise A

Set both questions for pairwork discussion and whole class feedback. Do not comment or correct at this point.

Elicit general words *'attractions/spin-offs/activities'* and write them on the board. Note that students may come up with more than one answer per question. Also note that with agriculture and forestry products and related activities, there is often common ground. Use the example of modern apple/fruit-picking where retired people provide casual labour to fruit farms for fun and profit, living in expensive motor homes. Also note that depending on the student's background and cultural knowledge, you can drill down into socio-economic groups using formal (A/B) or informal (rich) terms.

Language note

General words like *vehicle* are called hypernyms or superordinate terms. Words like *van* and *tractor* are called hyponyms or class members.

Answers

Model answers:

1 Spin-off: holiday cabins (1); game shooting (2); rock festivals (3).

Mainstream: fruit growing and picking (4); forestry: recreational (5); forestry: non-recreational products (6).

2 holiday cabins (families, middle and lower income)

game shooting (farming and farm communities and tourists, high disposable income)

rock festivals (young people and businesses – medium for sales or marketing)

fruit growing and picking (families, tourists, low and middle income, small and large agribusinesses)

forestry: recreational (businesses, families, tourists, sports enthusiasts, middle and lower income)

forestry: non-recreational products (businesses, adults, building trades)

Exercise B

The purpose of this exercise is to build sets of synonyms. This not only helps in understanding textual cohesion, but is useful for paraphrasing.

Set both questions for pairwork. Students should look for pairs of words/items. Tell them to use their dictionaries if necessary to check the grammatical information, and to note if they find other words with similar meanings.

Feed back with the whole class, building up a table on the board, and eliciting other words which can be used with a similar meaning.

Answers

Model answers:

Word 1	Part of speech	Word 2	Part of speech	Words with similar meanings/notes
activity	n (C)	task	n (C)	job
advertising	n (U)	promotion	n (C/U)	a *special promotion* = a special offer to promote a product
aim	n (C) v (T)	target	n (C) v (T)	objective, goal if *aim* is a transitive verb, it is followed by *at* or *for*
business	n (C/U)	company	n (C)	firm, organization business U = the activity
buy	v (T)	purchase	v (T)	get, acquire
consumer	n (C)	customer	n (C)	client (for services); shopper (for shops)
goods	pl n	products	pl n	can also use *items*, meaning things
main	adj	principal	adj	major, chief, key
meet	v (T/I)	satisfy	v (T)	use with *needs* or *requirements*
needs	pl n	requirements	pl n	
retail outlet	n (C)	shop	n (C)	*shop* can also be a verb

Exercise C

Set for individual work and pairwork checking. Feed back with the whole class. Discuss alternative ideas and decide whether they are acceptable. Check the meaning of any unknown words in the text (e.g., *mailing*, *sponsorship*). Note that the blue words are general purpose words frequent in academic contexts.

Answers

Model answers:

part	aspect(s)
known as	called
methods	means
kind	type
concern(s)	involve(d)

Exercise D

1 Set for pairwork. Feed back. Start the first column of the table as shown in the Answers section.

2 Do the first pair of words with the whole class as an example. Set the remainder for pairwork. Feed back, completing the second column of the table on the board.

3 Discuss with the whole class. Elicit a word or phrase which describes the whole set of words and add this to the table.

Answers

Possible answers:

Opposites	Other words	Word for set
careful, impulsive	cautious, risk-taking	personality
conventional, trendy	traditional, fashionable	style/lifestyle
elderly, young	old, middle-aged, teenage, in his/her 20s etc., child, adult	age
female, male	woman, man	gender/sex
low-income, wealthy	high/middle/average income, poor, rich, well-off	income
manual, professional	managerial, clerical, skilled/unskilled/semi-skilled, casual, part time, blue collar/white collar, unemployed, higher/lower professional/managerial, middle manager, housewife, retired	occupation/ type of job/socio-economic status
married, single	divorced, widowed, with/without children, parent	marital or family status

Language note

There are clearly a very large number of words which can be used to describe lifestyle or personality. The words in the previous table are just a small selection. A quick search on the Internet into marketing language will get you to websites where useful terminology information can be found.

Exercise E

Introduce the terms *market segmentation* and *target market*. Elicit or give definitions, e.g., *Market segmentation is the division of a market for a product into groups of customers who have similar needs or characteristics. Each of the groups is likely to want different types of the product. Each market 'segment' can thus be a* ***target market.***

Point out that the names for the word sets in Exercise D are actually the categories that are used to segment markets. Choose one of the spin-offs yourself and give a description of its target market (see table in Answers section, asking students to identify which spin-off you are talking about). Set the remainder for pairwork, telling students to use several of the categories in Exercise D for each agriproduct, adding other words or ideas if they need to.

Feed back as a class discussion, building the table in the Answers section if you wish. Note that the suggestions here may not fit certain countries or cultures. While the segmentation categories should apply universally, the descriptors can be changed to suit students' culture.

Answers

Possible answers:

Spin-off	Family, single, etc.	Income	Personality/lifestyle
1. Holiday cabins	families (young)	middle and lower	outdoors, sporty, health-conscious
2. Game shooting	adult, male	high	traditional, conventional
3. Rock festivals	predominantly young, m/f; also businesses	all: mainly young, working	single, active, travellers, social
4. Fruit growing and picking	families, older + grandchildren, m/f	low and middle	traditional, conventional
5. Forestry: recreational (sports)	professional, under 30, m/f	middle and lower	conventional/modern, healthy, successful
6. Forestry: non-recreational products	couples (teenage children), m/f; building trades; businesses	middle; business all size	conventional, home makers, practical; dependant on specific business

Exercise F

Introduce the idea of proportions and shares. Elicit or give a definition, e.g., *looking at Figure 1, the proportion of people using the forest has changed.* With the whole class, discuss what Figure 1 shows. Elicit some of the verbs and adverbs which students may need in order to discuss question 1. For example:

Go up	No change	Go down	Adverbs
rise increase grow improve	stay the same remain at … doesn't change is unchanged level off	fall decrease decline worsen drop drop off	slightly gradually steadily significantly sharply dramatically proportionately

Note: These verbs are generally used in an intransitive sense when describing trends.

1 Discuss with the whole class. The answer to this question should be one sentence for each graph.

2 Set for pairwork. Feed back, eliciting sentences from the student. Write correct sentences on the board, or display the model answers on an OHT.

Answers

Model answers:

1 Figure 1 shows the number of people visiting between 2005 and 2009 and the sorts of activities that attracted them. Figure 2 shows the developing economic impact that the development of Hadford Forest has had on employment levels in the area.

2 There is an increase in overall number of visitors. There is a diversification in the sort of activities that visitors engage in. There is an overall increase in the economic benefits in terms of jobs created. The new jobs are tending to be created in new recreation and education-related activities and not in traditional primary forestry activities.

Underline the verbs and adverbs. Ask students to make nouns from the verbs and adjectives from the adverbs. Alternatively, you could reproduce the following table (minus the noun and adjective forms) on the board, on an OHT or on a handout. The incomplete table is reproduced in the additional resources section (Resource 5B) to facilitate this.

Verbs	Nouns	Adverbs	Adjective
rise	a rise	gradually	gradual
increase	an increase	sharply	rise
grow	growth*	slightly	slight
improve	an improvement	markedly	marked
fall	a fall	significantly	significant
decrease	a decrease	rapidly	rapid
drop	a drop	steeply	steep
decline	a decline	steadily	steady

*usually (but not always) uncountable in this sense

Return to the original answer sentences and ask students to make sentences with the same meaning, using the nouns and adjectives in place of the verbs and adverbs. Note that when using the noun + adjective, sentences can be made using *There was* ... or *showed*. Do one or two examples orally, then ask students to write the remaining sentences. Feed back.

Model answers:

Figure 1 shows the number of people visiting Hadford Forest rising quite dramatically between 2005 and 2009.	There was a quite dramatic rise in the number of people visiting Hadford Forest between 2005 and 2009.
The number of people visiting to walk their dogs fell sharply as a proportion of the total and levelled off from 2008.	There was a sharp fall in the number of people walking their dogs as a proportion of the total and there was a levelling off of numbers in 2008.
If the trends continue, then overall visitor numbers will continue to increase steadily.	If the trends continue, then there will be a continued, steady increase in overall visitor numbers.
New activities will grow faster and by comparison, the number of traditional visitors will drop back proportionately.	There will be faster growth in new activities and by comparison, there will be a proportionate drop in traditional visitor numbers.

Exercise G

Ask students to look at Figure 1 and discuss in pairs the information it shows. Feed back. Elicit two types of sentence using a verb and a noun to express the changes. For example:

There was a dramatic <u>rise</u> in the number of visitors.

The number of visitors <u>rose</u> dramatically.

Set the text completion for individual work and pairwork checking.

Answers

Figure 1 shows the number of people visiting Hadford Forest <u>rising</u> quite <u>dramatically</u> between 2005 and 2009. The number of people visiting to walk their dogs <u>fell</u> sharply as a proportion of the <u>total</u> and levelled <u>off</u> from 2008. If the trends <u>continue</u>, then overall visitor numbers will continue to <u>increase</u> steadily. New activities will <u>grow</u> faster and <u>by</u> comparison, the number of traditional visitors will drop <u>back</u> proportionately.

Exercise H

Set for pairwork. Encourage lateral thinking and creativity. Feed back with the whole class.

Answers

Model answers:

The hotel could:

- promote outdoor activity breaks in cooperation with the education centre
- develop off-season study breaks for education establishments
- rebrand itself to highlight the links with the forest
- promote orienteering weekends
- promote themselves as providers of good-value lunches for walkers/cyclists

Closure

Listed below are some features, words and expressions associated with the spin-off activities. Ask students to work in small groups or pairs to decide which feature might be important when people make their choice of spin-off activity a) in the UK and b) in the students' own culture and community.

healthy
environmentally friendly
enhances status
a sense of freedom
tradition
best value
de-stressing

5.2 Listening

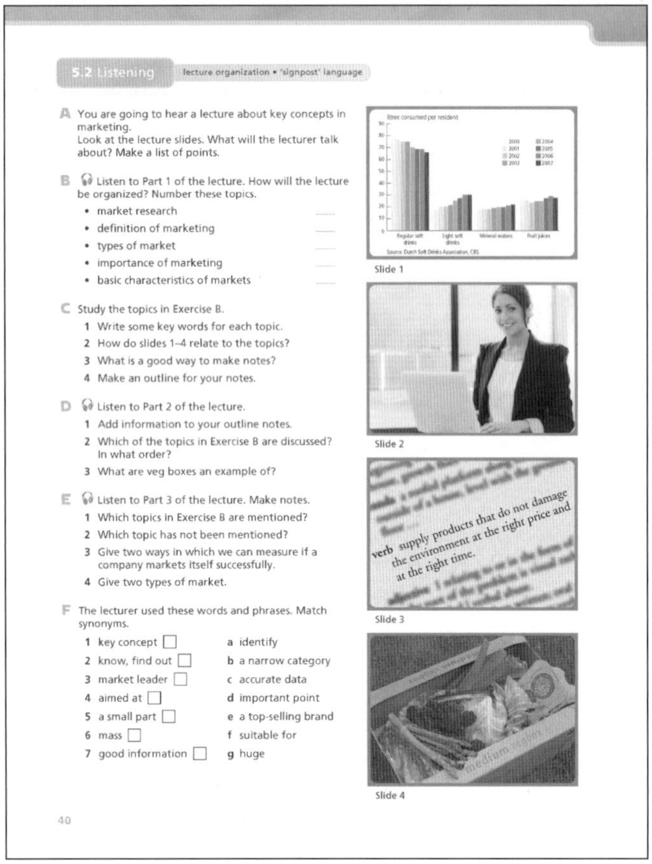

5.2 Listening lecture organization • 'signpost' language

A You are going to hear a lecture about key concepts in marketing.
Look at the lecture slides. What will the lecturer talk about? Make a list of points.

B Listen to Part 1 of the lecture. How will the lecture be organized? Number these topics.

- market research ___
- definition of marketing ___
- types of market ___
- importance of marketing ___
- basic characteristics of markets ___

C Study the topics in Exercise B.

1 Write some key words for each topic.
2 How do slides 1–4 relate to the topics?
3 What is a good way to make notes?
4 Make an outline for your notes.

D Listen to Part 2 of the lecture.

1 Add information to your outline notes.
2 Which of the topics in Exercise B are discussed? In what order?
3 What are veg boxes an example of?

E Listen to Part 3 of the lecture. Make notes.

1 Which topics in Exercise B are mentioned?
2 Which topic has not been mentioned?
3 Give two ways in which we can measure if a company markets itself successfully.
4 Give two types of market.

F The lecturer used these words and phrases. Match synonyms.

1 key concept ☐	a identify
2 know, find out ☐	b a narrow category
3 market leader ☐	c accurate data
4 aimed at ☐	d important point
5 a small part ☐	e a top-selling brand
6 mass ☐	f suitable for
7 good information ☐	g huge

General note

Read the *Skills bank – Signpost language in a lecture* at the end of the Course Book unit. Decide when, if at all, to refer students to it. The best time is probably at the very end of the lesson or the beginning of the next lesson, as a summary/revision.

Lesson aims

- improve comprehension through understanding of signposts and lexical cohesion
- deal with disorganization in lectures/fractured text

Further practice in:

- predicting content from own background knowledge and from the lecture introduction
- using the introduction to decide the best form of notes to use

Introduction

1 Review key vocabulary by writing a selection of words from Lesson 5.1 on the board and asking students to put the words in groups, giving reasons for their decisions.

2 Remind students about preparing for a lecture. If you wish, review Unit 1 *Skills bank – Making the most of lectures.*

Exercise A

Remind students that when lecturers begin their talks, they usually provide their listeners with an outline. Remind/tell students about the *signpost language* which speakers use at the beginning to list the areas they will cover. On the board, build the table below, eliciting suggestions from the students. Alternatively (or in addition), you could refer to the *Skills bank* at this point.

Sequencing words		Verbs
To start with, Firstly,	I'll	begin/start by …ing discuss examine consider mention talk about look at define give a(n) outline/overview/definition/ summary of … end/finish/conclude by …ing
Secondly, Then … After that,		
Finally,		

Language note

Speakers will usually avoid repeating words. So they would be unlikely to say *To start with, I'll start by … .*

Refer students to the lecture slides. Set the exercise for pairwork.

Ask students to feed back their possible lecture ideas to the whole class using the signpost language on the board to order their points. Accept any reasonable ideas. One possibility is given below.

Answers

Possible answer:

To start with, the lecturer will examine the market for soft drinks in the Netherlands. Then he/she will discuss market research and the concept of green marketing. He/she will finish by looking at marketing a new style and innovative agriproduct.

Methodology note

If students are new to the subject of marketing, they may only be able to make simple points about the slides, as in the model answer above. If they already know something about the subject, they may realize that the slides illustrate the concepts which the lecturer will discuss, i.e., Slide 1: market size, market share and market leaders; Slide 2: market research; Slide 3: definitions; and Slide 4: niche markets. These words will appear in the exercises that follow.

Exercise B

Tell students they are only going to hear the introduction to the lecture. Give students time to read the topics. Check that they understand the meaning. Remind them they will only hear the introduction once, as in a lecture. Tell them to listen out for the signpost language on the board. While they listen, they should number the topics from 1–5 in the order in which the lecturer will talk about them.

Play Part 1. Allow students to compare answers. Feed back. Ask students to say what signpost language they heard related to each topic. Confirm the correct answers.

Answers

market research – 3 (*After that, we're going to look at …*)

definition of marketing – 1 (*… to start with, we need to think about …*)

types of market – 5 (*Then I'll finish by mentioning …*)

importance of marketing – 2 (*And secondly, …*)

basic characteristics of markets – 4 (*… so next I'm going to discuss …*)

Transcript 1.20

Part 1

Good morning, everyone. This morning, I want to begin the topic of agribusiness marketing. In this first talk, I'm going to give you an overview of a few key concepts. We'll look at other aspects of food and agricultural products marketing in the next few lectures. Also, in your seminars and assignments, you'll be able to cover all the important points in more detail. OK, so … er … let's see – yes – to start with, we need to think about what marketing is. In other words, why do agribusinesses need to market themselves? And secondly, why is marketing such an important part of building a successful enterprise? After that, we're going to look at market research because agribusinesses, like all types of business, need good information on which to base decisions and build a marketing strategy. Part of this is about analyzing, so next I'm going to discuss the characteristics of markets. Then I'll finish by mentioning some different types of markets.

Exercise C

1 Set for pairwork. Divide the topics up among the pairs so that each pair concentrates on one topic. Feed back. Accept any reasonable suggestions.

2 Refer students to the lecture slides. Students should try to guess which of the topics each slide could refer to. Set for individual work and pairwork checking. Feed back but do not confirm or correct yet.

3 Elicit suggestions from the whole class. If you wish, refer students to Unit 1 *Skills bank*.

4 Set for individual work. Students should prepare an outline on a sheet of paper, preferably using either numbered points (with enough space between the points to allow for notes to be added) or a mind map/spidergram (see example below).

Answers

Possible answers:

1 Some key words are:

definition of marketing – *promotion, satisfy customer needs*

importance of marketing – *ensure company profits, help meet company objectives*

market research – *survey, focus group, interview, qualitative, quantitative*

basic characteristics of markets – *size, trends, growth, share, leader, brand, value, volume of sales, analysis*

types of market: any words from Lesson 5.1 concerned with market segmentation; also *niche* and *mass* (to be mentioned in the lecture)

2 Accept any reasonable answers with good justifications.

3/4 Example of spidergram:

Methodology note

There is no need to teach all the words given in the model answers for question 1. However, if students suggest words that others do not know, it would of course be reasonable to check/clarify meanings of such words at this point.

Exercise D

Tell students to use their outline from Exercise C to take notes. Which topics do they expect to hear in this section?

Play Part 2. Put students in pairs to compare their notes and discuss the questions.

Feed back. When it becomes clear that the lecturer did not actually stick to the plan in the introduction, say that this happens very often in lectures. Lecturers are human! Although it is a good idea to prepare outline notes, students need to be ready to alter and amend these. Discuss how best to do this. One obvious way is to use a non-linear approach such as a mind map or spidergram, where new topics can easily be added.

After checking answers to questions 2 and 3, build a complete set of notes on the board as a spidergram, as in the example in the Answers section.

Answers

Possible answers:

1 Example notes: see spidergram below.
2 Discussed first: importance of marketing; second: definition of marketing.
3 Creating new markets.

Transcript 1.21

Part 2

So let's consider a moment what marketing is and why it's important – why many people believe that marketing is potentially *the* critical factor in determining if an agribusiness will succeed or not. The Chartered Institute of Marketing talks about identifying, anticipating and satisfying customer requirements profitably in its definition. So the company must not only identify the customers' current needs correctly, but also predict or anticipate them. This enables the company to meet the customer's needs. The veg box distributor is an example I often think about from the area where I live. The company began as a conventional farming business selling products wholesale. When wholesale prices for vegetables dropped and profits started to fall a few years ago, they took a gamble and went into farm-gate sales. One or two local customers said that was great, but they wanted them to deliver their vegetables, and from that small beginning a major new retailing activity grew up. Anyway, er … to return to the main point, like those farmers, every company, ethical or conventional, wants to build sustainable, profitable income. And good marketing – predicting what people want, accurately, and then making sure the product or service meets the need – is what competitive, successful business is all about.

Exercise E

Ask students what they expect to hear about in the next part. Refer students to their outline again. Give them time to read the questions. Note that the final part of the lecture will be heard in Lesson 5.3, but there is no need to tell them this at this point. Play Part 3. Set the questions for pairwork. Students should use their notes to help them answer the questions.

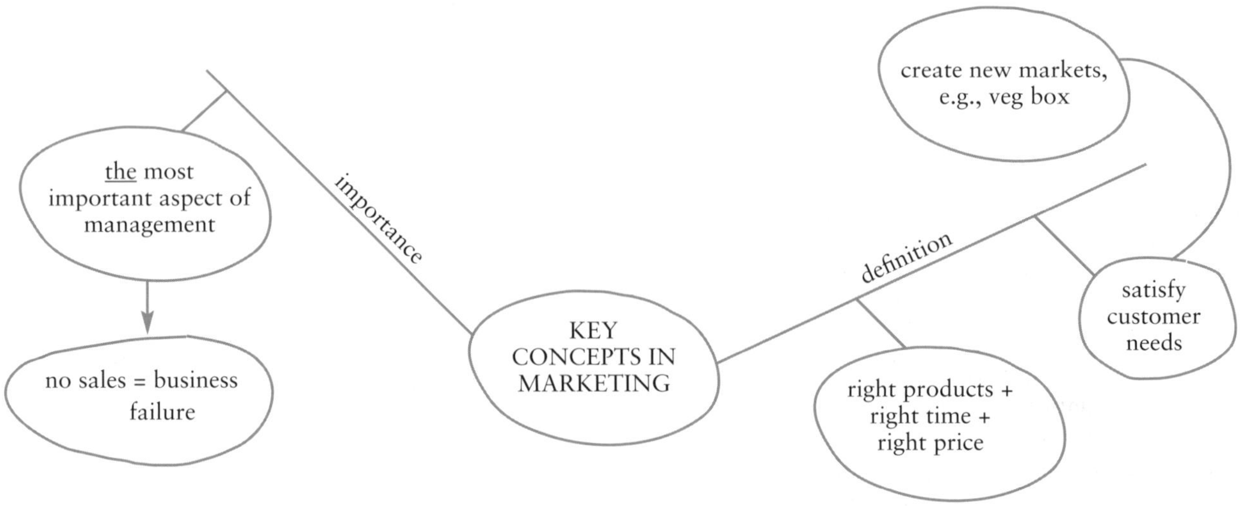

Feed back. Note that there is no need to build a set of notes on the board at this point – this will be done in Lesson 5.3. Ask students if they can remember what the lecturer was talking about when she lost her place (consumption of soft drinks) and exactly what she said to indicate that she had lost her place (*So, um ... where was I? Oh yes, ...*).

Answers

Model answers:

1 Basic characteristics of markets; types of market.
2 Market research is not mentioned explicitly, although data and trends are discussed.
3 By what it is worth (value) and by the quality of reputation.
4 Mass market; niche market.

Transcript 1.22

Part 3

OK, so, in theory, good marketing means knowing your customer so you can meet their needs. And in practice? Well, there are two things the marketing expert has to provide as a basis for ensuring their business meets customer needs correctly. First, they provide excellent and in-depth information about the market or markets for the company's products or services. And they need this so that, secondly, they can identify the correct *type* of market for their product. What do I mean by market 'types'? Well, 'mass' and 'niche' markets are two classic examples. Let's take an agribusiness, Cuckmere Valley Organics (CVO), and an agribusiness making a well-known brand of cheap lemonade. CVO makes organic soft drinks using traditional methods from a small rural village. The lemonade company produces industrial quantities of its product from a factory near Birmingham. The lemonade producer targets supermarkets and food chains: high volume, a *mass* market. CVO might also target supermarkets, but their marketing experts will probably have identified a different type of market: higher margins, small volume and the 'specialist foods' areas of the supermarkets – a *niche* market. Right then, but let's assume that CVO wants to grow their business by developing an export capability. Here is where the marketing team have to use that other expertise – data gathering and analysis. And their data on the Netherlands and Belgium indicates they are a potential market for low-calorie, healthy soft drinks. As the slide shows, consumption of soft drinks in the Netherlands reached around 30 litres per head of population in 2007. And trends towards healthier living are growing demand for low-calorie drinks – like CVO's. In fact, since 2000, the consumption of low-calorie, or light, drinks increased by nearly 75 per cent. So, er ... where was I? Oh yes, so if we are thinking niche markets, then the Netherlands looks a very interesting prospect because the data indicates an existing niche market there. And why interesting? Well, because there is existing demand for products similar to the organic producer, but the demand is growing and it could offer CVO the opportunity to move from a niche marketing operation to a mass marketing operation. And whatever niche companies may say about their intentions and principles, good businesses all tend to dream about becoming market leaders in a mass market – but I probably shouldn't say that!

Exercise F

This gives further practice in identifying words and phrases used synonymously in a particular context.

Set for individual work and pairwork checking.

Answers

1 key concept	d important point
2 know, find out	a identify
3 market leader	e a top-selling brand
4 aimed at	f suitable for
5 a small part	b a narrow category
6 mass	g huge
7 good information	c accurate data

Closure

Ask students to work in groups and list the agriproducts that they use regularly or are familiar with. Get them to feed back informally. Then ask them to think in their groups if they tend to select the product according to the brand. Then they should compare this with non-agriproducts such as computers, cars and jeans. Discuss and try to elicit comments and explanations on the basis of their observations.

Note: Students will need their lecture notes from this lesson in Lesson 5.3.

5.3 Extending skills

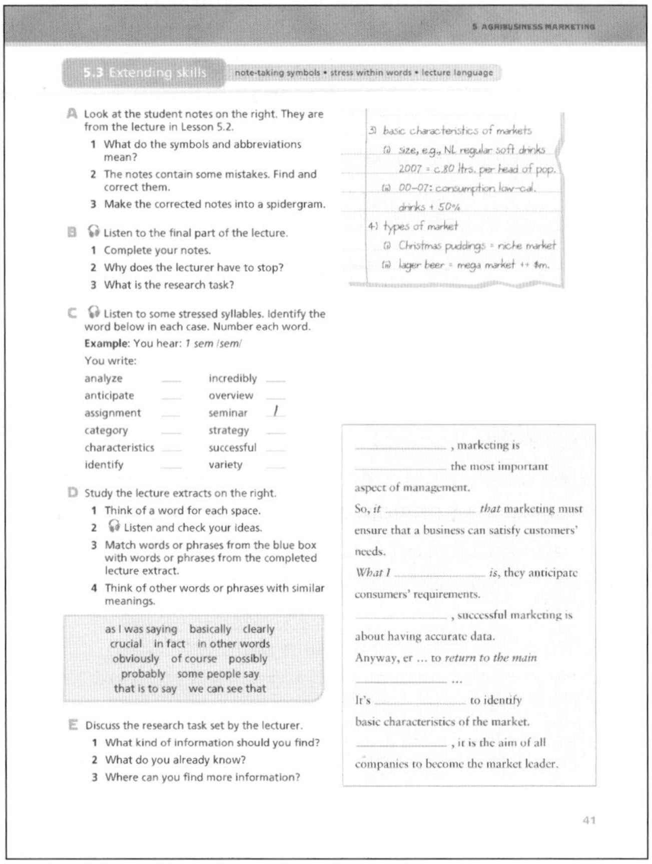
5 AGRIBUSINESS MARKETING

5.3 Extending skills — note-taking symbols • stress within words • lecture language

A Look at the student notes on the right. They are from the lecture in Lesson 5.2.

1 What do the symbols and abbreviations mean?
2 The notes contain some mistakes. Find and correct them.
3 Make the corrected notes into a spidergram.

3) basic characteristics of markets
(i) size, e.g., NL regular soft drinks 2007 = c.80 ltrs. per head of pop.
(ii) 00-07: consumption low-cal. drinks + 50%
4) types of market
(i) Christmas puddings = niche market
(ii) lager beer = mega market ++ $m.

B Listen to the final part of the lecture.

1 Complete your notes.
2 Why does the lecturer have to stop?
3 What is the research task?

C Listen to some stressed syllables. Identify the word below in each case. Number each word.

Example: You hear: *1 sem /sem/*

You write:

analyze	___	incredibly	___
anticipate	___	overview	___
assignment	___	seminar	1
category	___	strategy	___
characteristics	___	successful	___
identify	___	variety	___

D Study the lecture extracts on the right.

1 Think of a word for each space.
2 Listen and check your ideas.
3 Match words or phrases from the blue box with words or phrases from the completed lecture extract.
4 Think of other words or phrases with similar meanings.

as I was saying basically clearly crucial in fact in other words obviously of course possibly probably some people say that is to say we can see that

________, marketing is ________ the most important aspect of management.

So, *it* ________ *that* marketing must ensure that a business can satisfy customers' needs.

What I ________ *is*, they anticipate consumers' requirements.

________, successful marketing is about having accurate data.

Anyway, er … to *return to the main* ________ …

It's ________ to identify basic characteristics of the market.

________, it is the aim of all companies to become the market leader.

E Discuss the research task set by the lecturer.

1 What kind of information should you find?
2 What do you already know?
3 Where can you find more information?

41

Lesson aims

- use symbols in note-taking
- understand and use lecture language such as stance adverbials (*obviously, arguably*), restatement (*in other words …*) and other commentary-type phrases

Further practice in:

- stress within words
- asking for information
- formulating polite questions

Introduction

As in Unit 3, encourage students to ask you questions about the information in the lecture in Lesson 5.2 as if you were the lecturer. Remind them about asking for information politely. If they can't remember how to do this, you could tell them to revise the *Skills bank* for Unit 3.

Put students in pairs. Student A must ask Student B about the information in the lecture in Lesson 5.2 to help him/her complete the notes from the lecture. Then they reverse roles. Again, they can revise language for this in the *Skills bank* for Unit 3.

Exercise A

1 Revise/introduce the idea of using symbols and abbreviations when making notes. Ask students to look at the example notes and find the symbols and abbreviated forms. Do they know what these mean? If not, they should try to guess.

If you wish, expand the table in the Answers section below with more symbols and abbreviations that will be useful for the students. There is also a list at the back of the Course Book for students' reference.

2 Ask students to tell you what kind of notes these are (linear and numbered). Set the question for pairwork. Students will need to agree what the notes are saying and then make the corrections.

3 Set for individual work. Feed back with the whole class and build the spidergram in the Answers section on the board.

Answers

Model answers:

1

Symbol/abbreviation	Meaning
e.g.	(for) example
=	equals, the same as, is
c.	circa, around, approximately
ltrs.	litres
pop.	population
mega	huge
++	lots of
$	US dollars
m	million

2 Suggested corrections:

3) basic characteristics of markets

(i) size, e.g., NL soft drinks 2007 = <u>c.</u> 30 ltrs. per head of pop.

(ii) 00-07: consumption low-cal. drinks + <u>75</u>%

4) types of market

(i) <u>organic soft drinks</u> = niche market

ii) <u>lemonade</u> = <u>mass</u> market ++ $m.

3 Example note:

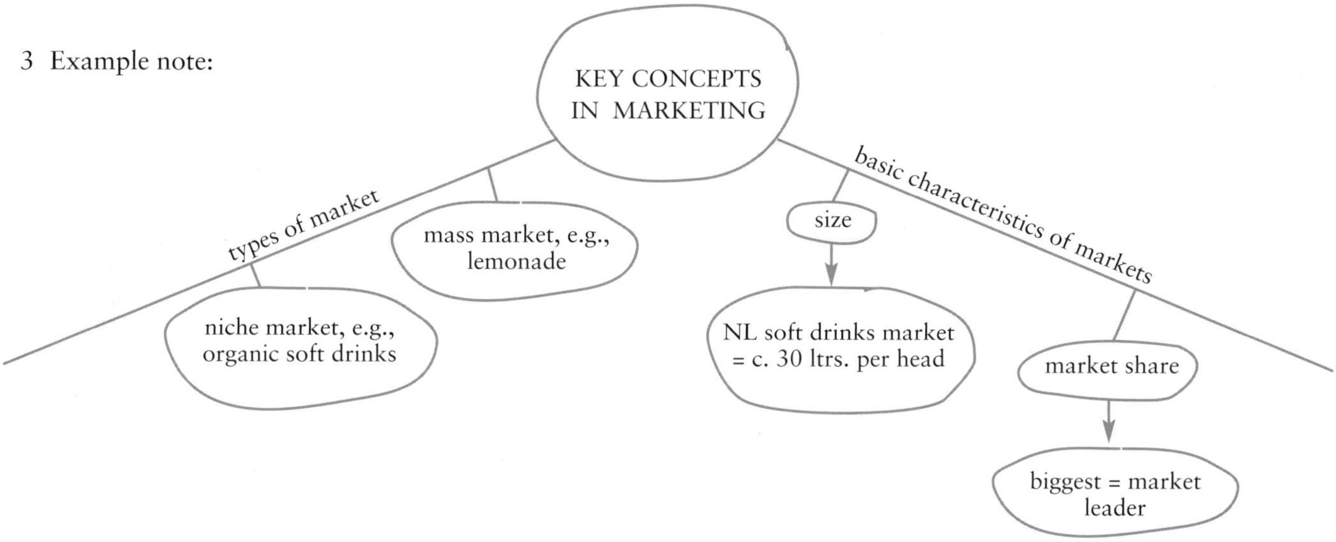

Language note

Some abbreviations are universal and some are personal. People often develop their own personal system of symbols and abbreviations. For example, *c.* for *circa* is used by many people, but *NED* is an example of a longer word abbreviated by the individual who wrote these notes.

Exercise B

Tell students they will hear the final part of the lecture. Give them time to read the questions. They should complete the final leg of the spidergram.

Play Part 4. Put students in pairs to compare their notes and discuss the questions. Feed back. For question 2, ask students if they can remember the exact words used by the lecturer (*oh, dear... sadly, I see that we've run out of time.*).

Answers

Model answers:

1

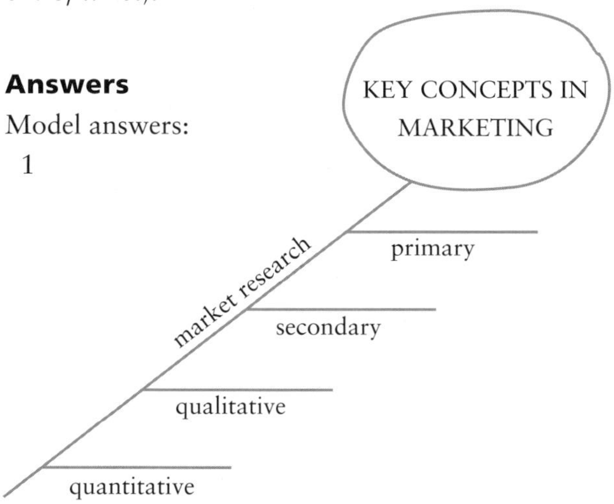

2 Because there is no more time.

3 The research task is to find out about the four different types of market research: *primary, secondary, qualitative* and *quantitative* research.

Transcript 1.23

Part 4

So, how does the marketer get the necessary information? By research, obviously. There are several ways to categorize market research. Let me see … one way is to distinguish between primary and secondary research. Another important distinction is between qualitative and quantitative research. However … oh, dear … sadly, I see that we've run out of time. This means that I'll have to ask *you* to do some research. I'd like you to find out what is meant by the four types of research I've just mentioned; that is, primary and secondary research, and qualitative and quantitative research. We'll discuss what you've found out next time I see you.

Exercise C

Remind students of the importance of stressed syllables in words (see the teaching notes for Unit 3.3, Exercise A). Play the recording, pausing after the first few to check that students understand the task.

Feed back, perhaps playing the recording again for each word before checking. Ideally, mark up an OHT of the words.

Answers

analyze	8
anticipate	7
assignment	3
category	11
characteristics	5
identify	10
incredibly	9
overview	2
seminar	1
strategy	4
successful	6
variety	12

Transcript 1.24

1 'seminar
2 'overview
3 a'ssignment
4 'strategy
5 characte'ristics
6 suc'cessful
7 an'ticipate
8 'analyze
9 in'credibly
10 i'dentify
11 'category
12 va'riety

Exercise D

This exercise gives students a chance to focus on some typical lecture language.

1 Set for pairwork. Students should try to think of a word for each of the blank spaces.

Note that they should *not* try to use the words from the box for this. Do not feed back at this point.

2 Tell students they will hear the sentences from the lecture and should fill in the missing words as they listen. There will be pauses at the end of each sentence but you will play the recording straight through without stopping (as a kind of dictation). Feed back with the whole class, playing the sentences again if necessary. Check the meanings and functions of the words and phrases. Point out the fixed phrases (in italics in the text) and encourage students to learn these. Ask students to repeat the sentences for pronunciation practice, making sure that the stress and intonation are copied from the model.

3 Set for individual work and pairwork checking. Students should check in their dictionaries for meanings or pronunciations of words from the box that they don't know. Feed back, building the first two columns of the table in the Answers section on the board.

4 Elicit suggestions from the whole class for a third column: 'Other similar words/phrases'.

If you wish, students can practise saying the sentences in question 2 but this time with words from questions 3 and 4.

After completing Exercise D, students can be referred to the *Vocabulary bank – Stance* and the *Skills bank – Signpost language in a lecture* for consolidation.

Answers

Model answers:

1/2 Actually, marketing is arguably the most important aspect of management. So, *it follows that* marketing must ensure that a business can satisfy customers' needs. *What I mean is*, they anticipate consumers' requirements. Fundamentally, successful marketing is about having accurate data. Anyway, er ... *to return to the main point*, it's essential to identify basic characteristics of the market. Naturally, it is the aim of all companies to become the market leader.

3/4

Word/phrase from the lecture	Words/phrases from the box	Other similar words/phrases
Actually	in fact	in reality
arguably	probably, possibly, some people say	perhaps
it follows that	we can see that	logically
What I mean is	that is to say, in other words	or, by that I mean, to put it another way
Fundamentally	basically	in essence, really
to return to the main point	as I was saying	
essential	crucial	important
Naturally	of course, obviously, clearly	certainly

Transcript 1.25

Actually, marketing is arguably *the* most important aspect of management.

So, it follows that marketing must ensure that a business can satisfy customers' needs ...

What I mean is, they anticipate consumers' requirements.

Fundamentally, successful marketing is about having accurate data.

Anyway, er ... to return to the main point, ...

It's essential to identify basic characteristics of the market ...

Naturally, it is the aim of all companies to become the market leader ...

Language note

There are three main categories of language here:

1 Stance markers. These are words or phrases that speakers use to show what they feel or think about what they are saying. Adverbs used like this are generally (though not always) positioned at the beginning of the sentence.

2 Phrases used to indicate a restatement. It is very important for students both to understand and to be able to use these, since speakers frequently need to repeat and explain their points.

3 Phrases used to show that the speaker has deviated from the main point and is now about to return to it. Again, this type of phrase is very common in lectures and discussions.

Exercise E

Remind students of the task set by the lecturer at the end of Part 4. Set the questions for pairwork discussion. Students should first list the sort of information they will need to find, then discuss and make notes on what they already know. Then they should compile a list of possible sources of information.

Feed back on all three questions with the whole class. Do not discuss further at this point, as the topic will be taken up in the next lesson.

If you wish, you can tell students to focus on only *one* of the types of research in preparation for Lesson 5.4 Exercise G, and to follow up one of their references, not forgetting to record the necessary bibliographical details. Since this will act as preparation for Exercise G, you need to make sure that there are equal numbers of students investigating each type of market research.

Answers

Possible answers:

1 Definitions, methods and anything else that is relevant.

2 Answers depend on the students.

3 Internet, library, subject textbooks, encyclopedias, etc.

Closure

Play a version of the game 'Just a minute'. Put students in groups of four. Give them an envelope in which they will find topics written on slips of paper. Students take turns to take a slip of paper from the envelope and then talk for one minute on the topic. Encourage them to use as many of the words and phrases from Exercises C and D as they can. Each person should talk for up to a minute without stopping. If they can talk for one minute they get a point. If they deviate from their topic or can't think of anything more to say, they have to stop. The person who has the most points is the winner.

Suggestions for topics follow. Or if you prefer, you can use other topics unrelated to agribusiness.

target markets
market leaders
holidays
socio-economic status
eco products
customers' needs
types of markets
green marketing
the four Ps
real food
advertising
market segmentation

5.4 Extending skills

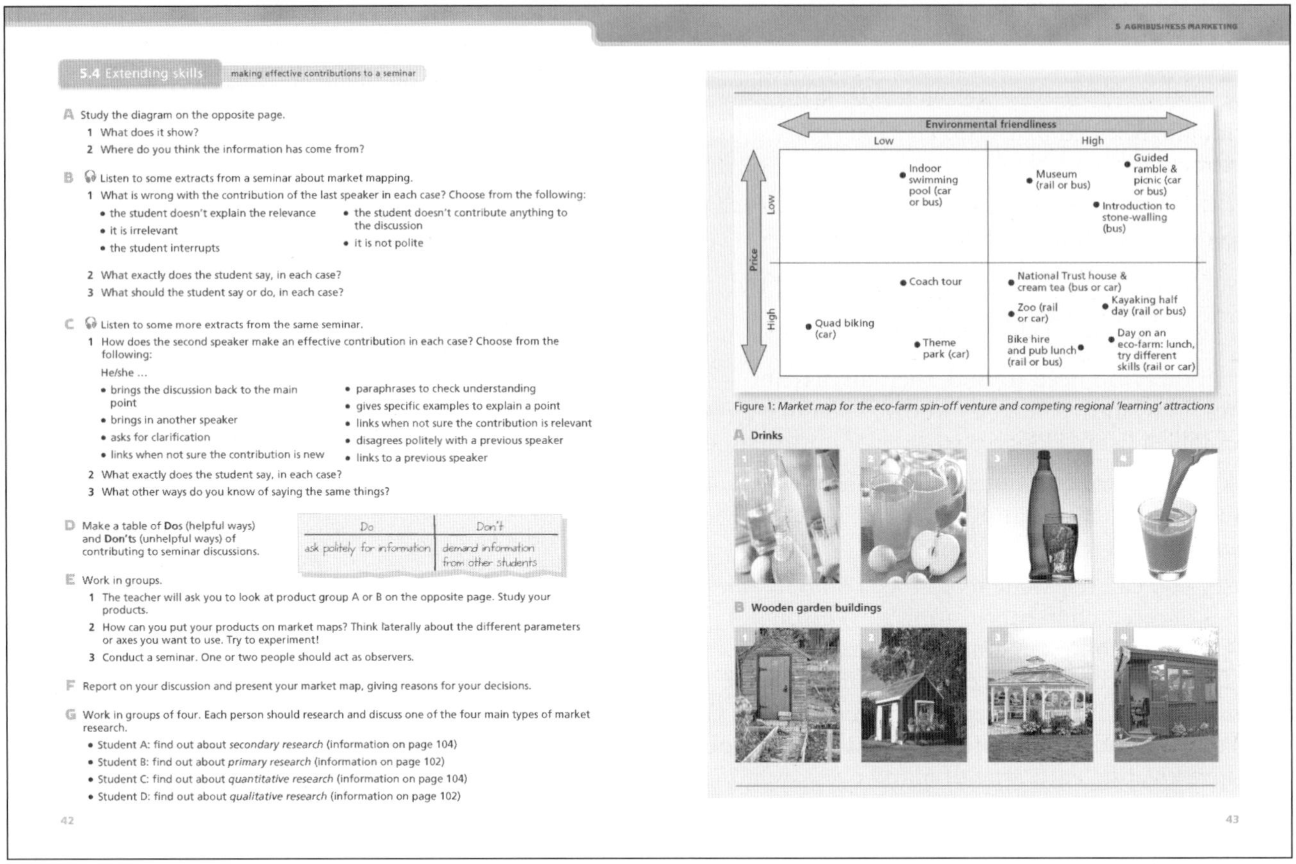
5 AGRIBUSINESS MARKETING

5.4 Extending skills — making effective contributions to a seminar

A Study the diagram on the opposite page.

1 What does it show?
2 Where do you think the information has come from?

B Listen to some extracts from a seminar about market mapping.

1 What is wrong with the contribution of the last speaker in each case? Choose from the following:
   - the student doesn't explain the relevance
   - it is irrelevant
   - the student interrupts
   - the student doesn't contribute anything to the discussion
   - it is not polite
2 What exactly does the student say, in each case?
3 What should the student say or do, in each case?

C Listen to some more extracts from the same seminar.

1 How does the second speaker make an effective contribution in each case? Choose from the following:
   He/she ...
   - brings the discussion back to the main point
   - brings in another speaker
   - asks for clarification
   - links when not sure the contribution is new
   - paraphrases to check understanding
   - gives specific examples to explain a point
   - links when not sure the contribution is relevant
   - disagrees politely with a previous speaker
   - links to a previous speaker
2 What exactly does the student say, in each case?
3 What other ways do you know of saying the same things?

D Make a table of **Dos** (helpful ways) and **Don'ts** (unhelpful ways) of contributing to seminar discussions.

Do	Don't
ask politely for information	demand information from other students

E Work in groups.

1 The teacher will ask you to look at product group A or B on the opposite page. Study your products.
2 How can you put your products on market maps? Think laterally about the different parameters or axes you want to use. Try to experiment!
3 Conduct a seminar. One or two people should act as observers.

F Report on your discussion and present your market map, giving reasons for your decisions.

G Work in groups of four. Each person should research and discuss one of the four main types of market research.

- Student A: find out about *secondary research* (information on page 104)
- Student B: find out about *primary research* (information on page 102)
- Student C: find out about *quantitative research* (information on page 104)
- Student D: find out about *qualitative research* (information on page 102)

42

Figure 1: *Market map for the eco-farm spin-off venture and competing regional 'learning' attractions*

A Drinks

B Wooden garden buildings

43

Lesson aims

- make effective and appropriate contributions to a seminar

Further practice in:

- speaking from notes
- reporting information

Introduction

Revise stance words and restatement/deviation phrases from the previous lesson. Give a word or phrase and ask students to give one with a similar meaning. Alternatively, give a sentence or phrase from the lecture in Lessons 5.2 and 5.3 and ask students to tell you the accompanying stance word or restatement phrase.

Exercise A

1 Tell students to look at the market map. Set for pairwork discussion.
2 Ask students to use what they know about marketing to speculate on the answer. Feed back, accepting any reasonable suggestions.

Answers

Possible answers:

1 It shows how a spin-off agribusiness compares with similar 'learning' attractions regionally using two criteria – price and environmental friendliness.
2 The information has probably come from some form of market research, for example by asking people, by reading or by observing. Some useful information might also have come from the attractions' website.

Subject and methodology note

Market mapping is a technique which is based on surveys of customers' attitudes and profiles. Two main variables are identified which enable products in a particular market to be distinguished from each other, and then the products in a particular market are plotted on the map according to the customers' attitudes and profiles. Companies can use the technique to help spot gaps in a market which could be filled. They can also see if their target market is being reached or if they need to reposition their brand with a new advertising campaign and a different image.

In this lesson, students will be asked to make market maps based on their own ideas rather than by using statistics from a survey.

Exercise B

In this exercise, students will hear examples of how *not* to contribute to a group discussion.

1/2 Allow students time to read the questions. Tell them they will hear five extracts. They should choose a different answer for each one. Set for individual work and pairwork checking. Play all the extracts through once.

Play the extracts a second time, pausing after each one. Students should write down the actual words, as in a dictation, then check in pairs. When students have completed questions 1 and 2, feed back with the whole class, maybe building up columns 1 and 2 of the table in the Answers section on the board.

3 Set for pairwork discussion. Feed back, adding a third column to the table on the board.

Answers

Model answers:
See table below.

Transcript 1.26

Extract 1

LECTURER: Right, Leila and Majed, what did you find out about market mapping?

LEILA: Well, first of all, we checked out the attractions in the region using the Internet. Then we visited the local tourist information centre.

MAJED: I went to the cinema yesterday evening.

Extract 2

LECTURER: And what else did you do?

LEILA: We talked to the manager. She was quite helpful.

MAJED: You're joking! She spent most of her time dealing with visitors.

Extract 3

LECTURER: Leila, can you give us an explanation of your market map?

LEILA: Well, yes, it has a vertical and a horizontal axis: attraction price versus how environmentally friendly it is.

LECTURER: What do the rest of you make of this? Evie, what about you?

EVIE: Well, erm … I'm not sure really.

Extract 4

LECTURER: Majed, can you explain how you decided where to place the different attractions on the map?

MAJED: Well, yes, we wanted a simple picture so we used the information the attractions put up on their sites and this told us clearly how they fitted the map.

JACK: So it's secondary.

Extract 5

LECTURER: What do you mean by 'secondary', Jack?

JACK: I mean it's an example of secondary research. They did two things – they asked someone for information and then …

EVIE: [interrupting] Actually, that's primary.

	Contribution is poor because	Exact words	How to improve
Extract 1	it is irrelevant	MAJED: I went to the cinema yesterday evening.	say something relevant: for example, something about the attractions they read about
Extract 2	it is not polite	MAJED: You're joking! She spent most of her time dealing with visitors.	use polite (tentative) language when disagreeing, e.g., *Actually, I'm not sure. I don't think she really wanted to talk to us.*
Extract 3	the student doesn't contribute anything to the discussion	EVIE: Well, erm … I'm not sure really.	be ready to contribute something when brought into the discussion by the lecturer or other students
Extract 4	the student doesn't explain the relevance	JACK: So it's secondary.	the comment is relevant to the topic but he doesn't explain why. He should say, for example, what he said later after the lecturer asked him to explain (i.e., *it's an example of secondary research*)
Extract 5	the student interrupts	EVIE: (interrupting) Actually, that's primary.	she should wait until the speaker has finished

Exercise C

1/2 This time, students will hear good ways of contributing to a discussion. Follow the same procedure as for 1 and 2 in Exercise B. This time, they need to listen for the second speaker.

Again, when students have completed 1 and 2, feed back with the whole class, maybe building up a table on the board. If you wish, students can look at the transcript at the back of the Course Book.

3 Ask the whole class for other words or phrases that can be used for the strategy and add a third column to the table as below.

Answers

Model answers:
See table below.

Transcript 1.27

Extract 6

LECTURER: Let's go back to this diagram for the moment to see how it can help with segmentation. First of all, tell us about the dimensions you chose.

LEILA: Well, price was easy, but the parents of a student here have a farm. They are trying out some radical new ideas. They gave us the idea of the environmental criteria. Didn't they, Majed?

MAJED: Absolutely. That seemed an interesting dimension – not the usual.

Extract 7

MAJED: The people who run the eco-farm are interested in the pricing, but they're also really concerned about how their business impacts locally on the environment.

JACK: Sorry, I don't follow. Could you explain why that's important for a business?

MAJED: Well, they depend on their local communities – they're stakeholders. It makes sense.

Extract 8

EVIE: But does this information help them improve profits?

LEILA: Well, if you look at the Sunday papers, for example, green tourism is big business, so I guess they'd say it does help profits, wouldn't they Majed?

Extract 9

MAJED: Yes, businesses need to be aware of what's going on in the world.

JACK: If I understand you correctly, you're saying that even if they're very specialist and target a niche market, these businesses could use the methods conventional companies use.

MAJED: Yes, that's right.

Extract 10

LECTURER: This is all very interesting, isn't it?

EVIE: Yes, but if we go back to the diagram, it's actually quite simple and logical.

LEILA: Correct!

	Helpful strategy	Exact words	Other ways to say it
Extract 6	brings in another speaker	LEILA: Didn't they, Majed?.	What do you think, Majed?/Isn't that right, Majed?
Extract 7	asks for clarification	JACK: Sorry, I don't follow. Could you explain …?	I don't quite understand. Could you say a bit more about …?
Extract 8	gives specific examples to explain a point	LEILA: Well, if you look at the Sunday papers, for example …	For instance, …
Extract 9	paraphrases to check understanding	JACK: If I understand you correctly, you're saying that …	So what you're saying is …
Extract 10	brings the discussion back to the main point	EVIE: Yes, but if we go back to the diagram, …	Thinking about …/If we can go back to … for a moment, …
Extract 11	disagrees politely with a previous speaker	MAJED: Actually, we discovered that …	I don't think I agree with that. In my opinion …
Extract 12	links to a previous speaker	EVIE: As Majed said earlier …	Going back to what Majed said a while ago …
Extract 13	links when not sure the contribution is new	JACK: I'm sorry. Has anybody made the point that …?	I don't know if this has been said already, but …
Extract 14	links when not sure the contribution is relevant	EVIE: I don't know if this is relevant, but …	I'm not sure if this is a little off the point, but …

Extract 11

Jack: I still think the whole idea is silly. Does anyone take a bus for a day out?

Majed: Actually, we discovered that more and more people do, especially older people.

Extract 12

Lecturer: So what do you think is the most important point about the diagram?

Evie: As Majed said earlier, they help to show a business where it is.

Extract 13

Lecturer: Any other ideas?

Jack: I'm sorry. Has anybody made the point that the information comes from the attractions themselves?

Lecturer: Yes, actually. Majed did say that earlier, but it's an important point.

Extract 14

Lecturer: So are there different things you could do with the map next time?

Evie: I don't know if this is relevant, but you could combine the price and environment and then look at visitor background, for example.

Lecturer: Yes, that's interesting. Perhaps that is another way to segment.

Exercise D

Set for group work. Tell students to brainstorm suggestions for more good and bad seminar strategies. They should think about what helps a seminar discussion to be successful. It may help to think about having seminar discussions in their own language, but they should also think about what is involved in having a seminar discussion in English. Aspects to consider include language, how to contribute to discussions and how to behave.

Feed back, making a list on the board.

Answers

Possible answers:

Do	Don't
prepare the topic beforehand	
ask politely for information	demand information from other students
try to use correct language	
speak clearly	mumble, whisper or shout
say when you agree with someone	get angry if someone disagrees with you
link correctly with previous speakers	
build on points made by other speakers	
make a contribution, even if you are not sure if it is new or relevant	stay silent, waiting for 'the perfect moment'
be constructive	be negative
give specific examples to help explain a point	be vague
listen carefully to what others say	start a side conversation
allow others to speak	dominate the discussion
paraphrase to check understanding	

Exercise E

Set students to work in groups of five or six. Give the groups a letter: A or B. Group As should look at the drinks, and Group Bs should look at the wooden garden buildings.

In each group, there should be one or two observers and three or four discussing. Groups should also appoint one person to take notes on the discussion, since they will have to present their solution to another group. During the discussion, they will need to identify two key variables (consisting of a pair of antonyms) for the market map. They will then need to decide where to place each product on the map and be able to justify their decisions. If you wish, give each group a large sheet of paper on which to draw their market map. While students are talking, you can listen in and note where students may need help with language, and where particularly good examples of language are used.

The students acting as observers for the discussion should use a checklist of things to watch for. One observer can concentrate on poor contributions and the other on good contributions. Sample checklists are provided in the additional resources section (Resource 5C) – students simply mark in each cell whenever the behaviour occurs.

Exercise F

For this exercise, an A and a B group can join together to make one larger group. Alternatively, if the groups are already large, divide each group in half and send one half plus one observer to another group, so that the new groups consist of 50% As and 50% Bs.

Before the groups report on their discussion, remind them about speaking from notes (see Unit 1 *Skills bank*).

First, the observers should give an overview of how the seminar discussion went and should highlight especially good practice. They can also report on poor contributions, but this needs to be done carefully and constructively (possibly without mentioning names), so that individuals are not embarrassed or upset.

Then the person who took notes should present the decisions of their group to the other group.

Finally, feed back to the whole class on what you heard as you listened in to the groups. Suggest improvements for words and phrases, and highlight good practice.

Exercise G

With the whole class, revise asking for information, opinions and clarification, and agreeing or disagreeing in a seminar. Remind students of the questions used by the lecturer in Exercise B. If necessary, play the seminar extracts again or refer students to the *Skills bank*. Remind students also about reporting information to people (see Unit 3 *Skills bank*).

For this discussion exercise, students will make use of the information at the back of the Course Book, plus the information they have already found on the four research types, as set in Lesson 5.3. Ideally, each student should already have found out information about one type of research (primary, secondary, qualitative, quantitative) and be able to give the source reference for their information, as set in Lesson 5.3.

Set students to work in groups of four. If students have done the research task in Lesson 5.3, each group should consist of students who have each found out about a different type of market research. The easiest way to manage this is to give students a letter: all the people who have researched primary research are As, secondary research are Bs, qualitative are Cs and quantitative are Ds. Make groups with one of each. Give each student the appropriate *discussion task cards* from the additional resources section (Resource 5D).

In their groups, students should do some more research *individually* on definitions, methods, advantages and disadvantages of the four categories of market research. They should work as follows:

As (who have already looked at primary research) can find out about secondary research

Bs (who have already looked at secondary research) can find out about primary research

Cs (who have already looked at qualitative research) can find out about quantitative research

Ds (who have already looked at quantitative research) can find out about qualitative research

Refer each student to the relevant page at the back of the Course Book.

Then each student should feed back to the group the information they have found (as given on the *discussion task card*).

Alternatively, the research activity can be done as a 'wall dictation' as follows. Make large copies of the research notes (Resource 5E), one type of research per page, and pin the sheets on the classroom walls. Each student should leave his/her seat and go to the wall to find the information he/she needs. Students should not write anything down: instead they should read and try to remember the information. Then they return to their group and tell them the information. If they forget something they can go back to the wall to have another look. (Alternatively, you can ask students to read the information, remember it and then fold up the paper – it's important that they do not simply read out the information but use it to inform their speaking.)

It is best to discuss each type of research in turn. For example, first, Student A goes to the wall to find out about secondary research and returns with the information. After hearing what A has to say, the other students can ask the student who has already researched the definition of secondary research (i.e., Student B): *Do you agree with X's view on Y?* Student B should give their definition and also the source reference for the information.

If students have *not* previously researched these topics, do *not* use the discussion task cards. Instead, use the wall dictation method or refer each student to the appropriate notes on market research at the back of the Course Book.

Closure

1 If you wish, refer students to the *Skills bank – Seminar language* for consolidation.

2 Focus on some of the vocabulary connected with research from Lessons 5.2 and 5.4. For example:

analyze (v)	*market research*
analysis	*methods*
attitudes	*observation*
axis	*overview*
beliefs	*qualitative*
category	*quantitative*
concept	*questionnaire*
data	*reliable*
define	*response rate*
definition	*sample*
dimension	*segment (n and v)*
examine	*statistical*
factual	*statistics*
focus group	*survey*
generalize	*trend*
horizontal	*verbal*
identify	*vertical*
interview (n and v)	

Extra activities

1 Work through the *Vocabulary bank* and *Skills bank* if you have not already done so, or as a revision of previous study.

2 Use the *Activity bank* (Teacher's Book additional resources section, Resource 5A).

A Set the crossword for individual work (including homework) or pairwork.

Answers

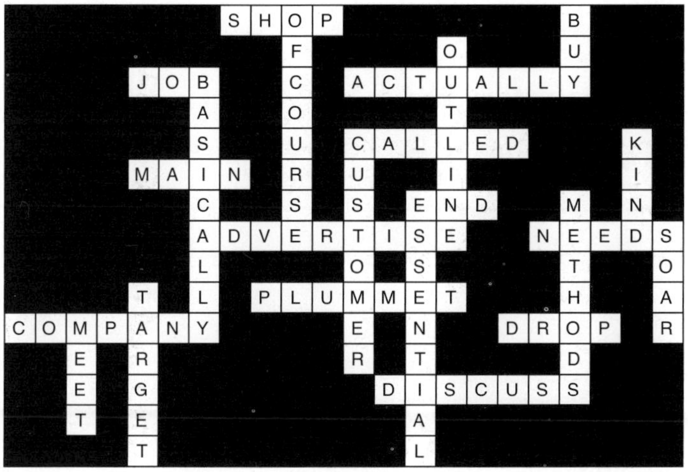

Students' words	Teacher's words
careful	impulsive
child	adult
easy	difficult
elderly	young
fall	rise
large	small
low	high
luxury	economy
male	female
poor	well-off
professional	manual
rigid	flexible
sharply	gradually
single	married
slightly	significantly
slow	rapid
traditional	modern
unskilled	skilled

B Students should select any six words from the box (listed in *Students' words* column). Call out words at random from the *Teacher's words* column, not forgetting to note which words you have called. Students cross out the antonym, if they have it on their card. When someone has managed to cross out all the words on his/her bingo card, he/she should call out *Bingo!* Check to see if all the words are correctly crossed out. If the student has made a mistake, they are out of the game. The first person to correctly cross out all the words on his/her bingo card is the winner.

An alternative is to put students in groups to play the game, with one student acting as the teacher. In this case, you will need to prepare a list of teacher's words for each group.

3 Students can choose a suitable product for which they can try to identify the target markets and/or segmentation. This can be anything they wish, but some suggestions are:

- examples of products that have already been discussed in Unit 5 (eco-farms, veg boxes, garden products)
- eco-tourism
- food and drink

Inspiration for products can be gained by looking at online shopping sites. Students can present their findings to the whole class or in groups.

6 PRODUCTS AND STRATEGIES

Unit 6 continues the theme of marketing. It focuses on the agriproduct, which is central to the marketing mix. Lesson 6.1 looks at a typical product life cycle in terms of costs, sales, cash flow and sustainability. The reading text in Lesson 6.2 discusses how a company can assess the potential of its products and markets, how it can analyze its opportunities and build a strategic marketing plan, and how it can extend the life of ageing products by rebranding and repackaging. The unit finishes with a case study of a marketing strategy in a developing market – fresh produce in major supermarkets.

Skills focus

Reading

- locating key information in complex sentences

Writing

- reporting findings from other sources: paraphrasing
- writing complex sentences

Vocabulary focus

- synonyms, replacement subjects, etc. for sentence-level paraphrasing

Key vocabulary

brand (n)	level (n)	profitable
cash flow	level off	profit margin
cost (n and v)	life cycle	psychological
decline (n and v)	life-span	rebrand
design (n and v)	market share	relaunch
established (adj)	matrix	repackage
evaluate	mature (adj and v)	sales figures
extend	maturity	saturate
extension	merchandise	saturation
feel-good factor	model (n, v and adj)	sponsor (adj)
identify	peak (n, v and adj)	stabilize
intangible	phase	stable (adj)
integrated	portfolio	tangible
introduce	position (n and v)	trial (n and v)
launch (n and adj)	profit (n)	unit cost

6.1 Vocabulary

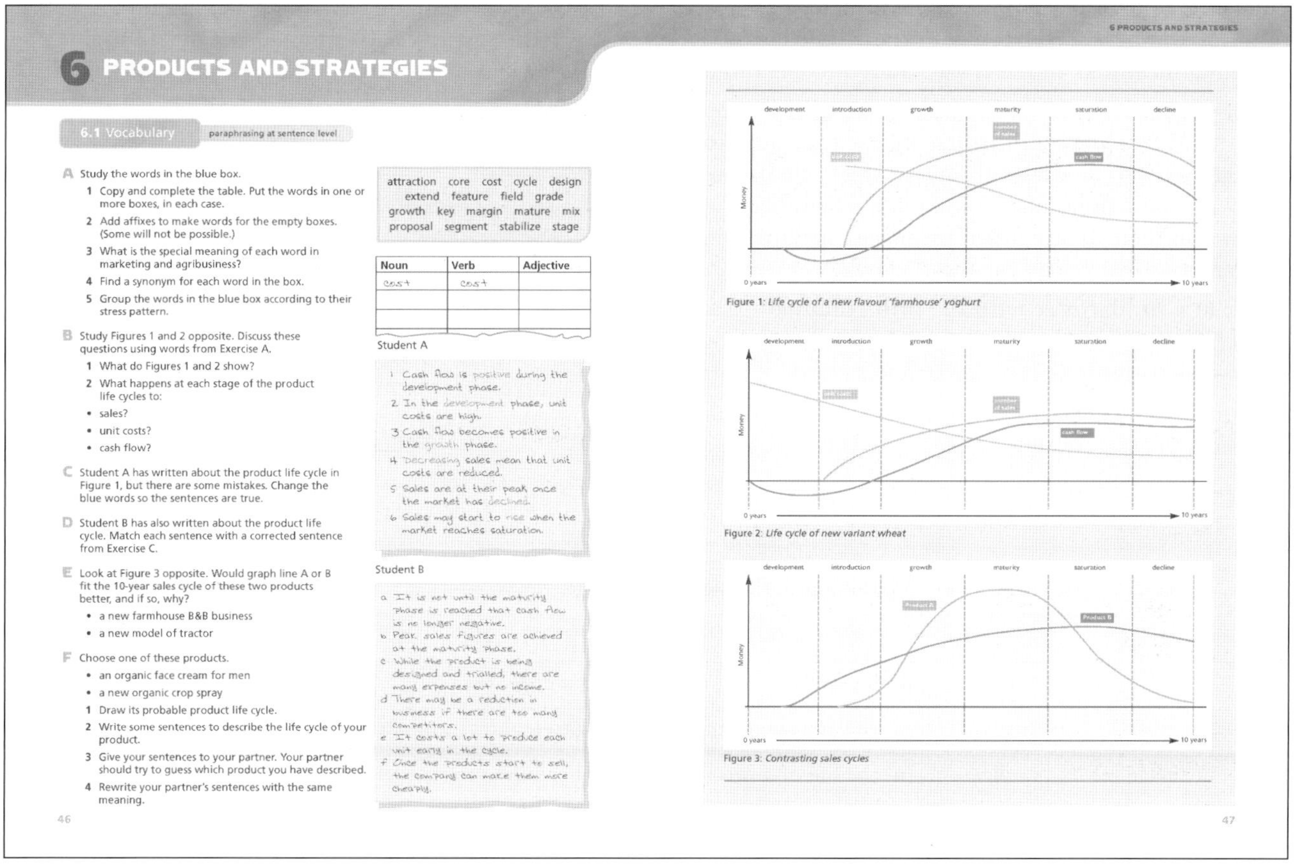

6 PRODUCTS AND STRATEGIES

6.1 Vocabulary — paraphrasing at sentence level

A Study the words in the blue box.
1 Copy and complete the table. Put the words in one or more boxes, in each case.
2 Add affixes to make words for the empty boxes. (Some will not be possible.)
3 What is the special meaning of each word in marketing and agribusiness?
4 Find a synonym for each word in the box.
5 Group the words in the blue box according to their stress pattern.

attraction core cost cycle design extend feature field grade growth key margin mature mix proposal segment stabilize stage

Noun	Verb	Adjective
cost	cost	

B Study Figures 1 and 2 opposite. Discuss these questions using words from Exercise A.
1 What do Figures 1 and 2 show?
2 What happens at each stage of the product life cycles to:
- sales?
- unit costs?
- cash flow?

C Student A has written about the product life cycle in Figure 1, but there are some mistakes. Change the blue words so the sentences are true.

D Student B has also written about the product life cycle. Match each sentence with a corrected sentence from Exercise C.

E Look at Figure 3 opposite. Would graph line A or B fit the 10-year sales cycle of these two products better, and if so, why?
- a new farmhouse B&B business
- a new model of tractor

F Choose one of these products.
- an organic face cream for men
- a new organic crop spray
1 Draw its probable product life cycle.
2 Write some sentences to describe the life cycle of your product.
3 Give your sentences to your partner. Your partner should try to guess which product you have described.
4 Rewrite your partner's sentences with the same meaning.

Student A

1 Cash flow is positive during the development phase.
2 In the development phase, unit costs are high.
3 Cash flow becomes positive in the growth phase.
4 Decreasing sales mean that unit costs are reduced.
5 Sales are at their peak once the market has declined.
6 Sales may start to rise when the market reaches saturation.

Student B

a It is not until the maturity phase is reached that cash flow is no longer negative.
b Peak sales figures are achieved at the maturity phase.
c While the product is being designed and trialled, there are many expenses but no income.
d There may be a reduction in business if there are too many competitors.
e It costs a lot to produce each unit early in the cycle.
f Once the products start to sell, the company can make them more cheaply.

46

Figure 1: Life cycle of a new flavour 'farmhouse' yoghurt

Figure 2: Life cycle of new variant wheat

Figure 3: Contrasting sales cycles

47

General note

Read the *Vocabulary bank* at the end of the Course Book unit. Decide when, if at all, to refer students to it. The best time is probably at the very end of the lesson or the beginning of the next lesson, as a summary/revision.

Lesson aims

- paraphrase at sentence level using passives, synonyms, negatives, replacement subjects

Further practice in:

- affixes
- words with different meanings in general English
- stress within words
- word sets – synonyms, antonyms

Introduction

1 Revise affixes, e.g., *re~, un~, in~, out~, ~ize, ~al, ~ty, ~ly, ~ion, ~ive*. Do this by dividing the class into small groups. Give each group one affix. Allow three or four minutes. The group which can list the most words is the winner.

2 Revise words describing graphs (from Unit 5). Draw a line graph on the board. The line should rise and fall, sharply and gradually, have a peak and a point where it levels off. Point to each part of the line and ask students to give you the appropriate verb and adverb. Alternatively, draw your own line graph and describe it. Students should try to draw an identical line graph from your description while you are talking.

Exercise A

1 Tell students to make a table with three columns and 18 rows in their notebooks. Go through the example in the Course Book. Set the exercise for individual work and pairwork checking. Tell students to use their dictionaries if they need to check meanings, grammatical category, etc. Feed back with the whole class, building the first three columns of the table in the Answers section on the board. Ask students to say what general meanings they can give for the words.

2 Refer to the example (*cost*) in the Course Book. Ask students to suggest a form of *cost* which is an adjective (*costly*). Set for pairwork. Students should try to fill as many empty boxes as possible with

words with appropriate affixes. They should continue to use their dictionaries to check meanings and spellings. Note that it is possible to use the past participle of a verb as an adjective if there is no other possibility. Feed back with the whole class, checking meanings of the words added to the table.

3 Add a fourth column on the board and give it the heading *Marketing/agribusiness meaning*. Underline or highlight the words as shown in the table below, and with the whole class, ask students to suggest (or find in their dictionaries) meanings specific to agribusiness/marketing for these words.

4 Work in a similar way with the fifth column, *Marketing/agribusiness synonym*. Limit the synonyms to those for the underlined words.

5 Set for pairwork. Feed back with the whole class, checking pronunciation.

Language note

Rules in language are made to be broken. The suffix *~ly* normally makes an adjective into an adverb, but there are common cases where this is not what happens. Here, the noun *cost* is made into an adjective *costly* by the addition of *~ly*. Other anomalies are:

hard – hardly

poor – poorly

kind – kindly

Note that *cost* is neutral, whereas *costly* is a stance adjective – it suggests that the writer or speaker thinks the price is too high. In this regard, *expensive* is not a true synonym since it does not have the same connotation. *Pricey*, however, does.

Answers

Model answers:

Noun	Verb	Adjective	Marketing/agribusiness meaning	Marketing/agribusiness synonym
attraction	attract	attractive	(n) a tourist or leisure venue	(n) draw
core		core	(adj) essential, e.g., *Marketing is a core skill.*	(adj) fundamental
cost	cost	costly	(n) the amount of money needed to make a product	(n) expense, overhead
cycle	cycle	cyclical	(n) the life of a product from its beginning to its end	(n) life-span
design	design	designed	(v) create the concept and/or look for a product	(v) develop
extension	extend	extended, extensive	(v) to make sales of a product last longer	(v) prolong
feature	feature	featured	(n) an element of a product or service	(n) element, benefit
field	field		(n) a business area, e.g., *She's skilled in the field of marketing.*	(n) area
grade	grade	graded	(v) to sort a fruit or vegetable product by size or quality	(v) size, classify
growth	grow	grown	(n) increase of, e.g., sales, production, income	(n) increase, improvement
key		key	(adj) essential, e.g., *This is a key decision.*	(adj) important
margin	marginalize	marginal	(n) the difference between cost and the selling price of a product	(n) difference
maturity	mature	mature	(adj) a product which has been available for a long time and is familiar to consumers	(adj) established
mix	mix	mixed	(n) a combination of elements in marketing	(n) combination
proposal	propose	proposed	(n) a formal suggestion for a business offering	(n) offer, plan
segment	segment	segmented	(v) to divide, e.g., *to segment the market*	(v) divide
stability	stabilize	stable	(v) numbers become stable; a market can stabilize = sales remain much the same	(v) level off
stage	stage	staged	(n) one of a number of different sales periods through which a product goes	(n) phase

5

one syllable	core, cost, field, grade, growth, key, mix, stage
Oo	cycle, feature, margin
oO	design, extend, mature, segment (v)*
oOo	attraction, proposal
Ooo	stabilize

*Stress pattern changes to oO when a noun

Language note

Some grammatical forms of the words may have particular meanings that are unlikely (though probably not impossible) in a business context, e.g., *field* (v), *stage* (v). Point this out if necessary but don't spend much time on these words.

Exercise B

1 Set for pairwork discussion. Students should refer to the words they have looked at in Exercise A to help describe the stages shown in the diagrams. Monitor but don't assist. Feed back with the whole class, checking that students can give the topic of the diagram, and the meanings of the vertical and horizontal axes. Elicit words which can be used from Exercise A (underlined below, including synonyms).

2 Set for pairwork discussion. Remind students about words they have already studied for describing trends in graphs. Feed back with the whole class, asking one or two students to describe the trends for each. Make sure that students use the present simple tense to talk about the stages as this is a *process*.

Answers

Model answers:

1 The diagrams show two product life cycles: of a new flavour yoghurt and a new-variant wheat. The vertical axis represents money/sales and the horizontal axis represents time. As can be seen, a product goes through a series of stages over time and the unit cost, cash flow and sales vary with each stage.

Development phase: Ideas for a product are developed. A product may be designed and tested or trialled.

Introduction phase: The new product is launched into a particular market segment.

Growth phase: People become aware of the product and start to buy it. So sales grow.

Maturity phase: The product is selling well, so competitors may copy it and start to sell their own version. The product has now become established, the market has stabilized and the margins are reasonable.

Decline phase: In the final phase, fewer and fewer people buy the product and margins will shrink until eventually the product is withdrawn, although this will happen at different times with different types of product.

2 **Sales:** In the development phase there are no sales, obviously. After introduction, sales are few. Then they start to increase until the market is mature, when they level off. When the product is in the decline phase, sales fall.

Unit costs: To begin with, the company may have a lot of expenses (for promotion, distribution, adapting or buying production machinery, etc.) so the unit costs are high. As sales increase, the unit costs start to fall.

Cash flow: In the beginning, cash flow is negative, which means that the company is spending more than it is earning. Although cash comes in from the introduction phase onwards, it takes some time before money from sales means that the company is earning more than it is spending and cash flow becomes positive. This happens during the maturity stage.

Note that companies often use marketing strategies to try to extend the life of a product, though this is not shown – it will be discussed in later lessons.

Exercise C

As well as requiring the use of antonyms, this exercise checks that students have understood the diagrams in Exercise B. Set for individual work and pairwork checking. Feed back with the whole class. A good way to do this is to use an OHT with blanks for the blue words (see additional resources section, Resource 6B).

Answers

1 Cash flow is negative during the development phase.

2 In the introduction phase, unit costs are high.

3 Cash flow becomes positive in the maturity phase.

4 Increasing sales mean that unit costs are reduced.

5 Sales are at their peak once the market has stabilized/matured.

6 Sales may start to fall/drop/decline when the market reaches saturation.

Exercise D

Introduce the idea of paraphrasing – or restating. Elicit from the students the main ways to do this at sentence level, namely:

- using different grammar
- using different words
- reordering the information

Write these points on the board. Also make the point very strongly that a paraphrase is not a paraphrase unless 90 per cent of the language is different. There are some words which must remain the same, but these are very few, and are likely to be words specific to the subject, such as *cash flow*. It is best to try to use all three of the above strategies, if possible.

Students should look carefully at the corrected sentences from Exercise C and then compare them with the paraphrases. The first step is to identify which sentences match. Set for individual work and pairwork checking. It may be helpful for the students if you reproduce the corrected sentences from Exercise C and the sentences in Exercise D on strips of paper so that they can move them around. Both sets of sentences are reproduced in the additional resources section (Resource 6C) to facilitate this.

Feed back with the whole class. A good way to do this is to reproduce the sentences on OHTs, with each sentence cut into a separate strip. Lay the sentences on the OHP one at a time, as you agree what is the correct match.

Once the sentences are correctly paired, ask students to locate the parts of each sentence which seem to match. They will need to look at the overall meaning of each phrase, using what they know about the subject, to make sure that the phrases are similar. Set for pairwork. Feed back with the whole group, using the OHT strips and highlighting the matching parts with coloured pens.

Answers

1 Cash flow is negative during the development phase.

c While the product is being designed and trialled, there are many expenses but no income.

2 In the introduction phase, unit costs are high.

e It costs a lot to produce each unit early in the cycle.

3 Cash flow becomes positive in the maturity phase.

a It is not until the maturity phase is reached that cash flow is no longer negative.

4 Increasing sales mean that unit costs are reduced.

f Once the products start to sell, the company can make them more cheaply.

5 Sales are at their peak once the market has stabilized/matured.

b Peak sales figures are achieved at the maturity phase.

6 Sales may start to fall/drop/decline when the market reaches saturation.

d There may be a reduction in business if there are too many competitors.

A final step is to discuss the changes that have been made in detail. Students should refer to the list of types of changes you have written on the board. Look at each paraphrase with the class and ask students what changes have been made. Be specific about the types of vocabulary or grammar changes. For example, in the first answer above, the paraphrase reorders the information, uses time clause/passive, uses replacement subject *there*, changes all vocabulary.

Exercise E

Put students in pairs to discuss Figure 3, and which product is most likely to have which type of sales cycle. Feed back with the whole group. If you wish, you could write some descriptive sentences on the board.

Answers

Model answer:

Graph line A shows a sales cycle that better fits a product/agriproduct such as a new model tractor where the expectation is a period of high sales followed by a sharp fall off when a new model is introduced. Graph line B shows the more gradual development of sales and a very gradual (if any) decline associated with a new tourism product in an assumed 10-year period.

Exercise F

1/2 Set for individual work. Go round and check the students' work, giving advice if necessary.

3 Set for pairwork. Once the students have guessed the products, feed back with the whole group.

4 Set for individual work or for homework. Tell students to try to follow the advice for paraphrasing in Exercise D, i.e., to reorder the information and to change vocabulary and grammar as far as possible. You may wish to refer students to the *Vocabulary bank* at this point to provide a reminder for grammar structures to use.

Closure

Discussion:

1 Can students think of other products with life cycles similar to those shown in Figures 1 and 2?

2 What can an agribusiness do to prolong the life of a product with a life cycle similar to the one in Figure 1?

6.2 Reading

6.2 Reading identifying subject–verb–object in long sentences • paraphrasing

A Discuss these questions.

1 What do companies need to know about their products?
2 How many ways of analyzing a product are there?

B Study Figure 1.

1 Imagine you are a mixed farm, dairy and arable. Give an example of a product that might be included in 'product development'.
2 What sort of agriproduct would you use a market penetration strategy for?
3 How could you diversify away from traditional dairy and arable activities and products to exploit new and more radical opportunities?
4 You currently sell your products using traditional wholesalers. In what ways could you innovate your marketing to increase your income and profitability selling the same range of products?

	Existing products	New products
Existing markets	market penetration	product development
New markets	market development	diversification

Figure 1: *Ansoff Matrix*

C Look at the photo, the title, the introduction and the first sentence of each paragraph on the opposite page.

1 What will the text be about?
2 Using your ideas from Exercise A, B and C1, write some research questions.

D Read the text. Does it answer your questions?

E Study the highlighted sentences in the text. Find and underline the subject, verb and object or complement in each sentence. See *Skills bank*

F Two students paraphrased part of the text.

1 Which part of the text are these paraphrases of?
2 Which paraphrase is better? Why?

Student A

Applecart has therefore demonstrated its commercial promise.

However, agri-tourism is fairly new and Applecart must not make judgements based on conventional tourism.

Luckily, Applecart weighed up the situation and took a positive approach.

Philippe and Anna consulted a marketing consultant who, from the beginning, showed them new methods of analyzing their situation.

Student B

Thus, there is no doubt that Applecart's business has the potential to turn a profit.

It would be dangerous for Applecart to base business decisions solely on examples from conventional tourism, given that agribusiness is quite a new field.

Applecart decided to meet the challenges of the market head on, after considering all the options.

A marketing specialist introduced Philippe and Anna to many new ways of looking at the market.

G Work in groups. Write a paraphrase of a different part of the text. See *Vocabulary bank*

48

Understand your product for business success!

Central to the marketing mix is the agriproduct itself. A company must understand the potential of its product in order to build a marketing strategy. There are many ways a product can be analyzed, with varying beneficial impacts. One way is by looking at the psychological benefits it offers. Another is by analyzing the product life cycle. A third could be by using analytical tools such as the Ansoff Matrix.

Applecart Farm is a traditional agribusiness that has diversified into agritourism products. Its decision was instinctive not analytical, but before talking about analysis, let us be clear what we mean by an 'agri-tourism product'. One definition is that it is a product or merchandise created by a mainstream agricultural business for the tourism market. Applecart remains a mixed farm with a beautiful farmhouse and outbuildings at its core. Its owners, Philippe and Anna, quite recently diversified into providing B&B packages. They have quickly established a good reputation with the family market for their B&B product. They offer a place to stay, offering double beds, en suite bathrooms, full English breakfasts – examples of functional or concrete features, delivering functional benefits: comfort and nourishment.

Applecart depends on the Internet for its marketing. Its website communicates not just traditional features, but the psychological benefits of a visit to Applecart. How does it achieve this? It has a video presentation of the farm which identifies and highlights the feel-good factors: images of home baking, sunshine, animals and friendly hosts. Applecart is an organic producer, which is advantageous and well highlighted. The basic information has been augmented to communicate intangible as well as tangible features: Applecart's functional features – nice rooms, healthy food, service and environment, are exploited to communicate something extra, the psychological benefits. This special mix of functional and psychological benefits and features appears to be an important characteristic of this growth sector and a key factor in visitors' decision to purchase.

Applecart has proven commercial potential, therefore. But having created its agriproduct and had some early success, it needs to be aware of factors that could affect future success, such as the product life cycle. Put simply, understanding the product life cycle enables an agribusiness to reduce risk, plan finances and manage cash flow. Tourism products generally tend to have quite long life cycles, with slow development and growth, and a long maturity and steady decline. But agritourism is relatively new and Applecart must not become complacent or make assumptions based on traditional tourism. The importance of this was highlighted when visitor numbers and inflow of funds began to fall recently. Fortunately, it assessed its situation and took a proactive approach. Philippe and Anna called in a marketing consultant who, from the start, introduced them to new ways of analyzing their situation. One of the tools the consultant used was the Ansoff Matrix, (reproduced in Figure 1, opposite).

An Ansoff Matrix assumes four strategies for achieving growth. The first strategy is market penetration, where the business targets growth with existing products in its current market segment to gain market share. The second is market development: creating growth by targeting existing products on new market segments. Product development and product diversification are the third and fourth strategies. If we look at our example, Applecart is currently upgrading its website to further optimize the psychological benefits of its existing B&B, as well as rebranding, repackaging and enhancing the product to target a new market – retired people in the 'shoulder periods' between the high and low seasons – before it is relaunched next year, with the changes delivering an expected 25% increase in visitors. It is diversifying and preparing to market its soft drinks and cheeses for the local market. It was established at the outset that successful companies need to understand their markets and their products. With minimal outside input, Applecart recognized it needed to understand all aspects of its product and market better to build a sustainable and profitable business. It has begun that process, establishing an effective marketing strategy using the three methods mentioned, and taking the first steps towards a product portfolio to improve the flow of funds and protect against the risk to a single-product business. So the message for sustainable business success is 'analyze and plan!' And Applecart heard it.

49

General note

Read the *Skills bank* at the end of the Course Book unit. Decide when, if at all, to refer students to it. The best time is probably after Exercise E, or at the very end of the lesson or the beginning of the next lesson, as a summary/revision.

Lesson aims

- identify the kernel SVO/C of a long sentence

Further practice in:

- research questions
- topic sentences
- paraphrasing

Introduction

Remind the class about techniques when using written texts for research. Ask:

What is it a good idea to do:

- *before reading?* (think of research questions)
- *while you are reading?* (look for topic sentences)
- *after reading?* (check answers to the research questions)

What words in a text signal the development of a topic in a new direction? (markers showing contrast such as *but*, *however*, *at the same time*, *on the other hand*, etc.)

If you wish, refer students to Unit 4 *Skills bank*.

Exercise A

Set for general discussion. Allow students to debate differences of opinion. Encourage them to give examples if they can. Do not correct or give information at this point, as these topics will be dealt with in the text.

Exercise B

1/2 Set for pairwork discussion. Feed back with the whole class. Accept any reasonable answers. If students don't know, don't tell them, but say they may find out in the text.

3 Set for pairwork. Collate key points on the board.

4 Discuss with the whole class. Accept any sensible suggestions. Encourage lateral thinking.

Answers

Model answers:

1 milk/cheese/yoghurt/ice-cream
2 well-established products with a clear market, e.g., commodity products such as cereals where marketing tends to focus on growing existing market share
3 add value by processing raw materials/leasing land for eco-communities/wind farming
4 keep the same product, but innovate the places and sectors targeted, e.g,. retailers, farmers markets, exhibitions and shows

Exercise C

1 Set for individual work. Elicit ideas, but do not confirm or correct.
2 Set for individual work and pairwork checking.

Exercise D

Set for individual work. Feed back with the whole class.

Exercise E

Draw a table with the headings from the Answers section on the board. If you wish, students can also draw a similar table in their notebooks. Explain that in academic writing, sentences can seem very complex. This is often not so much because the sentence structure is highly complex in itself, but that the subjects and objects/complements may consist of clauses or complex noun phrases. Often the verb is quite simple. But in order to fully understand a text, the grammar of a sentence must be understood. Subject + verb + object or complement is the basic sentence structure of English. Students need to be able to locate the subjects, main verbs and their objects or complements.

Elicit from the students the subject, main verb and object for the first sentence. Ask students for the *head word* of each subject, main verb and object (underlined in the table in the Answers section). Write them in the table on the board. Using high-speed questioning, get students to build the whole phrase that constitutes the subject/main verb/object/complement.

Example 1:

Put simply, understanding the product life cycle enables an agribusiness to reduce risk, plan finances and manage cash flow.

What is this sentence about? = understanding

What does understanding do? = it enables

What does it enable? = an agribusiness

Which agribusiness? = an = any agribusiness

Write these head words in the table on the board.

Then elicit the remaining words and add to the table:

Understanding what exactly? = the product life cycle

Give me more information about how this will enable an agribusiness = (it will enable it) to reduce risk, plan finances and manage cash flow

Example 2:

This example shows how to deal with *is* + complement.

So the message for sustainable business success is 'analyze and plan!'

What is this sentence about in general? = business success

More particularly? = a message for businesses to hear

What is the main verb in this sentence? = is

*So what **is** the message?* = 'analyze and plan'

Why? = to be sustainable and successful

The idea is that students should be able to extract something which contains the kernel even if it does not make complete sense without the full phrase.

Ask students to identify the prepositional/adverbial phrase in the fourth sentence (*Put simply* …). Point out that this part contains information which is extra to the main part of the sentence. The sentence can be understood quite easily without it.

Set the remainder of the exercise for individual work followed by pairwork checking. Finally, feed back with the whole class.

You may wish to refer students to the *Skills bank – Finding the main information.*

Answers

Subject	Verb	Object/complement
A company	must understand	the potential of its product in order to build a marketing strategy.
One definition	is	that it is a product or merchandise created by a mainstream agricultural business for the tourism market.
This special mix of functional and psychological benefits and features	appears to be	an important characteristic of …
… understanding the product life cycle	enables	an agribusiness to reduce …
So the message for sustainable business success	is	'analyze and plan!'

Exercise F

Set for individual work and pairwork checking. Make sure that students identify the original phrases in the text first (they appear in paragraph 3 of the main text) before looking at the paraphrases.

Feed back with the whole class. A good way to demonstrate how Student A's text contains too many words from the original is to use an OHT and highlight the common words in colour. (A table giving the sentences plus commentary is included in the additional resources section – Resources 6D and 6E.) Check that students are able to say which parts of the paraphrase match with the original, and which structures have been used.

Answers

1 The first part of paragraph 3.

2 Student B's paraphrase is better, because it uses fewer words from the original text and uses different sentence structures.

Language note

It is important that students understand that when paraphrasing, it is not sufficient to change a word here and there and leave most of the words and the basic sentence structure unchanged. This approach is known as 'patch-writing' and is considered to be plagiarism. It is also important when paraphrasing not to change the meaning of the original – also quite hard to do.

Exercise G

Refer students to the *Vocabulary bank* at this stage. Review paraphrasing skills with the whole class before starting this exercise.

Divide the text into parts. For example, each paragraph can be divided into two so that there are eight different sections (though of course you should not use paragraph 3). Give each section to different students to work on. Alternatively, you could choose one part of the text for all students to work on, for example the second part of paragraph 4. This can be done in class or if you prefer as individual work/homework.

If students are doing the work in class in groups or pairs, a good way to provide feedback is to get them to write their paraphrase on an OHT. Show each paraphrase (or a selection) to the class and ask for comments. Say what is good about the work. Point out where there are errors and ask for suggestions on how to improve it. Make any corrections on the OHT with a different coloured pen.

Closure

1 Divide the class into two teams. Write the five topic sentences from the reading text on strips, or photocopy them from the additional resources section (Resource 6F). One team chooses a topic sentence and reads it aloud. The other team must give the information triggered by that topic sentence. Accept only the actual paragraph content.

2 Dictate the following to the class:

Think of an agribusiness sector or product line that:

... is known for innovation.

... has developed through different stages.

... is strong on marketing.

... has not changed over many years.

... has migrated to new markets.

Students work in pairs to think of sectors or products which fit the descriptions. The first pair to find a sector or product for each category are the winners. Feed back with the whole class.

6.3 Extending skills

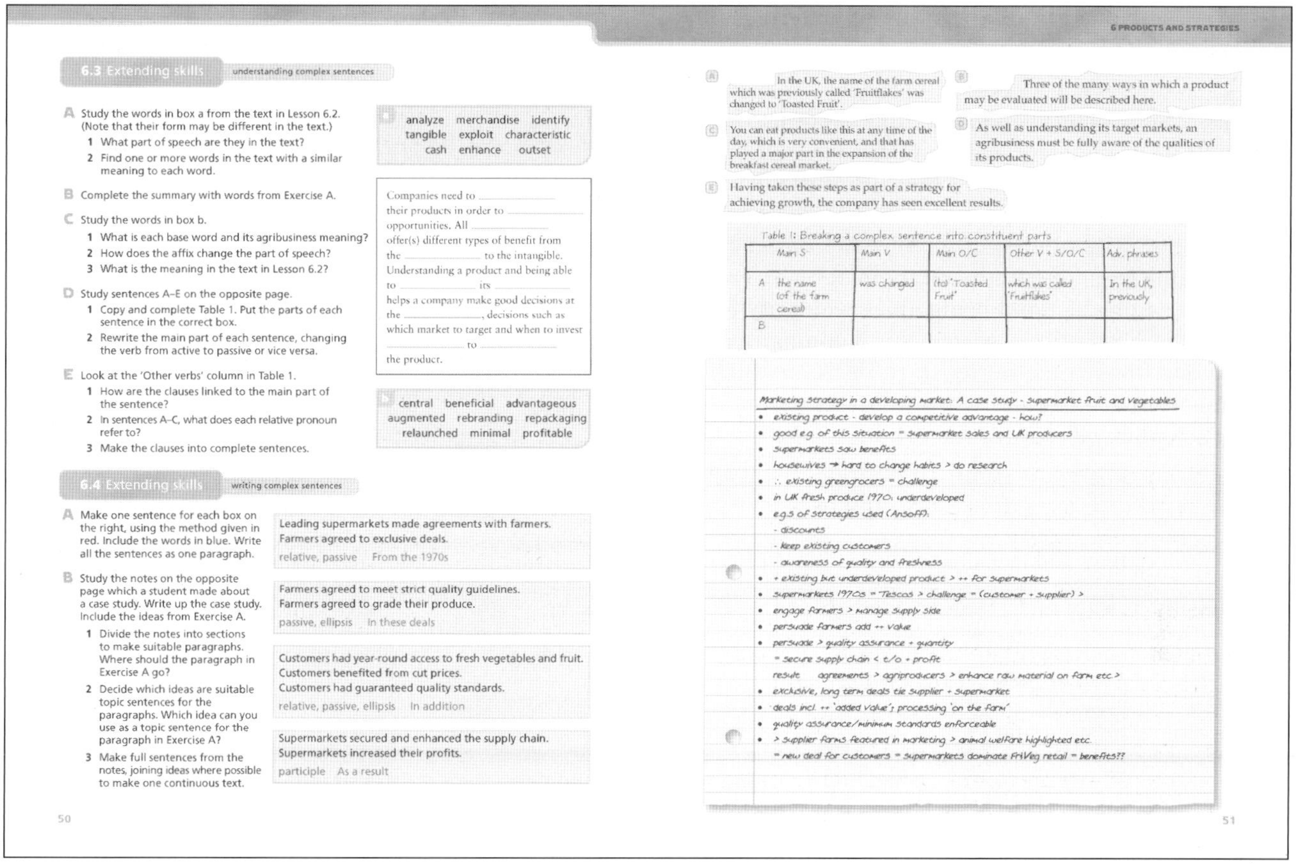

6.3 Extending skills understanding complex sentences

A Study the words in box a from the text in Lesson 6.2. (Note that their form may be different in the text.)
1 What part of speech are they in the text?
2 Find one or more words in the text with a similar meaning to each word.

a
analyze merchandise identify
tangible exploit characteristic
cash enhance outset

B Complete the summary with words from Exercise A.

Companies need to ______ their products in order to ______ opportunities. All ______ offer(s) different types of benefit from the ______ to the intangible. Understanding a product and being able to ______ its ______ helps a company make good decisions at the ______, decisions such as which market to target and when to invest ______ to ______ the product.

C Study the words in box b.
1 What is each base word and its agribusiness meaning?
2 How does the affix change the part of speech?
3 What is the meaning in the text in Lesson 6.2?

b
central beneficial advantageous
augmented rebranding repackaging
relaunched minimal profitable

D Study sentences A–E on the opposite page.
1 Copy and complete Table 1. Put the parts of each sentence in the correct box.
2 Rewrite the main part of each sentence, changing the verb from active to passive or vice versa.

E Look at the 'Other verbs' column in Table 1.
1 How are the clauses linked to the main part of the sentence?
2 In sentences A–C, what does each relative pronoun refer to?
3 Make the clauses into complete sentences.

6.4 Extending skills writing complex sentences

A Make one sentence for each box on the right, using the method given in red. Include the words in blue. Write all the sentences as one paragraph.

Leading supermarkets made agreements with farmers.
Farmers agreed to exclusive deals.
relative, passive From the 1970s

Farmers agreed to meet strict quality guidelines.
Farmers agreed to grade their produce.
passive, ellipsis In these deals

Customers had year-round access to fresh vegetables and fruit.
Customers benefited from cut prices.
Customers had guaranteed quality standards.
relative, passive, ellipsis In addition

Supermarkets secured and enhanced the supply chain.
Supermarkets increased their profits.
participle As a result

B Study the notes on the opposite page which a student made about a case study. Write up the case study. Include the ideas from Exercise A.
1 Divide the notes into sections to make suitable paragraphs. Where should the paragraph in Exercise A go?
2 Decide which ideas are suitable topic sentences for the paragraphs. Which idea can you use as a topic sentence for the paragraph in Exercise A?
3 Make full sentences from the notes, joining ideas where possible to make one continuous text.

50

6 PRODUCTS AND STRATEGIES

A In the UK, the name of the farm cereal which was previously called 'Fruitflakes' was changed to 'Toasted Fruit'.

B Three of the many ways in which a product may be evaluated will be described here.

C You can eat products like this at any time of the day, which is very convenient, and that has played a major part in the expansion of the breakfast cereal market.

D As well as understanding its target markets, an agribusiness must be fully aware of the qualities of its products.

E Having taken these steps as part of a strategy for achieving growth, the company has seen excellent results.

Table 1: Breaking a complex sentence into constituent parts

	Main S	Main V	Main O/C	Other V + S/O/C	Adv. phrases
A	the name (of the farm cereal)	was changed	(to) 'Toasted Fruit'	which was called 'Fruitflakes'	In the UK, previously
B					

Marketing strategy in a developing market: A case study - supermarket fruit and vegetables
- existing product - develop a competitive advantage - how?
- good e.g. of this situation = supermarket sales and UK producers
- supermarkets saw benefits
- housewives → hard to change habits > do research
- ∴ existing greengrocers = challenge
- in UK fresh produce 1970: underdeveloped
- e.g.s of strategies used (Ansoff):
  - discounts
  - keep existing customers
  - awareness of quality and freshness
- + existing but underdeveloped product > ++ for supermarkets
- supermarkets 1970s = Tescos > challenge = (customer + supplier) >
- engage farmers > manage supply side
- persuade farmers add ++ value
- persuade > quality assurance + quantity
  = secure supply chain < t/o + profit
  result agreements > agriproducers > enhance raw material on farm etc. >
- exclusive, long term deals tie supplier + supermarket
- deals incl. ++ 'added value'; processing 'on the farm'
- quality assurance/minimum standards enforceable
- > supplier farms featured in marketing > animal welfare highlighted etc.
  = new deal for customers = supermarkets dominate F/Veg retail = benefits??

51

Lesson aims

- study sentence structure in more detail
- identify the main information in:
  - an active sentence
  - a passive sentence
  - a complex sentence with participles
  - a complex sentence with embedded clauses

Further practice in:

- vocabulary from Lesson 6.2

Introduction

Ask students to see how many phrases or compound nouns they can make with the word *market* or *marketing*. Tell students to brainstorm a list in pairs. Feed back with the whole class.

Possible answers: *market characteristics, market share, market leader, market segment, target market, market segmentation, market size, market trends, growth market, agri-tourism market, gap in the market, marketing mix, marketing strategy, marketing approach*, etc.

Exercise A

Ask students to study the words in the box and to find the words in the text. Set for individual work and pairwork checking. Tell students *not* to use their dictionaries to begin with but to use what they know to guess meanings and parts of speech. If necessary, they should use dictionaries when checking in pairs. Deal with any common problems with the whole class.

Answers

Model answers (paragraph numbers are given in brackets):

Word	Part of speech	Similar meaning
analyze (1)	v (T)	assess (3)
merchandise (1)	n (U)	products (1, 3, 4)
identify (2)	v (T)	be aware of (3)
tangible (2)	adj	concrete (2)
exploit (2)	v (T)	optimize (4)
characteristic (2)	n (C)	aspect (4)
cash (3)	usually n (U); here part of compound *cash flow*, so qualifying	funds (3)
enhance (4)	v (T)	improve (4)
outset (4)	n (U)	start (3)

Exercise B

Set for individual work and pairwork checking. Students should make use of all the words they have discussed in Exercise A (i.e., the synonyms as well as the words in the box). Feed back with the whole class.

Answers

Model answers:

Companies need to analyze/assess their products in order to exploit/optimize opportunities. All merchandise/products offer(s) different types of benefit from the tangible/concrete to the intangible. Understanding a product and being able to identify/be aware of its characteristics/aspects helps a company make good decisions at the outset/start, decisions such as which market to target and when to invest funds/cash to enhance/improve the product.

Language note

The use of words as synonyms often depends on the context. For example, although the base meanings of *identify* and *be aware of* are not exactly synonymous, they could both be used in the third and final sentence of the summary text with very little difference in meaning.

Exercise C

Set for pairwork. Feed back with the whole class. Note that not all the base words have specifically agribusiness meanings. Tell students to explain the meaning in agribusiness terms as far as possible.

Answers

Model answers (paragraph numbers are given in brackets):

Word	Base and meaning	Effect of affix	Meaning in text
central (Intro)	centre (n, C) – the middle	*~al* = adjective ending	most important
beneficial (Intro)	benefit 1. (n, C) – something good or useful that a product has 2. (v, T) – to have a good effect on; be useful	*~(ci)al* = adjective ending	helpful, useful, good
advantageous (2)	advantage (n, C) – something that helps to bring a good result, especially compared to other people/companies	*~ous* = adjective ending	desirable, beneficial
augmented (2)	augment (v, T) – to make something bigger	*~ed* = past participle ending, makes an adjective	something extra added to a core product
rebranding (4)	brand (n, C) – the name which a business uses to identify a product as theirs. The brand also represents all the values and qualities which the business wants to associate with its product	*re~* = do something again *~ing* = gerund form indicating the process of creating a brand	changing or making over the brand/image to freshen, stimulate interest, reposition in the market
repackaging (4)	package 1. (n, C) – parcel 2. (v, T) – make into a package for selling	*re~* = do something again *~ing* 1. = noun ending: the materials used to make the package 2. = gerund ending: the process of making a package	giving the product new features a new image so that they appear new and fresh (to a new target market segment)
relaunched (4)	launch (v, T) – put a new product on the market; begin selling a product	*re~* = do something again *~ed* = past simple ending	start to sell their B&B service again as if it was new
minimal (4)	minimum (adj) – the smallest possible; can only be used in front of the noun, i.e., *The minimum amount* not *The amount is minimum*; sometimes a noun with *of*: *a minimum of …*	*~al* = adjective ending; difference *minimum*: only an adjective, and can be used after *be*: *The amount is minimal.*	very little
profitable (4)	profit (n, C/U) – money from business activity	*~able* = adjective ending	produces profits

Exercise D

1 Copy the table headings from the Answers section onto the board and complete the example with the students. Tell them that when they look at the 'Other verbs' column they may well find several, and should number each verb and subject/object/complement section separately. Point out that the order of each part of the sentence is not reflected in the table: The table is just a way to analyze the sentences.

Set the rest of the sentences for individual work and pairwork checking. Feed back with the whole class. Draw their attention to the 'main' parts of the sentence: it is very important in reading that they should be able to identify these. Notice also that the main parts can stand on their own and make complete sentences.

2 Set for individual work. If the clause is active, it should be changed to passive, and vice versa.

Answers

Model answers:

1 See table below.

2 Possible answers:

A (The company) changed the name of the farm cereal to 'Toasted Fruit'.

B I/the author will describe three (of the many) ways (to evaluate products).

C Products like this can be eaten at any time of the day.

D The qualities of (its) products must be fully understood (by an agribusiness/agribusinesses).

E Excellent results were seen (by the company).

Exercise E

This exercise involves looking carefully at the dependent clauses in sentences A–E.

1 Say that these clauses have special ways to link them to the main part of the sentence. Do this exercise with the whole class, using an OHT of the table in Exercise D, and a highlighter pen to mark the relevant words. (A version of the table without underlining is included in the additional resources section – Resource 6G.) Go through the clauses, asking students what words or other ways are used to link the clauses to the main part of the sentence.

2 Set for individual work and pairwork checking. Students should look at each sentence and identify the antecedents of the relative pronouns. You could ask them to use a highlighter pen or to draw circles and arrows linking the words.

3 Students must be able to get the basic or kernel meaning of the clause. Take sentence A as an example and write it on the board. Point out that the relative pronouns and other ways of linking these clauses to the main clause will need to be changed or got rid of. Students should aim to write something that makes good sense as a complete sentence. They can break a sentence into shorter sentences if necessary.

Set the remaining clauses for individual work. Feed back with the whole class. Accept anything that makes good sense.

	Main subject	Main verb	Main object/complement	Other verbs + their subjects + objects/complements	Adverbial phrases
A	the name (of the farm cereal)	was changed	(to) 'Toasted Fruit'.	which* was called 'Fruitflakes'	In the UK previously
B	Three of the many ways	will be described		in which a product may be evaluated	here.
C	You	can eat	products like this	1. … which is very convenient 2. that has played a major part in the expansion of the breakfast cereal market.	at any time of the day
D	an agribusiness	must be	aware (of the qualities of its products.)	As well as understanding its target markets, …	fully
E	the company	has seen	excellent results.	Having taken these steps as part of a strategy for achieving growth, …	

*underlined text = means by which dependent clause is joined to main clause

Answers

1 See table in Exercise D above. Sentences A–C use relative clauses. D and E use participle clauses (*As well as understanding*; *Having taken*).

2 A *which* = the name of the farm cereal

B *which* = ways

C 1. *which* = the ability to eat products at any time of the day

2. *that* = the ability to eat products at any time of the day

3 Possible answers:

A The farm cereal was previously called 'Fruit Flakes'.

B A product may be evaluated in many ways.

C 1. Products like this are very convenient.

2. Products like this have played a major part in the expansion of the breakfast cereal market.

D An agribusiness must (also) be aware of its target markets.

E The company took these steps as part of a strategy for achieving growth.

Language note

A dependent clause contains a verb and a subject and is a secondary part of a sentence. It is dependent because it 'depends' on the main clause. A main clause can stand by itself as a complete sentence in its own right (usually). A dependent clause always goes with a main clause and cannot stand by itself as a sentence in its own right.

Dependent clauses are typically joined to main clauses with certain types of words: for example, relative pronouns (e.g., *who, which*, etc.), linking adverbials (e.g., *if, when, before, although, whereas*, etc.), words associated with reporting speech (e.g., *that*, a *Wh~* word such as *what* or *why*), and so on.

Some dependent clauses are non-finite; that is, they don't have a 'full verb' but a participle form (e.g., *having finished, opening*) and the subject may not be stated.

For more on this, see a good grammar reference book.

Closure

Write the following underlined beginnings and endings of words on the board or dictate them. Ask students to give the complete word. Accept alternatives and other parts of speech.

bene(fit)
(profit)able
ob(ject)
supp(ort)
(aug)ment
(serv)ice
(analy)sis
(custo)mer
(advanta)geous
(strat)egy
(succ)ess

6.4 Extending skills

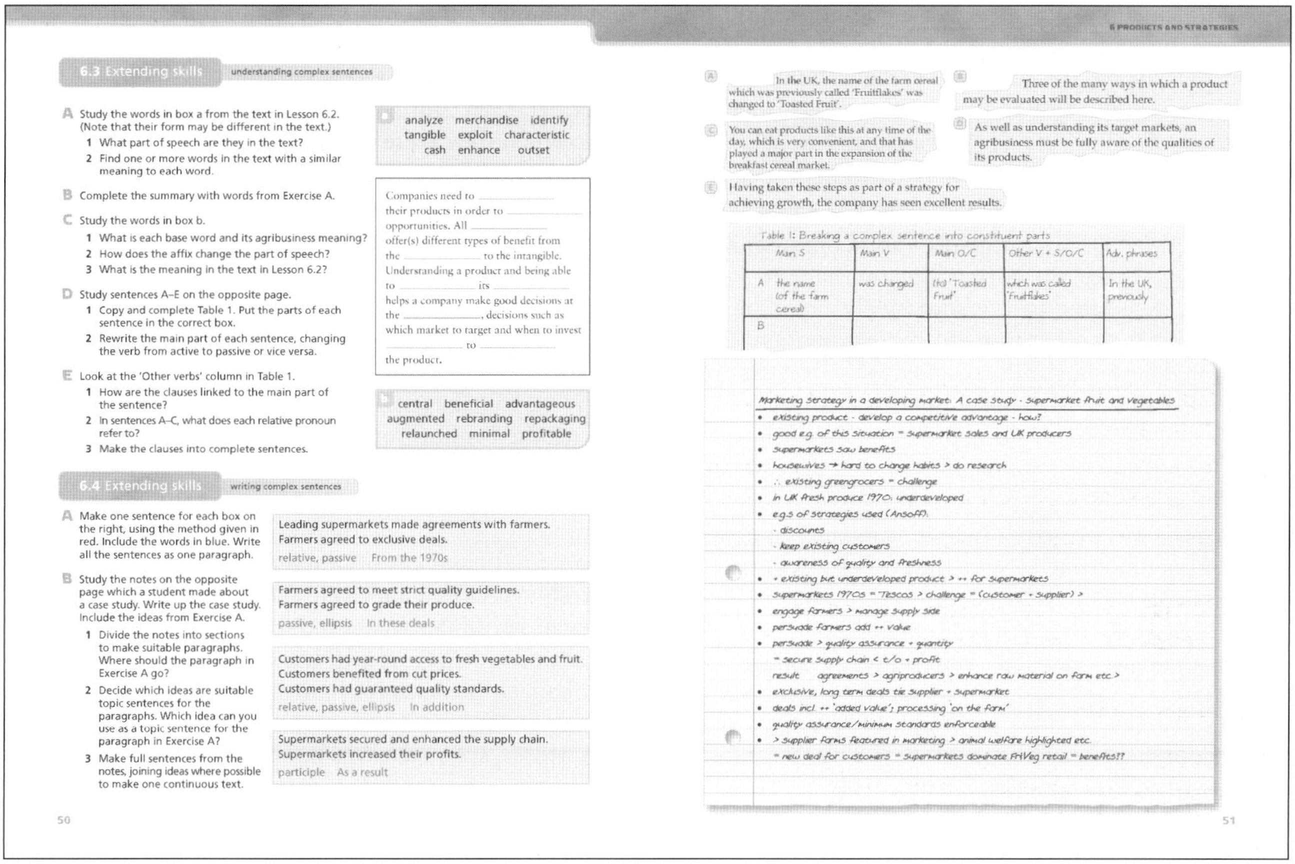

6 PRODUCTS AND STRATEGIES

6.3 Extending skills understanding complex sentences

A Study the words in box a from the text in Lesson 6.2. (Note that their form may be different in the text.)
1 What part of speech are they in the text?
2 Find one or more words in the text with a similar meaning to each word.

a: analyze merchandise identify tangible exploit characteristic cash enhance outset

B Complete the summary with words from Exercise A.

Companies need to ________ their products in order to ________ opportunities. All ________ offer(s) different types of benefit from the ________ to the intangible. Understanding a product and being able to ________ its ________ helps a company make good decisions at the ________, decisions such as which market to target and when to invest ________ to ________ the product.

C Study the words in box b.
1 What is each base word and its agribusiness meaning?
2 How does the affix change the part of speech?
3 What is the meaning in the text in Lesson 6.2?

D Study sentences A–E on the opposite page.
1 Copy and complete Table 1. Put the parts of each sentence in the correct box.
2 Rewrite the main part of each sentence, changing the verb from active to passive or vice versa.

E Look at the 'Other verbs' column in Table 1.
1 How are the clauses linked to the main part of the sentence?
2 In sentences A–C, what does each relative pronoun refer to?
3 Make the clauses into complete sentences.

b: central beneficial advantageous augmented rebranding repackaging relaunched minimal profitable

6.4 Extending skills writing complex sentences

A Make one sentence for each box on the right, using the method given in red. Include the words in blue. Write all the sentences as one paragraph.

Leading supermarkets made agreements with farmers.
Farmers agreed to exclusive deals.
relative, passive From the 1970s

B Study the notes on the opposite page which a student made about a case study. Write up the case study. Include the ideas from Exercise A.
1 Divide the notes into sections to make suitable paragraphs. Where should the paragraph in Exercise A go?
2 Decide which ideas are suitable topic sentences for the paragraphs. Which idea can you use as a topic sentence for the paragraph in Exercise A?
3 Make full sentences from the notes, joining ideas where possible to make one continuous text.

Farmers agreed to meet strict quality guidelines.
Farmers agreed to grade their produce.
passive, ellipsis In these deals

Customers had year-round access to fresh vegetables and fruit.
Customers benefited from cut prices.
Customers had guaranteed quality standards.
relative, passive, ellipsis In addition

Supermarkets secured and enhanced the supply chain.
Supermarkets increased their profits.
participle As a result

50

A In the UK, the name of the farm cereal which was previously called 'Fruitflakes' was changed to 'Toasted Fruit'.

B Three of the many ways in which a product may be evaluated will be described here.

C You can eat products like this at any time of the day, which is very convenient, and that has played a major part in the expansion of the breakfast cereal market.

D As well as understanding its target markets, an agribusiness must be fully aware of the qualities of its products.

E Having taken these steps as part of a strategy for achieving growth, the company has seen excellent results.

Table 1: Breaking a complex sentence into constituent parts

	Main S	Main V	Main O/C	Other V + S/O/C	Adv. phrases
A	the name (of the farm cereal)	was changed	(to) 'Toasted Fruit'	which was called 'Fruitflakes'	In the UK, previously
B					

Marketing strategy in a developing market: A case study - supermarket fruit and vegetables
- existing product - develop a competitive advantage - how?
- good e.g. of this situation = supermarket sales and UK producers
- supermarkets saw benefits
- housewives → hard to change habits > do research
- ∴ existing greengrocers = challenge
- in UK fresh produce 1970: underdeveloped
- e.g.s of strategies used (Ansoff):
  - discounts
  - keep existing customers
  - awareness of quality and freshness
- + existing but underdeveloped product > ++ for supermarkets
- supermarkets 1970s = Tescos > challenge = (customer + supplier) >
- engage farmers > manage supply side
- persuade farmers add ++ value
- persuade > quality assurance + quantity
  = secure supply chain < t/o + profit
  result agreements > agriproducers > enhance raw material on farm etc >
- exclusive, long term deals tie supplier + supermarket
- deals incl ++ 'added value'; processing 'on the farm'
- quality assurance/minimum standards enforceable
- > supplier farms featured in marketing > animal welfare highlighted etc.
  = new deal for customers = supermarkets dominate Fr/Veg retail = benefits??

51

Lesson aims

- write complex sentences:
  - with passives
  - joining with participles
  - embedding clauses
  - adding prepositional phrases

Further practice in:

- writing topic sentences
- expanding a topic sentence into a paragraph

Introduction

Ask students to think about and discuss the following questions:

1 Is the approach to marketing agribusinesses that supply fresh produce more likely to be product-led or market-led? (In other words, are these companies more likely to develop products in response to perceived market needs (= market-led) or are they more likely to develop new products and then try to promote them (= product-led)?) (Fresh produce examples include: frozen peas, chopped vegetables, fruit salads, etc.)

2 Why?

3 What are some of the marketing strategies a fresh vegetable producer might use to create competitive advantage?

Exercise A

Set for individual work and pairwork checking. If necessary, do the first box with the whole class. Make sure students understand that they should write the four sentences as a continuous paragraph.

Feed back with the whole class. Accept any answers that make good sense. Point out where the phrases in blue act as linkers between the sentences to make a continuous paragraph.

Answers

Possible answer:

From the 1970s onwards, agreements were made by farmers with the leading supermarkets in deals which were exclusive. In these deals, it was agreed by the farmers that they would meet strict quality standards and grade their produce. Customers benefited from cut prices and year-round access to fresh vegetables and fruit and, in addition, were rewarded by quality standards which were guaranteed. As a result, having secured and enhanced the supply chain, supermarkets increased their profits.

Exercise B

In this exercise, students are required to use all they have practised about sentence structure as well as revise what they know about topic sentences and paragraphing.

Set for pairwork. Do not feed back after each question but allow students to work through the questions, proceeding to write up the whole text. They will need to decide where is the best place for the paragraph in Exercise A, and should also add this to their text. Students can change the wording and add extra phrases to help the flow of the text, as long as the sense remains the same.

If possible, pairs should write their text on an OHT. Select two or three OHTs for display and comment by the whole class. Make any corrections on the text yourself with a coloured pen. Alternatively, circulate the transparencies to other pairs to correct and comment on. These pairs then display the corrected work and explain why they have made the corrections.

Answers

1/2 Possible answers:

Paragraph divisions are given below, with the possible topic sentences underlined. Note that other answers may be possible.

- existing product – develop a competitive advantage – how?
- good e.g. of this situation = supermarket sales and UK producers
- supermarkets saw benefits
- housewives → hard to change habits > do research
- ∴ existing greengrocers = challenge
- in UK fresh produce 1970: underdeveloped

- e.g.s of strategies used (Ansoff):
  – discounts
  – keep existing customers
  – awareness of quality and freshness
- + existing but underdeveloped product > ++ for supermarkets

---

- supermarkets 1970s > challenge = (customer + supplier) >
  – *engage* farmers > manage supply side
  – *persuade* farmers add ++ value
  – *persuade* > quality assurance + quantity
  = secure supply chain < t/o + profit

---

result → agreements > agriproducers > enhance raw material on farm etc.>
– exclusive, long term deals tie supplier + supermarket
– deals incl. ++ 'added value'; processing 'on the farm'
– quality assurance/minimum standards *enforceable*
– > supplier farms featured in marketing > animal welfare highlighted etc.
= new deal for customers = supermarkets dominate Fr&Veg retail = benefits??

3 How does a company introduce an existing product and then convert customers? An example is the supermarkets in the 1970s in the UK and the fresh fruit and vegetable market. This had been dominated by specialist greengrocers. The supermarkets saw the opportunity provided by new technologies and methods in the supply chain to sell to their mass market. Their traditional customers were not easy to convince.

The supermarkets which wanted to further consolidate their positions in the market had to plan their strategies very carefully. Using the Ansoff model, their aim was to penetrate further a market in which they were already present and grow the sales of fruit and vegetables. Their strategy was implemented using a combination of discounted prices and campaigns to raise awareness of the benefits that their products offered such as reliable quality and freshness.

Companies like Tesco provided a good model for how to develop this area of their business. From the 1970s onwards, agreements were made by farmers with the leading supermarkets in deals which were exclusive. In these deals, it was agreed by the farmers that they would meet strict quality standards and grade their produce. Customers benefited from cut prices and year-round access to fresh vegetables and fruit and, in addition, were rewarded by quality standards which were guaranteed. As a result, having secured and enhanced the supply chain, supermarkets increased their profits.

So the real challenge for the supermarkets was to engage the supply side and persuade them to do some of the work for them. Their aims were achieved by offering very substantial and secure contracts to the farmers which tied them not only to delivering standard, graded produce but also tied them to prices which enabled the supermarkets to take on the competition in the high streets. The new arrangements helped establish quality systems. And as a result of this they secured the aim of substantially increasing market penetration and enriching the experience of their existing customer base. This in turn has delivered customer loyalty and increased turnover and profits for the supermarkets. But is it a virtuous circle?

Closure

Give students some very simple three- or four-word SVO/C sentences from the unit (or make some yourself) and ask them to add as many phrases and clauses as they can to make a long complex sentence. Who can make the longest sentence?

For example:

Cash flow becomes positive.

→ It is not until the maturity phase, when sales are at their peak and demand for the product is established, that **cash flow becomes positive** and the agribusiness may have extra funds to start investing in a new product which they will develop over the next few years … (48 words)

Extra activities

1 Work through the *Vocabulary bank* and *Skills bank* if you have not already done so, or as a revision of previous study.

2 Look at Figure A. What advantages does a portfolio of products have over a single one?

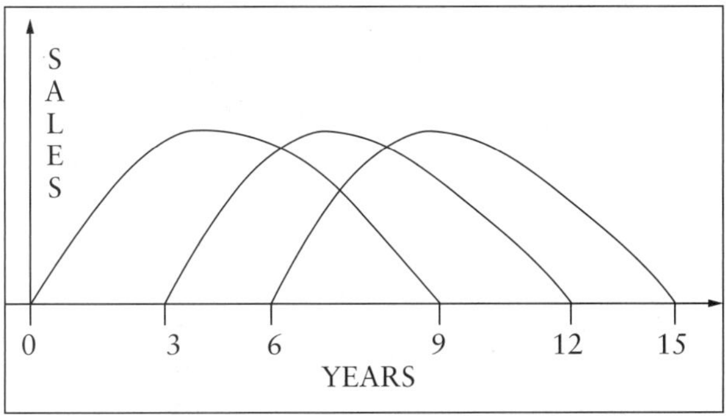

Figure A: *Strategy based on product portfolio*

Suggested answer: Figure A shows a business/marketing strategy based on a portfolio of products. The advantages include: as one product declines, there is another to replace it. Similarly, if one product is compromised – by a health scare, for example – the agribusiness may survive by promoting others in the range. The strategy also better ensures continuity for employment, investment and cash flow, although it may take longer to achieve overall business profitability.

3 Use the *Activity bank* (Teacher's Book additional resources section, Resource 6A).

A Set the wordsearch for individual work (including homework) or pairwork.

Answers

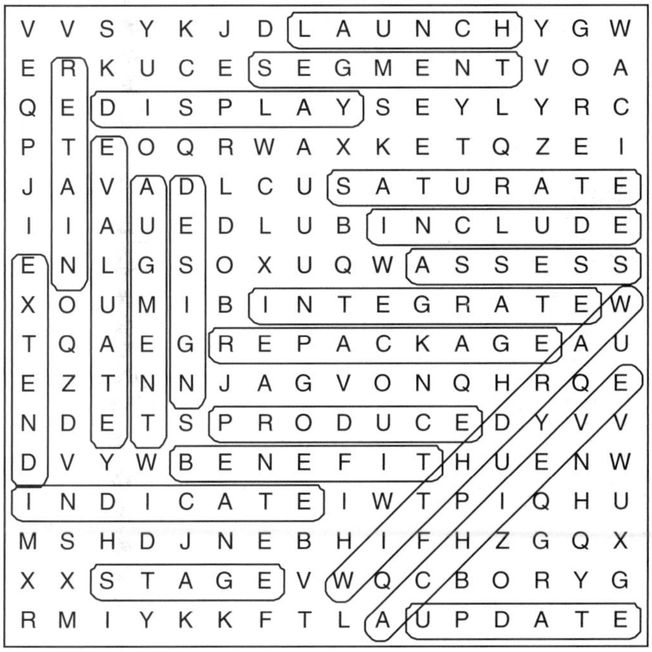

Verb	Noun
achieve	achievement
assess	assessment
augment	augmentation
benefit	benefit
design	design
display	display
evaluate	evaluation
extend	extension
include	inclusion
indicate	indication

Verb	Noun
integrate	integration
launch	launch
produce	production
repackage	repackaging
retain	retention
saturate	saturation
segment	segment
stage	stage
update	update
withdraw	withdrawal

B Students work in pairs or small groups and try to think of word pairs. They should be able to explain the meaning.

Alternatively, photocopy (enlarged) the words from the additional resources section (Resource 6H) and cut up into cards. Put the A and B words into separate envelopes. Put students into groups of four. Make one set of A and one set of B words for each group. Give one pair in each group the A words and the other pair the B words. Each pair takes it in turns to pick a word from their envelope. The other pair must try to find a word from their own envelope which can go with it.

Accept all reasonable word pairs. Possible pairs are:

A	B
added	value
analytical	tools
buying	habits
cash	flow
competitive	edge
customer	services
decline	phase
discounted	prices
extension	strategy
functional	benefits

A	B
good	practice
life	cycle
market	segments
marketing	strategy
mature	market
portfolio	strategy
quality	assurance
quality	standards
shoulder	period

7 THE EFFECTIVE PRODUCTION MANAGER

In this unit, the focus is on effective production management. The skills, techniques and qualities required for success are all explored. The first listening extract discusses concepts such as complexity and interrelationships and explains tools including Critical Path Analysis (CPA) and Ishikawa diagrams and their application in scheduling production activities. The second listening extract uses the idea of an experiment in eco-community development to explore the role of the production manager as project leader.

Skills focus

Listening

- understanding speaker emphasis

Speaking

- asking for clarification
- responding to queries and requests for clarification

Vocabulary focus

- compound nouns
- fixed phrases from agribusiness
- fixed phrases from academic English
- common lecture language

Key vocabulary

See also the list of fixed phrases from academic English in the *Vocabulary bank* (Course Book page 60).

adaptive management
adjust
allocate
brainstorming
co-housing
concept
controlled market
cooperative
Critical Path Analysis (CPA)
distribution
empathic skills
fishbone (Ishikawa) diagram
forklifts
fossil fuels
free market
greenhouse
helicopter
helicoptering
infrastructure
intervene
intervention
interventionist
Lifecycle Assessment (LCA)
micro-manager
minimize
mutual
occupation
oversee
pallet
premium quality
production manager
quality assurance
reflective skills
resilience
scheduling
Site of Special Scientific Interest (SSSI)
storage
systems thinking
whole-farm strategy

7.1 Vocabulary

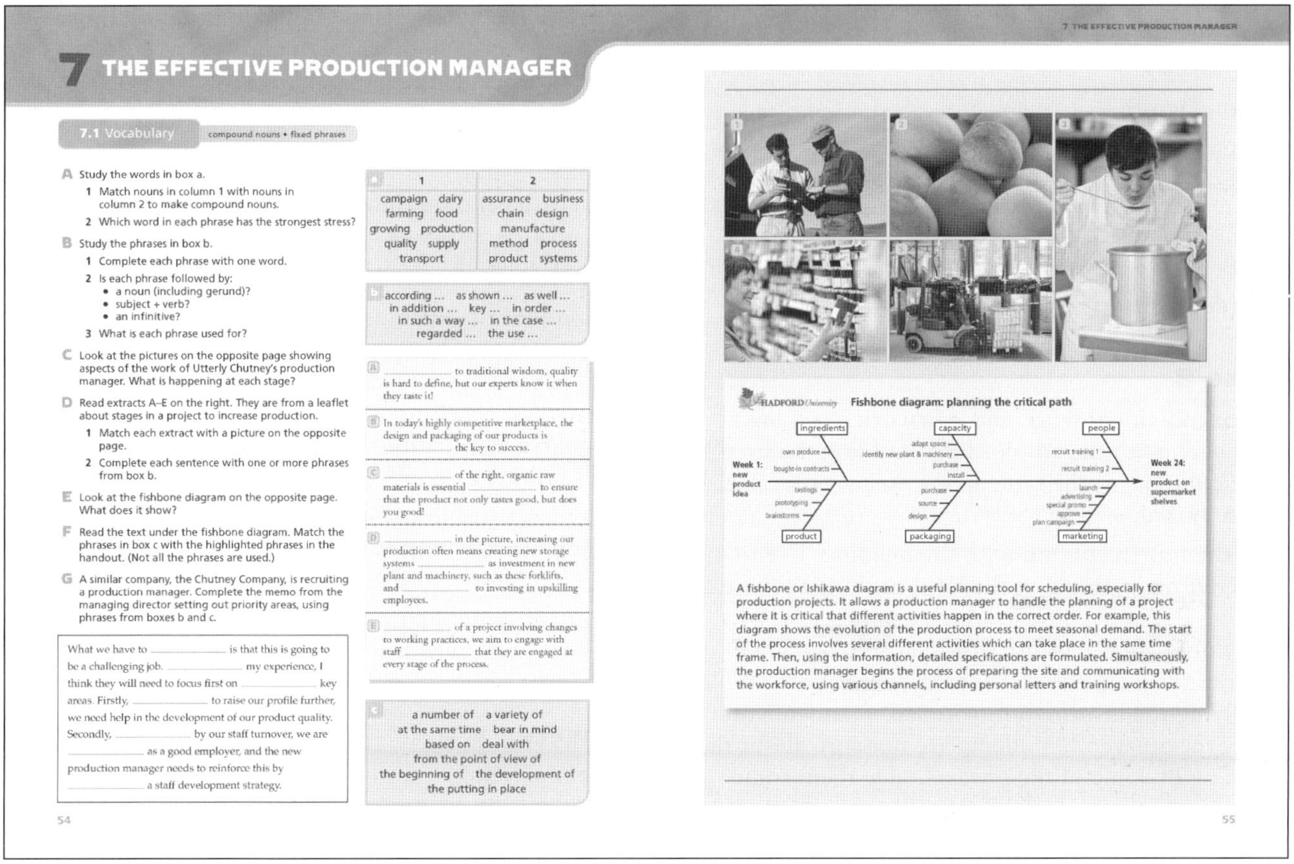

7 THE EFFECTIVE PRODUCTION MANAGER

7.1 Vocabulary compound nouns • fixed phrases

A Study the words in box a.
1 Match nouns in column 1 with nouns in column 2 to make compound nouns.
2 Which word in each phrase has the strongest stress?

B Study the phrases in box b.
1 Complete each phrase with one word.
2 Is each phrase followed by:
- a noun (including gerund)?
- subject + verb?
- an infinitive?

3 What is each phrase used for?

C Look at the pictures on the opposite page showing aspects of the work of Utterly Chutney's production manager. What is happening at each stage?

D Read extracts A–E on the right. They are from a leaflet about stages in a project to increase production.
1 Match each extract with a picture on the opposite page.
2 Complete each sentence with one or more phrases from box b.

E Look at the fishbone diagram on the opposite page. What does it show?

F Read the text under the fishbone diagram. Match the phrases in box c with the highlighted phrases in the handout. (Not all the phrases are used.)

G A similar company, the Chutney Company, is recruiting a production manager. Complete the memo from the managing director setting out priority areas, using phrases from boxes b and c.

What we have to ______ is that this is going to be a challenging job. ______ my experience, I think they will need to focus first on ______ key areas. Firstly, ______ to raise our profile further, we need help in the development of our product quality. Secondly, ______ by our staff turnover, we are ______ as a good employer, and the new production manager needs to reinforce this by ______ a staff development strategy.

54

a

1	2
campaign dairy farming food growing production quality supply transport	assurance business chain design manufacture method process product systems

b

according ... as shown ... as well ... in addition ... key ... in order ... in such a way ... in the case ... regarded ... the use ...

A ______ to traditional wisdom, quality is hard to define, but our experts know it when they taste it!

B In today's highly competitive marketplace, the design and packaging of our products is ______ the key to success.

C ______ of the right, organic raw materials is essential ______ to ensure that the product not only tastes good, but does you good!

D ______ in the picture, increasing our production often means creating new storage systems ______ as investment in new plant and machinery, such as these forklifts, and ______ to investing in upskilling employees.

E ______ of a project involving changes to working practices, we aim to engage with staff ______ that they are engaged at every stage of the process.

c

a number of a variety of at the same time bear in mind based on deal with from the point of view of the beginning of the development of the putting in place

7 THE EFFECTIVE PRODUCTION MANAGER

HADFORD University **Fishbone diagram: planning the critical path**

A fishbone or Ishikawa diagram is a useful planning tool for scheduling, especially for production projects. It allows a production manager to handle the planning of a project where it is critical that different activities happen in the correct order. For example, this diagram shows the evolution of the production process to meet seasonal demand. The start of the process involves several different activities which can take place in the same time frame. Then, using the information, detailed specifications are formulated. Simultaneously, the production manager begins the process of preparing the site and communicating with the workforce, using various channels, including personal letters and training workshops.

55

General note

Read the *Vocabulary bank* at the end of the Course Book unit. Decide when, if at all, to refer your students to it. The best time is probably at the very end of the lesson or the beginning of the next lesson, as a summary/revision.

Lesson aims

- understand and use some general academic fixed phrases
- understand and use fixed phrases and compound nouns from the discipline

Introduction

1 Revise some noun phrases (noun + noun, adjective + noun) from previous units. Give students two or three minutes to make word stars with a base word, trying to find as many possible combinations as they can (preferably without having to look at dictionaries).

For example:

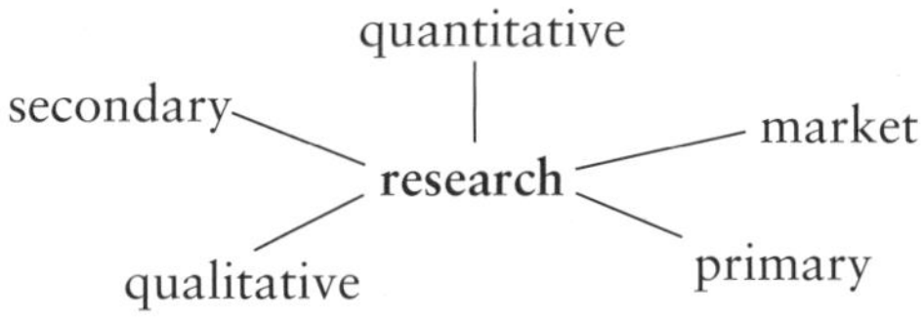

Other base words which could be used are *company, job, agribusiness, management*. If they are stuck for ideas, tell them to look back at previous units.

2 Introduce the topic of the lesson by looking at the pictures on page 55. Discuss what students know about agribusiness production processes. What is the difference between the production of, e.g., chutney, face cream and ginger beer?

Exercise A

Set for individual work and pairwork checking. Feed back with the whole class, making sure that the stress pattern is correct. Ask students to suggest other fixed phrases which could be made using the words in column 2.

Answers

Model answers:

cam'paign design
'dairy product
'farming business
'food manufacture
'growing method
pro'duction process
'quality assurance
su'pply chain
'transport systems

Exercise B

1/2 Set for individual work and pairwork checking. Feed back with the whole class, building the first three columns of the table in the Answers section on the board.

3 Add the fourth column with the heading 'Use to …'. Give an example of the kind of thing you are looking for, i.e., a phrase which can describe why you would choose to use this fixed phrase. Elicit suggestions from the students to complete the table, supplying the information yourself if students don't know the answer. If students are not sure about the meaning of some of the phrases, give them some example sentences and tell them that you will look further at how they are used shortly. Leave the table on the board as you will return to it.

Answers

Model answers:

Phrase		Followed by …	Use to …
according	to	noun	cite/refer/identify agent
as shown	in/by	noun/gerund	indicate a diagram or table
as well	as	noun/gerund	add information
in addition	to	noun/gerund	add information
key	to	noun/gerund	highlight/empahsis
in order	to	infinitive	give the purpose for doing something
in such a way	that*	subject + verb	give the result of doing something
in the case	of	noun/gerund	mention something
regarded	as	noun	give the reputation of something
the use	of	noun	refer to the use of something

**as to* is also possible after *in such a way*, although in this exercise, one word is required

Exercise C

Set for pairwork. Students should try to identify what each picture represents. One pair can describe each picture to the whole class. On the board, build up as many key words to describe the process as students can come up with. If students don't know some important words, tell them they will meet them shortly.

Answers

Answers depend on the students.

Exercise D

Explain that the information from the leaflet goes with the pictures they have just discussed. Each extract (A–E) goes with one picture. Students should first read the extracts, checking words they can't guess in the dictionary. They should not pay attention to the spaces at this point.

1 Set for pairwork. Feed back with the whole class. Add any key words which might have been useful in Exercise C to the board.

2 Set for individual work. Refer back to the table in Exercise B, which will help students to choose the correct phrase. Feed back with the whole class.

Answers

Model answers:

Picture	Extract
1	E <u>In the case</u> of a project involving changes to working practices, we aim to engage with staff <u>in such a way</u> that they are engaged at every stage of the process.
2	C <u>The use</u> of the right, organic raw materials is essential <u>in order</u> to ensure that the product not only tastes good, but does you good!
3	A <u>According</u> to traditional wisdom, quality is hard to define, but our experts know it when they taste it!
4	B In today's highly competitive marketplace, the design and packaging of our products is <u>regarded</u> as the key to success.
5	D <u>As shown</u> in the picture, increasing our production often means creating new storage systems <u>as well</u> as investment in new plant and machinery, such as these forklifts, and <u>in addition</u> to investing in upskilling employees.

If you wish, ask students to return to the table in Exercise B and write one sentence for each of the fixed phrases to show their meaning. If you can put this into the context of a production process which students are very familiar with, such as a recipe, so much the better.

Exercise E

Introduce the fishbone chart. If students have not seen one before, you can explain that it is called an Ishikawa diagram, and that it is a highly important tool in project management. It shows one way to plot a critical path – which analyzes the stages in a production process and makes sure that they interrelate correctly – and ensure the correct ordering of activities.

Set for pairwork discussion. Feed back with the whole group, making sure that students understand the concept behind the chart. Do not correct or confirm students' views of the content at this point.

Subject note

Ishikawa diagrams were introduced in the 1960s by one of the leading characters in modern management. His name was Kaoru Ishikawa and he pioneered quality management processes in the Kawasaki shipyards. The name *fishbone* has been widely adopted due to the diagrams' resemblance to a piscine skeleton.

Exercise F

Set for individual work and pairwork checking. Students should use their dictionaries if they are not sure of the meaning of the phrases. Note that some phrases can be used for the same thing – it is a good idea to use a different word to avoid repetition. Ask students to say which sentence goes with which part of the chart.

Answers

Model answers:

A fishbone or Ishikawa diagram is a useful planning tool, especially (*for*) from the point of view of production projects. It allows a production manager to (*handle*) deal with the planning of a project where it is critical that (*different*) a number of/a variety of activities happen in the correct order. For example, this diagram shows (*the evolution of*) the development of the production process to meet seasonal demand. (*The start of*) The beginning of the process involves (*several*) a number of/a variety of different activities which can take place in the same time frame. Then, (*using*) based on the information, detailed specifications are formulated. (*Simultaneously*) At the same time, the production manager begins the process of preparing the site and communicating with the workforce, using (*various*) a number of/a variety of channels, including personal letters and training workshops.

Language note

The fixed phrases here are used in a situation which describes a series of chronological stages. However, the same words can be used when writing or talking in more general abstract academic terms, for example when introducing an essay or lecture or piece of research. This use of these words will be covered later in the unit.

Exercise G

Set for pairwork. Feed back with the whole class.

Answers

What we have to bear in mind is that this is going to be a challenging job. Based on my experience, I think they will need to focus first on a number of key areas. Firstly, in order to raise our profile further, we need help in the development of our product quality. Secondly, as shown by our staff turnover, we are regarded as a good employer, and the new production manager needs to reinforce this by putting in place a staff development strategy.

Closure

Tell students to cover the text and then describe:

- the typical main stages of a production process, e.g., chutney manufacture
- what a typical fishbone diagram looks like and what it includes

7.2 Listening

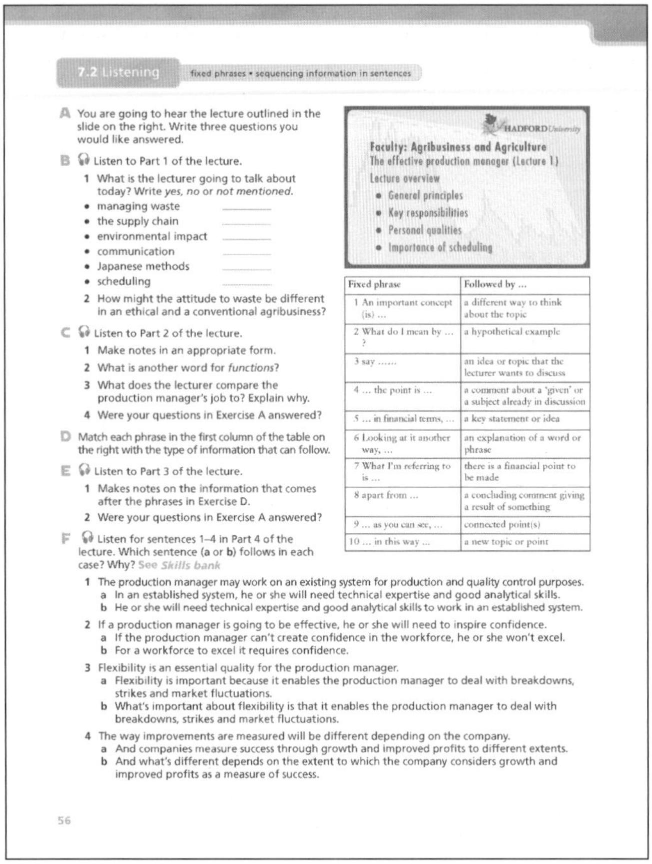

7.2 Listening fixed phrases • sequencing information in sentences

A You are going to hear the lecture outlined in the slide on the right. Write three questions you would like answered.

B Listen to Part 1 of the lecture.

1 What is the lecturer going to talk about today? Write *yes*, *no* or *not mentioned*.
- managing waste ______
- the supply chain ______
- environmental impact ______
- communication ______
- Japanese methods ______
- scheduling ______

2 How might the attitude to waste be different in an ethical and a conventional agribusiness?

C Listen to Part 2 of the lecture.

1 Make notes in an appropriate form.
2 What is another word for *functions*?
3 What does the lecturer compare the production manager's job to? Explain why.
4 Were your questions in Exercise A answered?

D Match each phrase in the first column of the table on the right with the type of information that can follow.

E Listen to Part 3 of the lecture.

1 Makes notes on the information that comes after the phrases in Exercise D.
2 Were your questions in Exercise A answered?

F Listen for sentences 1–4 in Part 4 of the lecture. Which sentence (**a** or **b**) follows in each case? Why? See *Skills bank*

1 The production manager may work on an existing system for production and quality control purposes.
  a In an established system, he or she will need technical expertise and good analytical skills.
  b He or she will need technical expertise and good analytical skills to work in an established system.

2 If a production manager is going to be effective, he or she will need to inspire confidence.
  a If the production manager can't create confidence in the workforce, he or she won't excel.
  b For a workforce to excel it requires confidence.

3 Flexibility is an essential quality for the production manager.
  a Flexibility is important because it enables the production manager to deal with breakdowns, strikes and market fluctuations.
  b What's important about flexibility is that it enables the production manager to deal with breakdowns, strikes and market fluctuations.

4 The way improvements are measured will be different depending on the company.
  a And companies measure success through growth and improved profits to different extents.
  b And what's different depends on the extent to which the company considers growth and improved profits as a measure of success.

HADFORD University
Faculty: Agribusiness and Agriculture
The effective production manager (Lecture 1)
Lecture overview
- General principles
- Key responsibilities
- Personal qualities
- Importance of scheduling

Fixed phrase	Followed by ...
1 An important concept (is) ...	a different way to think about the topic
2 What do I mean by ... ?	a hypothetical example
3 say	an idea or topic that the lecturer wants to discuss
4 ... the point is ...	a comment about a 'given' or a subject already in discussion
5 ... in financial terms, ...	a key statement or idea
6 Looking at it another way, ...	an explanation of a word or phrase
7 What I'm referring to is ...	there is a financial point to be made
8 apart from ...	a concluding comment giving a result of something
9 ... as you can see, ...	connected point(s)
10 ... in this way ...	a new topic or point

56

Lesson aims

- improve comprehension through recognition of fixed phrases and what follows them in terms of words/type of information
- understand how information can be sequenced in different ways within a sentence, e.g., for emphasis (see *Skills bank*)

Further practice in:

- understanding fractured text

General note

Read the *Skills bank – 'Given' and 'new' information in sentences* at the end of the Course Book unit. Decide when, if at all, to refer students to it. The best time, as before, is probably at the very end of the lesson or the beginning of the next lesson, as a summary/revision. Alternatively, use the *Skills bank* in conjunction with Exercise F.

Introduction

Review key vocabulary by writing a selection of words from Lesson 7.1 on the board and asking students to put them into phrases of two or more words.

Exercise A

Remind students about preparing for a lecture. If you wish, review Unit 1 *Skills bank – Making the most of lectures*. Remind students that, when they begin their talks, lecturers usually provide their listeners with an outline in order to aid comprehension. Elicit from the students the kinds of signpost words lecturers might use (e.g., *To start with, ..., Firstly, ..., I'll begin/start by ...ing*, etc.). If necessary, refer students to Unit 5.

Refer students to the lecture slide. Tell them to look at the title and bullet points and to list ideas/make questions for each bullet point. At this stage, do not explain any words from the slide, or allow students to check in their dictionaries, as the meanings will be dealt with in the lecture. Set the exercise for pairwork.

Feed back with the whole class: Ask several students to read out their questions. Write some of the questions on the board.

Exercise B

Tell students they are going to hear the introduction to the lecture – not the whole thing. Give students time to read questions 1 and 2. Remind them they will only hear the recording once. Play Part 1. Allow students to compare their answers.

Feed back. Confirm the correct answers. Note that 'scheduling' is mentioned on the slide, but not in the introduction, so we have no idea if this will be covered or not.

Answers

1

waste	yes
the supply chain	no
environmental impact	yes
communication	not mentioned
Japanese methods	not mentioned
scheduling	not mentioned

2 A conventional agribusiness may look at the cost of waste, whereas the ethical business may focus on minimizing environmental impact.

Transcript 2.1

Part 1

Good morning, everyone. What I'm going to talk about today is one of the key questions for an agribusiness: What makes an effective production manager? To put it another way: What does a typical job description for an effective production manager involve? Bearing in mind, of course, that the production manager's job is different depending on the type of company. The large traditional food manufacturer may look at things like waste from the point of view of minimizing cost. In the case of the small ethical food producer, which we'll be looking at an example of later, the production manager's priority may be minimizing environmental impact. What I mean is that a production manager for a small ethical company may focus on waste in the whole supply chain. Anyway, we're not going to look at supply chains now, I mean today.

OK, now, um … in later lectures, we will be considering how companies, um … a variety of companies, manage costs at the same time as … I mean in such a way as to reflect their principles. Today, however, we're going to deal with what the effective production manager does.

Exercise C

Refer students to the first point on the lecture slide ('general principles'). Ask students to suggest an appropriate type of notes. The key word here is *principles*, which should instantly trigger the idea of bullet points (see Unit 1).

Give students time to read the questions. Ask if they remember the model of business given in Unit 1. Can they suggest the other key words which accompany *functions*? (i.e., *responsibility* and *activity*). Put the first few letters of some of the words on the board to help students if needed. Write these on the board. Play Part 2.

Put students in pairs to compare their notes and discuss the questions. With the whole class, ask students how many answers to their questions in Exercise A they heard.

Build the notes from the Answers section on the board, at the same time checking the answers to questions 2 and 3.

Language note

Native English speakers often use the pronoun *they* to mean *he* or *she* when the sex of a person is not known. This is quite acceptable in spoken English but should be avoided in formal writing.

Answers

Model answers:

1

Functions of an effective production manager

<u>Plan and schedule existing activities</u>

- raw materials
- delivery services, etc.

<u>Plan and schedule new production projects</u>

- budget
- increased capacity
- new products

<u>Allocate resources</u>

- human resources
- equipment

<u>Track production</u>

- daily/weekly/monthly/annually

<u>Track costs</u>

- labour/hr
- material
- distribution

2 Roles.

3 Film producer: they are at the centre of activities and responsible for making sure those activities are monitored and controlled instantly, in real time.

4 Answers depend on students' questions.

Transcript 2.2

Part 2

As we have seen, the production manager's job can be very different depending on the type of agribusiness manufacturer they work for. As we know, producing food products or other agriproducts is not the same as managing an agriservice business such as an IT consultancy. I'll be talking about a day in the life of a production manager later, but their general responsibilities include scheduling things like materials and delivery, and planning budgets and capacity for current and new products and services, as well as allocating resources such as human resources and equipment. And, of course, they monitor costs such as production, labour, materials and distribution. The production manager's job in both manufacturing and service agribusinesses is similar in some ways to the job of a … er … film producer, for example. He or she plans and schedules new film projects, is responsible for budgets and resources, and manages teams and day-to-day activities, tracking progress and monitoring labour and material costs. In other words, he or she makes the instant decisions and adjustments to the process which ensure that everything from raw material supplies to distribution is working effectively within budget.

Feed back with the whole class. Deal with sentences 1 and 2 first. Tell students that all the sentences are correct, but sentence **a** 'sounds better' when it comes after the first sentence. This is because of the way that sentences go together and the way in which information is organized in a sentence. Draw the table on the previous page on the board. Show how the underlined words in the second sentence have been mentioned in the first sentence. In the second sentence the underlined words are 'old' or 'given' information. When sentences follow each other in a conversation (or a piece of writing), usually the 'given' information comes in the first part of a sentence.

Now look at sentences 3 and 4. These are different. The normal choice would be the **a** sentences. However, here the speaker wanted to emphasize the idea of 'important' and 'different'. So a *Wh~* cleft sentence structure was used, which changes the usual order of information. Show this on the table as on the previous page. This 'fronting' of information has the effect of special focus for emphasis.

Language note

In English, important information can be placed at the beginning or at the end of a sentence. There are two types of important information. The first part of the sentence contains the topic and the second part contains some kind of information or comment about the topic. Usually the comment is the more syntactically complicated part of the sentence.

Once a piece of text or a piece of conversation (i.e., a piece of discourse) has gone beyond the first sentence, a 'given'/'new' principle operates. Information which is 'given', in other words that has already been mentioned, goes at the beginning of the sentence. Normally speaking, information which is new goes at the end of the sentence. So in the second sentence of a piece of discourse, an aspect of the comment from the previous sentence may become the topic. Thus the topic of the second sentence, if it has already been mentioned in the previous sentence, is also 'given'. Of course, the given information may not be referred to with exactly the same words in the second sentence. Other ways to refer to the given information include reference words (*it, he, she, this, that, these, those,* etc.) or vocabulary items with similar meanings.

Information structure is covered in the *Skills bank* in the Course Book unit.

Transcript 🎧 2.4

Part 4

Now … er … let me see … oh dear, we're running short of time. Anyway, never mind, I want to say something about the *personal* qualities required of an effective production manager.

The production manager may work on an existing system for production and quality control purposes. In an established system, he or she will need technical expertise and good analytical skills. The production manager will need to be patient and not make recommendations until he or she has gathered enough information. In the process, he or she will have to develop good, trusting relationships with the people working for him or her. If a production manager is going to be effective, he or she will need to inspire confidence. If the production manager can't create confidence in the workforce, he or she won't excel.

Flexibility is an essential quality for the production manager. What's important about flexibility is that it enables the production manager to deal with breakdowns, strikes and market fluctuations. And if the production manager doesn't have a flexible approach, he or she won't adjust to changing situations. He or she may be based on the production floor, but it will be important not to get too involved in micro-managing the production process.
Micro-managing – taking care of small everyday details – is the responsibility of lower management levels. The production manager needs to have the ability to 'helicopter' – to see the whole picture.

Their main target is to ensure improvements in the performance of the business. The way improvements are measured will be different depending on the company. And what's different depends on the extent to which the company considers growth and improved profits to be a measure of success. Organizations like the chutney business will certainly put a high value on the quality of their products and their reputation as a responsible company. The production manager will need to be able to interpret the policies of their company and synthesize the results of his or her analysis in effective day-to-day strategies. The production manager gets satisfaction from tangible results. And to achieve these, he or she will need to communicate with colleagues and managers at all levels and in different departments and also with suppliers and customers.

I suppose one of the main qualities that every production manager needs is the ability to take criticism. There will always be times when they're going to make unpopular decisions. People will often resent the things they want to do at first, but if the production manager is good at the job, he or she will win them over. It's about a very practical sort of leadership – the most important quality a production manager brings to the role.

Closure

Ask students to look at the list of the qualities that they might associate with an effective production manager and try to rank them as either a. *essential*, b. *useful* or c. *non-essential*.

analytical
assertive
calm under pressure
experienced
financially astute
good communicator
good empathic skills
persuasive
quick thinker
reflective
strategic thinker
strong leader

7.3 Extending skills

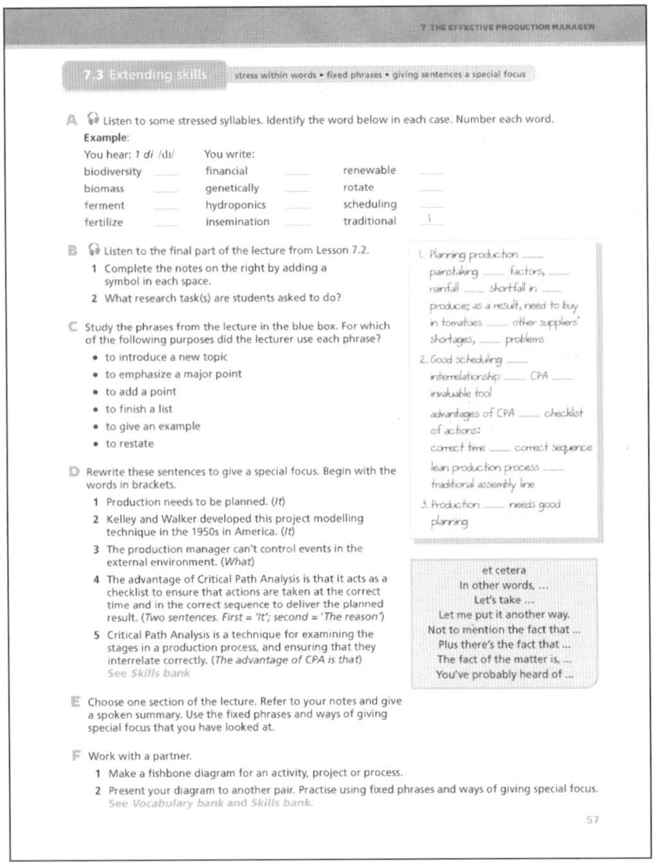

7 THE EFFECTIVE PRODUCTION MANAGER

7.3 Extending skills   stress within words • fixed phrases • giving sentences a special focus

A Listen to some stressed syllables. Identify the word below in each case. Number each word.

Example:

You hear: *1 di* /dɪ/   You write:

biodiversity	____	financial	____	renewable	____
biomass	____	genetically	____	rotate	____
ferment	____	hydroponics	____	scheduling	____
fertilize	____	insemination	____	traditional	1

B Listen to the final part of the lecture from Lesson 7.2.

1 Complete the notes on the right by adding a symbol in each space.

2 What research task(s) are students asked to do?

C Study the phrases from the lecture in the blue box. For which of the following purposes did the lecturer use each phrase?

- to introduce a new topic
- to emphasize a major point
- to add a point
- to finish a list
- to give an example
- to restate

D Rewrite these sentences to give a special focus. Begin with the words in brackets.

1 Production needs to be planned. (*It*)

2 Kelley and Walker developed this project modelling technique in the 1950s in America. (*It*)

3 The production manager can't control events in the external environment. (*What*)

4 The advantage of Critical Path Analysis is that it acts as a checklist to ensure that actions are taken at the correct time and in the correct sequence to deliver the planned result. (*Two sentences. First = 'It'; second = 'The reason'*)

5 Critical Path Analysis is a technique for examining the stages in a production process, and ensuring that they interrelate correctly. (*The advantage of CPA is that*)

See *Skills bank*

et cetera
In other words, …
Let's take …
Let me put it another way.
Not to mention the fact that …
Plus there's the fact that …
The fact of the matter is, …
You've probably heard of …

E Choose one section of the lecture. Refer to your notes and give a spoken summary. Use the fixed phrases and ways of giving special focus that you have looked at.

F Work with a partner.

1 Make a fishbone diagram for an activity, project or process.

2 Present your diagram to another pair. Practise using fixed phrases and ways of giving special focus.

See *Vocabulary bank* and *Skills bank*.

57

Lesson aims

- extend knowledge of fixed phrases commonly used in lectures
- give sentences a special focus (see *Skills bank*)

Further practice in:

- stress within words

Introduction

As in Units 3 and 5, tell students to ask you questions about the information in the lecture in Lesson 7.2 as if you were the lecturer. Remind them about asking for information politely. If they need to revise how to do this, tell them to look back at the *Skills bank* for Unit 3.

Exercise A

Remind students of the importance of stressed syllables in words (see the teaching notes for Unit 3, Lesson 3, Exercise A). Play the recording, pausing after the first few to check that students understand the task.

Feed back, perhaps playing the recording again for each word before checking. Ideally, mark up an OHT of the words. Finally, check students' pronunciation of the words.

Answers

biodiversity	4
biomass	10
ferment	9
fertilize	2
financial	11
genetically	7
hydroponics	3
insemination	6
renewable	8
rotate	12
scheduling	5
traditional	1

Transcript 2.5

1 tra'ditional
2 'fertilize
3 hydro'ponics
4 biodi'versity
5 'scheduling
6 insemi'nation
7 ge'netically
8 re'newable
9 fer'ment
10 'biomass
11 fi'nancial
12 ro'tate

Exercise B

Write these words on the board and ask students to say what symbols you can use for them when taking notes. Put the symbols on the board.

is, means	=
because	∵
for example	e.g.
leads to*	→
also; lots	+
therefore, so	∴
and	&
a list	numbers or bullet points
or	/

*the arrow has a wide range of possible meanings, including *made, invented, produced, did, causes, results in*, etc.

Tell students they will hear the final part of the lecture. Ask them to read through the notes. Remind them also to listen for their research task. Play Part 5.

Put students in pairs to compare their symbols. Feed back with the whole class, if possible using an OHT of the notes. Discuss acceptable alternatives, e.g., *lean production process & traditional assembly line* instead of *lean production process / traditional assembly line.*

Answers

Model answers:

1
   1. Planning production = painstaking ∵ factors, e.g., = rainfall → shortfall in produce; as a result, need to buy in tomatoes + other suppliers' shortages, ∴ problems
   2. Good scheduling = interrelationship &/+ CPA = invaluable tool advantages of CPA = checklist of actions: correct time + correct sequence lean production process / traditional assembly line
   3. Production ∴ needs good planning

2 They must research what an agribusiness needs in order to innovate successfully.

Transcript 🎧 2.6

Part 5

I'm going to finish by looking back at the factory environment and a vital part of the management process, which is scheduling.

Now, planning production is a very painstaking process. The reason is that planning decisions are based on a range of different factors – not to mention the fact that many are outside the production manager's control. Let's take chutney makers as an example. If they have a shortfall in produce because of a prolonged period without rain, they may have to buy in tomatoes. The production manager can't control that. Plus there's the fact that other suppliers may also be experiencing a shortage, which will cause additional problems because it may increase demand and exacerbate shortages.

OK. Where was I? Oh, yes … So scheduling means working out the key production timetable in relation to other processes. You've probably heard of 'Critical Path Analysis'. Kelley and Walker developed this project modelling technique in the 1950s in America. Critical Path Analysis is a technique for examining the stages in a production process, and ensuring that they interrelate correctly. The advantage of Critical Path Analysis is that it acts as a checklist to ensure that actions are taken at the correct time and in the correct sequence to deliver the planned result. It doesn't matter whether you're managing a lean production process – a more modern and creative system – or the traditional assembly line pioneered by Henry Ford. You need a clear framework for both, and although it was more than 50 years ago that Critical Path was introduced, it is still an invaluable tool and used widely today.

Good planning must permeate every part of the manufacturing agribusiness's activities and processes. That's the message. Let me put it another way … production needs to be planned.

Oh, I nearly forgot to mention your research topics. OK, well, what's very important for any business is innovation – the ability of a business to think laterally and find new directions. So I'd like you to find out what an agribusiness needs in order to achieve successful innovation.

Exercise C

Set for pairwork. Feed back with the whole class. If necessary, play the relevant sections again. Ask for other phrases which have similar meanings, particularly from Lesson 7.3, and also from Unit 5. Build the table in the Answers section on the board. Accept any suitable words or phrases for the third column.

Answers

Model answers:

Use	Fixed phrase	Other phrases
to introduce a new topic	You've probably heard of …	Now, an important concept is …
to emphasize a major point	The fact of the matter is, …	Actually, … In fact, … The point is that …
to add points	Not to mention the fact that … Plus there's the fact that …	also, and, too
to finish a list	et cetera	and so on
to give an example	Let's take …	For example, … e.g., … Let's look at an example of this. For instance, …
to restate	Let me put it another way. In other words, …	What I mean is … That is to say … By that I mean … To put it another way, …

Language note

The phrases in the table are appropriate in speaking. Many are not suitable for written language, for which different phrases should be used.

Exercise D

Students need to decide which word(s) should receive the particular focus and then try to rewrite the sentences. Depending on the class, they can work in pairs or individually first.

Feed back with the whole class. Take each sentence in turn. Ask for suggestions as to which aspect could receive special emphasis. Accept any reasonable answers. If you wish, replay Part 5 of the lecture for students to check their answers. Note that:

- sentences 1, 2 and the first part of 4 use an *It* construction to give the special focus
- sentence 3 uses a *Wh~* cleft sentence already seen in Lesson 7.2
- sentences 4 and 5 introduce new, general words (often found in academic contexts) and use *is* plus *that* clauses

Answers

Model answers:

1 Production needs to be planned (*It*)

It is necessary to plan production.

2 Kelley and Walker developed this project modelling technique in the 1950s in America. (*It*)

It was Kelley and Walker who developed this project modelling technique in the 1950s in America.

3 The production manager can't control events in the external environment. (*What*)

What the production manager can't control is events in the external environment.

4 The advantage of Critical Path Analysis is that it acts as a checklist to ensure that actions are taken at the correct time and in the correct sequence to deliver the planned result. (*Two sentences. First = 'It'; second = 'The reason'*)

It's an advantage of Critical Path Analysis that it acts as a checklist to ensure actions are taken at the correct time and in the correct sequence. The reason managers use it is to deliver the planned result.

5 Critical Path Analysis is a technique for examining the stages in a production process, and ensuring that they interrelate correctly. (*The advantage*)

The advantage of Critical Path Analysis is that it enables a production manager to examine the stages in a production process and ensure that they interrelate correctly.

After completing Exercises C and D, students can be referred to the *Vocabulary bank* and the *Skills bank* for consolidation and preparation for Exercise E.

Exercise E

Set the initial preparation for individual work. Students can refer to their notes in Lesson 7.2 (Exercises C and E) or the notes for completion in Lesson 7.3 (Exercise B). They should think about how they can use the phrases they have looked at, and ways of giving special focus/emphasis. (Note: They should not write out exactly what they are going to say in complete sentences and then read!)

Put students in pairs to give their oral summaries to each other, preferably pairing students who have chosen different sections to summarize.

Go around the class noting any problems or especially good examples of language use. You may wish to choose one or two individuals to give their summary to the whole class.

With the whole class, feed back any language or other difficulties which you noticed.

Exercise F

1 Set for pairwork. Suggest simple activities like making a cup of tea or a sandwich or writing an essay. Students should first list all the different processes and then decide how to order them and which processes overlap. They should make a fishbone diagram and put the activities in it.

2 Put the pairs in groups of four to present their diagrams to each other.

Closure

Dictate some words for which students have learnt note-taking symbols or abbreviations such as *and, minus, approximately, less than, results in, therefore, because, etc., as, since, for example*. Students should write the symbol or abbreviation.

Remind them of the list of symbols and abbreviations at the back of the Course Book.

7.4 Extending skills

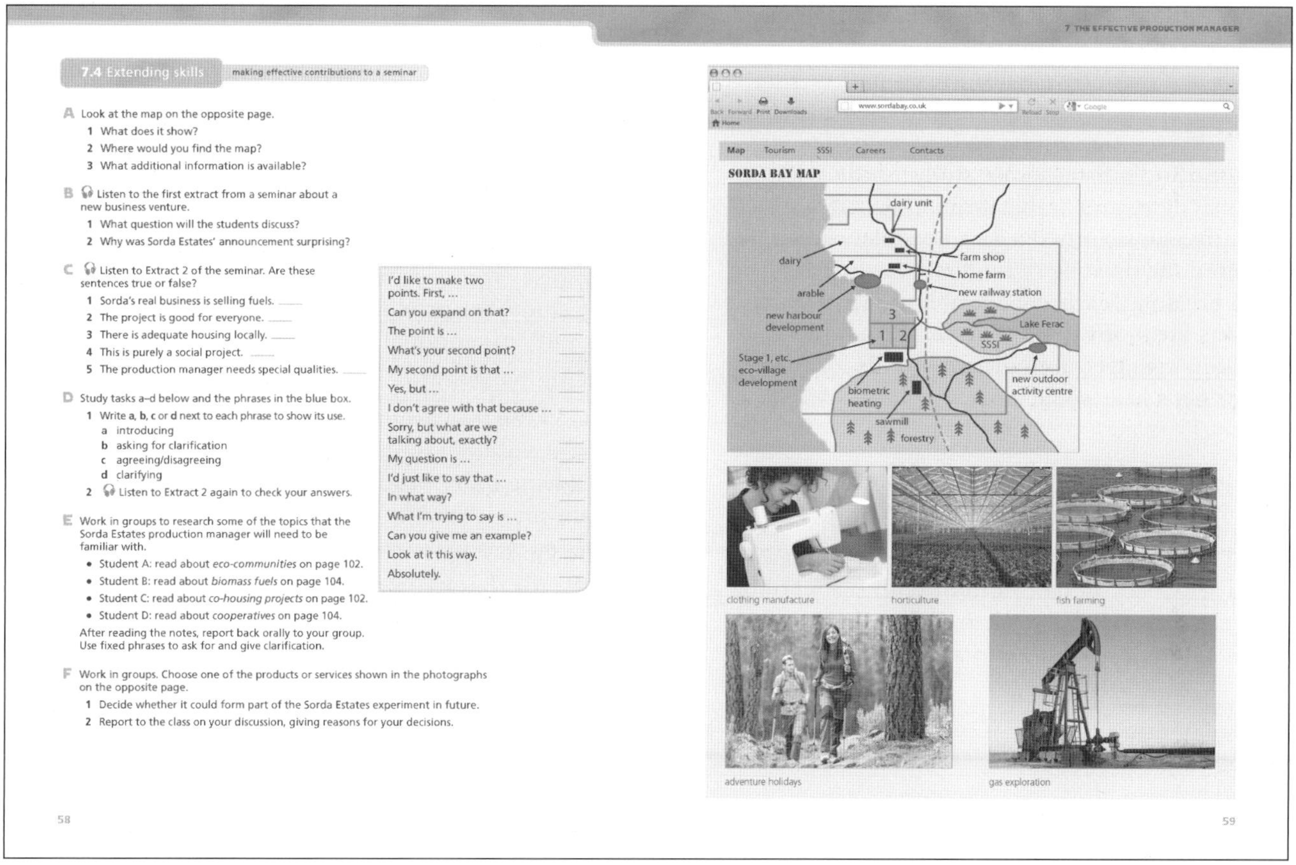

Lesson aims

- make effective contributions to a seminar:

  using pre-organizers – *I'd like to make two points; I don't agree with that because …*

  responding to queries by clarifying – *What I'm trying to say is … What I meant was …*

Introduction

Revise phrases from the previous lessons. Give a word or phrase and ask students to give one with a similar meaning. Ask for phrases from the previous lesson which can be used to:

- introduce a new topic
- emphasize a major point
- add a point
- finish a list
- give an example

Exercise A

Set for pairwork discussion. Feed back.

Answers

Possible answers:

1 It shows Sorda Bay and the various business types and developments in the area.
2 It is taken from the project website.
3 Information about Sorda Bay, Jobs, SSSI (Site of Special Scientific Interest), etc.

Exercise B

Allow students time to read the two questions. Play Part 1 once only. Check answers in pairs. Feed back with the whole class.

Answers

Model answers:

1 What motivated Sorda to diversify and develop a new agribusiness?
2 Because the new agribusiness was so different from their existing, traditional farming activities.

Transcript 🎧 2.7

Extract 1

Now, as we said earlier, a production manager needs to be able to think laterally, come up with new solutions and learn new skills. This is part of what makes a production manager effective, particularly in a forward-looking agribusiness. I suggested you look at the Sorda Estates case study. Sorda is a traditional dairy and arable farming business which six months ago surprised a lot of people when it announced the opening of the first phase of a ground-breaking, carbon-neutral housing project. What was the motivation to do this? And does the production manager need different skills for the new venture? Let's have some ideas.

🎧 Exercise C

Allow students time to read the questions. Play Extract 2 straight through once while they mark the answers true or false. Check in pairs and/or with the whole class. Check any unknown vocabulary, such as *elaborate on, 'win-win' situation.*

Answers

1	true	
2	true	
3	false	There is not adequate affordable housing locally.
4	false	Sorda's real business is to make profit from selling biomass fuels to the new houses.
5	true	

Transcript 🎧 2.8

Extract 2

(Note: The underlining relates to Exercise D)

JACK: Well. I'd like to make two points. First, I think it's important to point out that the real business for Sorda is selling biomass fuels to generate energy for the new houses.

LEILA: Can you expand on that, Jack?

JACK: Certainly, Leila. Sorda are obviously going to profit, short term, from building and selling the new houses. But they'll benefit long term from supplying the fuel needed.

LEILA: So?

JACK: So the point is that it's a 'win-win' situation for them.

LECTURER: OK. So, what's your second point, Jack?

JACK: I was coming to that! My second point is that they're also developing a range of dairy products. They're going to have a source of low-cost housing to help attract the workers they'll need to employ for the new dairy business.

LEILA: Yes, but the workers could have found housing anyway.

MAJED: Actually, I don't agree with that, Leila, because, from looking at the background to the project, affordable housing is a big problem in that area.

EVIE: Sorry, but what are we talking about, exactly? Is this a social project or a business?

LEILA: Well, I guess my question is: Does it need to be one or the other? I'm not sure. Anyway, I'd just like to say that, according to what I've read in the case study, the production manager would have to be pretty exceptional.

EVIE: In what way?

LEILA: Well, the original farm business – the mixed farm – was very traditional. Even when they started producing dairy products, it wasn't a complex agribusiness.

EVIE: I'm not sure what you're getting at. Does he, or she for that matter, have to build houses?

LEILA: No, but what I'm trying to say is, even if you look at the planning of the new project, there are a lot of different, interrelated elements that he or she needs some knowledge of.

EVIE: Nope, I'm still not clear. Can you give me an example, Leila?

LEILA: OK. Look at it this way. Sorda decide that they are going to replace a brownfield site – some redundant farm buildings – with houses. But the estate includes areas of special scientific interest. And they're going to make this a carbon-neutral eco-project. That means, for one thing, they're going to build the new 'village' round a biomass boiler and source the wood chips it uses from their own low-grade timber. And then they'll use the same technology in a new, purpose-built agrimanufacturing unit.

MAJED: OK, I'm with you now! So the production manager does need a good range of skills then!

LECTURER: Absolutely. And the Estates Board are interested in developing the food business as part of a cooperative. The production manager has to understand all these elements and make sure they are all planned for.

MAJED: Yes, that's right, but I'd just like to add something. I said earlier that housing is a problem and employment is also a problem and so is fuel poverty – people are having real trouble paying for their fuel in rural areas. So the project *is* complicated, but if the production manager is effective, they will have the advantage of working on a scheme which is really coherent. I mean, it doesn't just solve one problem and create others.

Exercise D

Check the meaning of 'introducing' phrases. This means a phrase to use before your main statement to announce that you are going to say something. It may also signal how much you are going to say, or how important you think what you are going to say is.

1 Set for individual work and pairwork checking. Feed back.

2 Play Extract 2 from Exercise C. Ask students to tell you to stop when they hear each phrase (underlined in the transcript above). Check what kind of phrase they think it is. Get students to repeat the phrase to copy the intonation.

If you wish, ask students to suggest other phrases that could be used in the same way.

Answers

Model answers:

I'd like to make two points. First, ...	a
Can you expand on that?	b
The point is ...	d
What's your second point?	b
My second point is that ...	a
Yes, but ...	c
I don't agree with that because ...	c
Sorry, but what are we talking about, exactly?	b
My question is ...	d
I'd just like to say that ...	a
In what way?	b
What I'm trying to say is ...	d
Can you give me an example?	b
Look at it this way.	d
Absolutely.	c

Exercise E

With the whole class, revise asking for information. Remind students of the questions used by the lecturer in Unit 5, Lesson 3 (see Unit 5 *Skills bank*). Remind students also about reporting information to people (see Unit 3 *Skills bank*).

Set students to work in groups of four. Each student should choose one aspect of Sorda's business activities and turn to the relevant page to make notes on the information. When everyone is ready they should feed back to their group, giving an oral report on the information. It's important that they do not simply read aloud the information, but use it to inform their speaking.

Alternatively, the research activity can be done as a 'wall dictation' as follows. Use Resource 7C in the additional resources section. Make large A3 (or A4) size copies of the topic information (one type of research per page) and pin the sheets on the classroom walls. Each student should leave his/her seat and go to the wall to find the information he/she needs. Students should not write anything down: instead, they should read and try to remember the information. Then they return to their group and tell them the information. If they forget something they can go back to the wall to have another look.

Circulate, encouraging students to ask for clarification and to use the appropriate phrases when giving clarification. Note where students are having difficulty with language and where things are going well. When everyone has finished, feed back to the class on points you have noticed while listening in to the discussions.

Exercise F

Split students into five groups, allocating one product or service to each group. Encourage students to think about what they already know about their product or service and to share information within their group.

Alternatively, you could have a 'pyramid discussion'. Choose one product or service for the whole class to debate and put students in pairs to discuss it. After a short while, the pair should join together with another pair. This group of four should then come to an agreement on whether the product or service would be suitable for Sorda Estates. The group of four should then join another group of four. One or two people from each group of eight should then present the decision and the reasons for the decision to the class. Finally, the whole class should try to reach agreement on the product or service, taking a vote if necessary.

Remind students about agreeing and agreeing, and about good and bad ways to contribute to seminar discussions (refer to Unit 5 if necessary).

While the representatives are presenting their group decisions, you should occasionally interrupt with a wrong interpretation so that students are forced to clarify their statements. Or you could ask for clarification.

Closure

Ask students to think of innovative agribusiness projects – encourage them to find projects that work in the context of their own countries or communities. Ask them to *describe* and *evaluate* the characteristics of each one according to the benefits and challenges they offer for:

- the economy
- transport systems
- workforce
- economy
- tourism
- quality of life

Extra activities

1 Work through the *Vocabulary bank* and *Skills bank* if you have not already done so, or as a revision of previous study.

2 Use the *Activity bank* (Teacher's Book additional resources section, Resources 7A).

A Set the crossword for individual work (including homework) or pairwork.

Answers

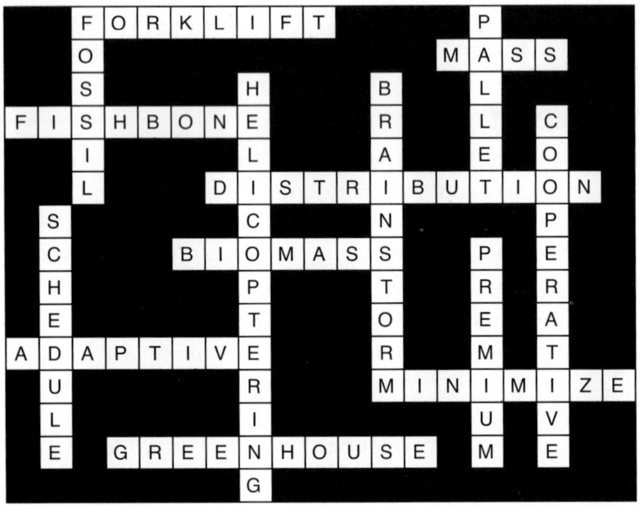

B Make some statements about what you're going to do after the class and ask students to transform them into *Wh~* cleft sentences. For example:

I'm going to play tennis after the class.

→ *What you're going to do after the class is play tennis.*

I might go jogging this evening.

→ *What you might do this evening is go jogging.*

Put students in pairs to practise.

8 RURAL ECONOMIES AND FINANCIAL MANAGEMENT

This is the first of two units on financial management. This unit looks at how different internal and external financial and management strategies can affect agribusinesses. Concepts such as free market and controlled market economies are examined. The reading text discusses financial management techniques such as whole-farm strategy and other 'systems thinking' techniques, as well as interventionist financial strategies such as economic subventions.

Skills focus

Reading

- understanding dependent clauses with passives

Writing

- paraphrasing
- expanding notes into complex sentences
- recognizing different essay types/structures:
  descriptive
  analytical
  comparison/evaluation
  argument
- writing essay plans
- writing essays

Vocabulary focus

- synonyms
- nouns from verbs
- definitions
- common 'direction' verbs in essay titles (*discuss*, *analyze*, *evaluate*, etc.)

Key vocabulary

apportion	financial management	market trading
business environment	finite	micro-economic
carrying capacity	market economy	organism
catering	geared	per head
command economy	growth	rapid
control	increase	regulation
creative sector	infinite	subvention
decrease	interrupt	supply and demand
economic policy	inverse	utility
economic unit	macro-economic	24/7
exploit	marginal	

8.1 Vocabulary

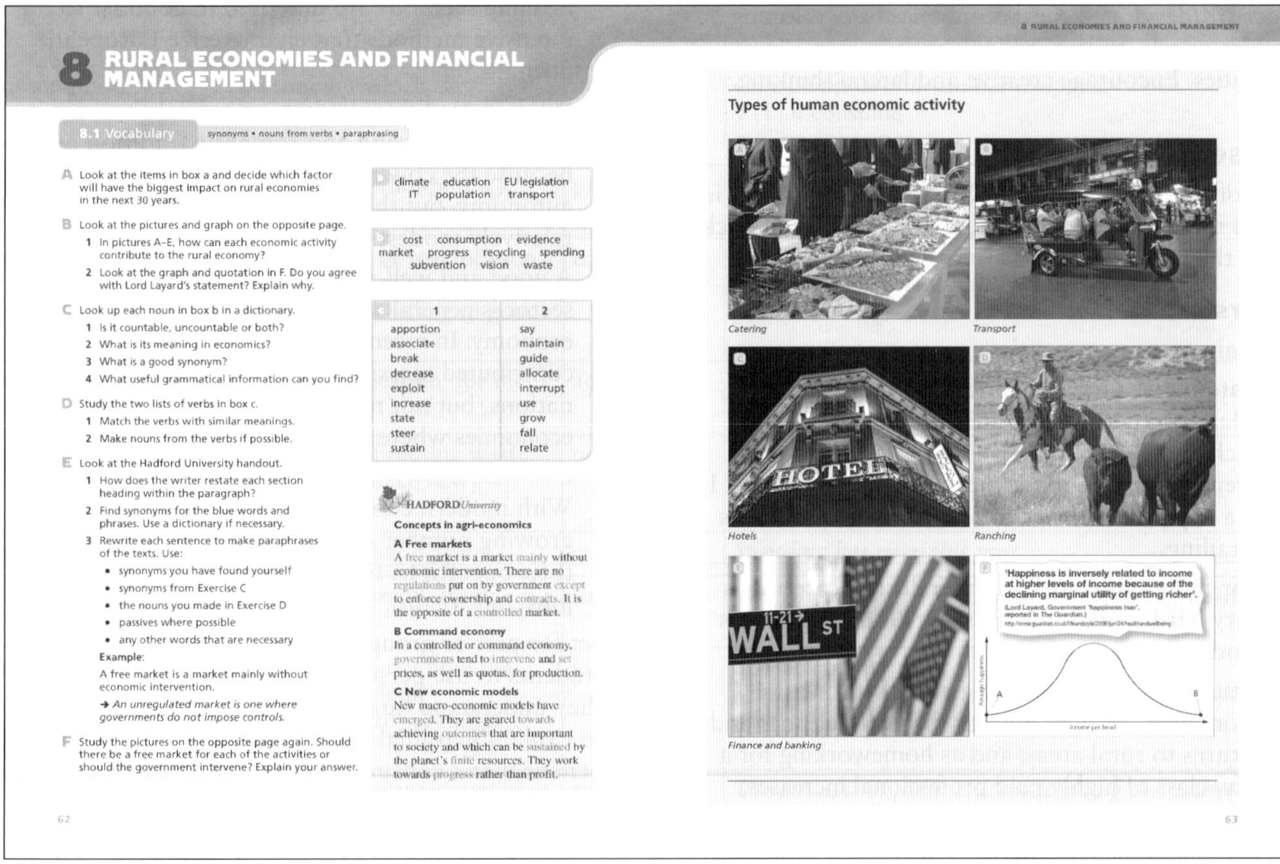

8 RURAL ECONOMIES AND FINANCIAL MANAGEMENT

8.1 Vocabulary synonyms • nouns from verbs • paraphrasing

A Look at the items in box a and decide which factor will have the biggest impact on rural economies in the next 30 years.

B Look at the pictures and graph on the opposite page.

1 In pictures A–E, how can each economic activity contribute to the rural economy?
2 Look at the graph and quotation in F. Do you agree with Lord Layard's statement? Explain why.

C Look up each noun in box b in a dictionary.

1 Is it countable, uncountable or both?
2 What is its meaning in economics?
3 What is a good synonym?
4 What useful grammatical information can you find?

D Study the two lists of verbs in box c.

1 Match the verbs with similar meanings.
2 Make nouns from the verbs if possible.

E Look at the Hadford University handout.

1 How does the writer restate each section heading within the paragraph?
2 Find synonyms for the blue words and phrases. Use a dictionary if necessary.
3 Rewrite each sentence to make paraphrases of the texts. Use:
   - synonyms you have found yourself
   - synonyms from Exercise C
   - the nouns you made in Exercise D
   - passives where possible
   - any other words that are necessary

   Example:
   A free market is a market mainly without economic intervention.
   → *An unregulated market is one where governments do not impose controls.*

F Study the pictures on the opposite page again. Should there be a free market for each of the activities or should the government intervene? Explain your answer.

a: climate education EU legislation IT population transport

b: cost consumption evidence market progress recycling spending subvention vision waste

c:

1	2
apportion	say
associate	maintain
break	guide
decrease	allocate
exploit	interrupt
increase	use
state	grow
steer	fall
sustain	relate

HADFORD University

Concepts in agri-economics

A Free markets
A free market is a market mainly without economic intervention. There are no regulations put on by government except to enforce ownership and contracts. It is the opposite of a controlled market.

B Command economy
In a controlled or command economy, governments tend to intervene and set prices, as well as quotas, for production.

C New economic models
New macro-economic models have emerged. They are geared towards achieving outcomes that are important to society and which can be sustained by the planet's finite resources. They work towards progress rather than profit.

62

8 RURAL ECONOMIES AND FINANCIAL MANAGEMENT

Types of human economic activity

A *Catering*

B *Transport*

C *Hotels*

D *Ranching*

E *Finance and banking*

F 'Happiness is inversely related to income at higher levels of income because of the declining marginal utility of getting richer'.
(Lord Layard, Government 'happiness tsar', reported in *The Guardian*.)
http://www.guardian.co.uk/lifeandstyle/2008/jun/24/healthandwellbeing

63

Lesson aims

- extend knowledge of synonyms and word sets (enables paraphrasing at word level)
- make nouns and noun phrases from verbs (enables paraphrasing at sentence level)

Further practice in paraphrasing at sentence level with:

- passives
- synonymous phrases
- negatives
- replacement subjects

Introduction

Revise ways of paraphrasing sentences. Write the following sentences on the board and ask students to say what changes have been made to the paraphrased sentences.

Original sentence: *If we take Applecart Farm as an example, they have established a good reputation among families …*

Paraphrase: *For example, a favourable impact has been made by Applecart Farm in the families segment of the market …*

(answer: change in word order, active to passive, use of synonyms)

Original sentence: *The farm currently produces a range of specialist cheeses.*

Paraphrase: *There are a number of distinctive cheeses offered by Applecart.*

(answer: change in word order, use of synonyms, active to passive, replacement subject)

8.4 Extending skills

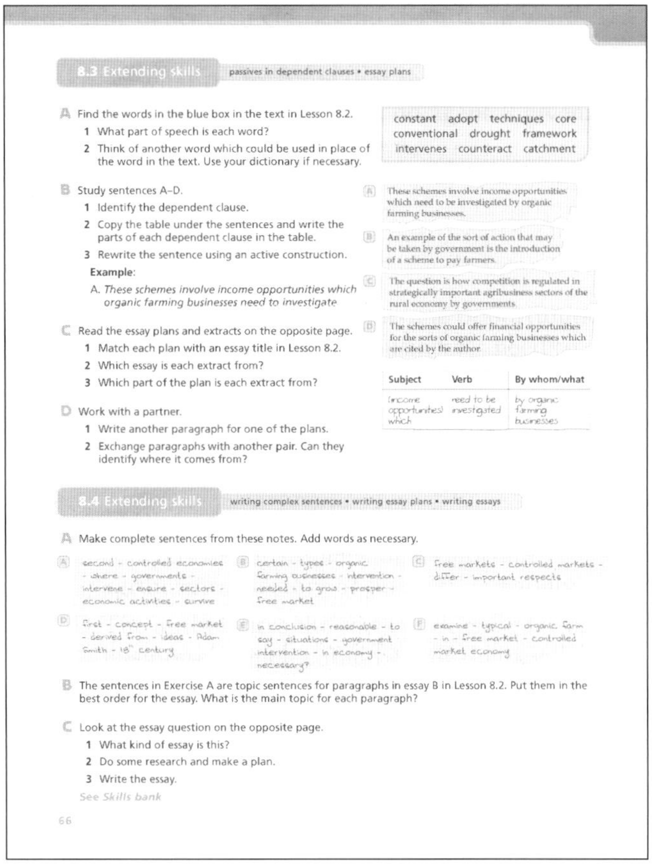

8.3 Extending skills — passives in dependent clauses • essay plans

A Find the words in the blue box in the text in Lesson 8.2.

1 What part of speech is each word?

2 Think of another word which could be used in place of the word in the text. Use your dictionary if necessary.

constant adopt techniques core conventional drought framework intervenes counteract catchment

B Study sentences A–D.

1 Identify the dependent clause.

2 Copy the table under the sentences and write the parts of each dependent clause in the table.

3 Rewrite the sentence using an active construction.

Example:

A. *These schemes involve income opportunities which organic farming businesses need to investigate*

A These schemes involve income opportunities which need to be investigated by organic farming businesses.

B An example of the sort of action that may be taken by government is the introduction of a scheme to pay farmers.

C The question is how competition is regulated in strategically important agribusiness sectors of the rural economy by governments.

D The schemes could offer financial opportunities for the sorts of organic farming businesses which are cited by the author.

Subject	Verb	By whom/what
(income opportunities) which	need to be investigated	by organic farming businesses

C Read the essay plans and extracts on the opposite page.

1 Match each plan with an essay title in Lesson 8.2.

2 Which essay is each extract from?

3 Which part of the plan is each extract from?

D Work with a partner.

1 Write another paragraph for one of the plans.

2 Exchange paragraphs with another pair. Can they identify where it comes from?

8.4 Extending skills — writing complex sentences • writing essay plans • writing essays

A Make complete sentences from these notes. Add words as necessary.

A second – controlled economies – where – governments – intervene – ensure – sectors – economic activities – survive

B certain – types – organic farming businesses – intervention – needed – to grow – prosper – free market

C free markets – controlled markets – differ – important respects

D first – concept – free market – derived from – ideas – Adam Smith – 18th century

E in conclusion – reasonable – to say – situations – government intervention – in economy – necessary?

F examine – typical – organic farm – in – free market – controlled market economy

B The sentences in Exercise A are topic sentences for paragraphs in essay B in Lesson 8.2. Put them in the best order for the essay. What is the main topic for each paragraph?

C Look at the essay question on the opposite page.

1 What kind of essay is this?

2 Do some research and make a plan.

3 Write the essay.

See *Skills bank*

66

Lesson aims

- expand notes into complex sentences
- make an essay plan
- write an essay

Further practice in:

- writing topic sentences
- expanding a topic sentence into a paragraph
- writing complex sentences with passives
- identifying required essay type

Introduction

Remind students about complex and compound sentences – that is, sentences with more than one clause. Remind students that academic texts typically consist of sentences with several clauses. Give the following simple sentences (or make your own) and ask students to add some more clauses to them:

Leadership skills are key skills.
Controlled markets suit some agribusinesses.
Financial managers need to be aware of the external business environment.
An EU directive on the environment can offer opportunities.
An organic farm is a typical modern agribusiness.
Climate change can cause droughts.

Exercise A

Set for individual work and pairwork checking. Remind students that they should try to make sentences in a good 'academic' style. Also remind them to use passives where necessary/possible, and to look out for ways of making dependent clauses, such as relative pronouns, linking words, etc. They will also need to pay attention to making correct verb tenses.

Feed back with the whole class.

Answers

Example answers:

A Secondly, controlled economies are economies where governments intervene to ensure that certain sectors and economic activities survive.

B For certain types of organic farming businesses, economic intervention is needed if they are to grow and prosper in a free market.

C Free markets and controlled markets differ in certain important respects.

D Firstly, the concept of the free market is derived from the ideas put forward by Adam Smith in the 18th century.

E In conclusion, is it reasonable to say that there are situations where government intervention in the economy is necessary?

F Let's examine a typical organic farm in a free market and in a controlled market economy.

Exercise B

Set for individual work. Feed back with the whole class. Point out how this argument essay is organized by giving opinions and opposing opinions (= counter arguments) and providing evidence in support of both. (See *Skills bank* for an alternative approach to arguments.)

If you wish, you could take this exercise further, asking students to build on the topic sentences by suggesting what ideas could follow the topic sentence in each paragraph. For this, they will need to refer to ideas in the text. Note that disadvantages of economic intervention are not fully discussed in the text, so ideas for this paragraph would need to be researched further. A Web search is a good place to start for this.

Answers

Model answers:

Topic sentences	Paragraph topic
B For certain types of organic farming businesses, economic intervention is needed if they are to grow and prosper in a free market.	introduction
D Firstly, the concept of the free market is derived from the ideas put forward by Adam Smith in the 18th century.	definition and examples of free markets
A Secondly, controlled economies are economies where governments intervene to ensure that certain sectors and economic activities survive.	definition and examples of controlled markets
C Free markets and controlled markets differ in certain important respects.	establish the key idea
F Let's examine a typical organic farm in a free market and in a controlled market economy.	presentation of evidence for and against free market
E In conclusion, is it reasonable to say that there are situations where government intervention in the economy is necessary?	conclusion

Exercise C

Discuss question 1 with the whole class. Set the research and planning (question 2) for group work, and the writing for individual work (this could be done at home). Students can do web searches to find more information.

Answers

1 Model answer:

This essay is largely analytical since it requires students to explain why certain skills are applicable to the situation. It also asks students to think critically about issues with the business and what skills are required to manage these successfully.

2 Possible essay plan:

- Introduction: short background to William Humphries and the reason for the project = need to diversify; aims of essay.
- Examination of the parameters
  - What is hemp?
    cultivation issues
    market
  - What is a conservation area?
    possible impact on project?
- Financial manager:
  - General comments
  - Personal qualities
    team player?
    sense of humour?
    assertive?
    leadership?
    diligent?
  - Skills and qualifications
    analytical skills
    flexibility
    appropriate qualifications
  - Experience
    range of different environments and businesses
    successful planning
    evidence of successful innovation projects
- Conclusion: relate the challenge to priority skills, etc.

Closure

Ask students if they can remember a word from the unit …

	Example(s)
beginning with *c*	controlled
beginning with *i*	intervention
ending with *y*	economy
ending with *s*	agribusiness
with two syllables	framework, market, rural
with three syllables	adaptive, organic
with four syllables	economy, political
which is a verb	benefit, sustain
which is an uncountable noun	water, importance
which is an adverb	correctly
which goes together with another word	organic farming
which is difficult to pronounce	initiative (students' answers will vary)

Accept all reasonable answers.

Extra activities

1 Work through the *Vocabulary bank* and *Skills bank* if you have not already done so, or as a revision of previous study.

2 Use the *Activity bank* (Teacher's Book additional resources section, Resource 8A).

A Set the wordsearch for individual work (including homework) or pairwork. Establish that the words are uncountable in the context in which they are used in the unit, although some can be countable in other contexts.

Answers

Q	S	J	L	T	U	I	I	L	T	Y	Y	G	B	J	L
W	I	F	C	M	A	I	N	T	E	N	A	N	C	E	E
Q	G	L	W	O	W	Z	J	G	Q	L	V	F	H	S	A
F	N	E	V	T	E	L	Q	R	O	G	V	O	P	S	R
A	I	X	R	L	X	M	G	R	D	Q	R	O	A	N	N
R	F	I	H	T	M	H	T	U	Z	D	Q	O	O	I	I
M	I	B	A	R	C	N	A	T	U	R	E	I	W	K	N
I	C	I	P	J	O	L	M	H	F	Z	T	O	A	T	G
N	A	L	P	C	L	U	U	K	C	C	F	W	Q	L	H
G	N	I	I	A	D	A	P	T	A	B	I	L	I	T	Y
P	C	T	N	D	T	J	B	F	O	C	T	P	P	K	D
S	E	Y	E	R	S	R	S	U	C	G	B	R	V	N	H
X	S	U	S	T	A	I	N	A	B	I	L	I	T	Y	W
C	H	F	S	U	T	D	W	E	A	L	T	H	R	E	R
S	D	W	K	A	X	I	M	P	O	R	T	A	N	C	E
X	E	P	S	E	C	O	N	O	M	I	C	S	I	A	H

B Set the spelling exercise for individual work and pairwork checking. If students are having difficulty, give them the first letter of the word.

Answers

Jumbled word	Correct spelling
sngiiro	origins
ngbeni	benign
nogrica	organic
nactfisngii	significant
kratem	market
iterflirez	fertilizer
toilxep	exploit
lolbag	global
rocam	macro
mehp	hemp

3 Check word stress by writing the following words on the board *without* stress markings. Students have to mark the stress and pronounce the words correctly.

a'daptable
di'rective
'factor
a'ware
'concept
e'conomy
su'stainable

4 Remind students of how to give definitions (see Lesson 8.2). Then select five or six familiar items (e.g., iPod, laptop, sunglasses, pen, mobile phone) and ask students to think of definitions (e.g., it's something that you use to listen to music; you need these when it is sunny; etc.).

This can also be done the other way round by giving the definitions and asking students to guess the word; once they get the idea students can come up with items, questions and definitions themselves. Other forms for definitions can include:

This is a place where ...
This is a company which ...
If you want to buy an (X), you need to go to ...

Other categories which can be used to practise both the language of definition and general business and cultural knowledge include:

- brand or company names, e.g., Marks & Spencer (well known in the UK for food and for quality); also use global multinationals
- familiar places in the town or college where the students are studying
- famous people
- movies and TV programmes

An alternative is the Weakest Link TV quiz show format, e.g., *What 'A' is a well-known brand of computer?* (Apple.)

9 MANAGING FINANCIAL ACCOUNTS

Continuing the theme of financial management, this unit focuses on two aspects: financial accounting and management accounting. The key differences between these two types of accounting are analyzed. Useful financial terms such as *balance sheet, profit and loss account, cash flow, profit margin, turnover,* etc., are defined. Reference is also made to treasury.

Skills focus

Listening

- using the Cornell note-taking system
- recognizing digressions in lectures

Speaking

- making effective contributions to a seminar
- referring to other people's ideas in a seminar

Vocabulary focus

- fixed phrases from agribusiness
- fixed phrases from academic English

Key vocabulary

See also the list of fixed phrases from academic English in the Vocabulary bank (Course Book page 76).

accountancy	equity	operating costs
accounting	expenditure	overdraft
accounts	expense	net profit
annual	finance	profit margin
asset	financial	profitability
audit	forward looking	profitable
balance sheet	functions	record keeping
baseline figures	green accountancy	revenue
borrowings	income	SAGE
budget (n and v)	interest	share
cash flow	investment	shareholder
collate	investor	stake
core business	legal obligations	standard
cost-benefit analysis	lender	statutory duties
credit	liabilities	turnover
debt	liquid	under licence
dividend	liquidity	variable
document	Management Information System (MIS)	
downstream		

9.1 Vocabulary

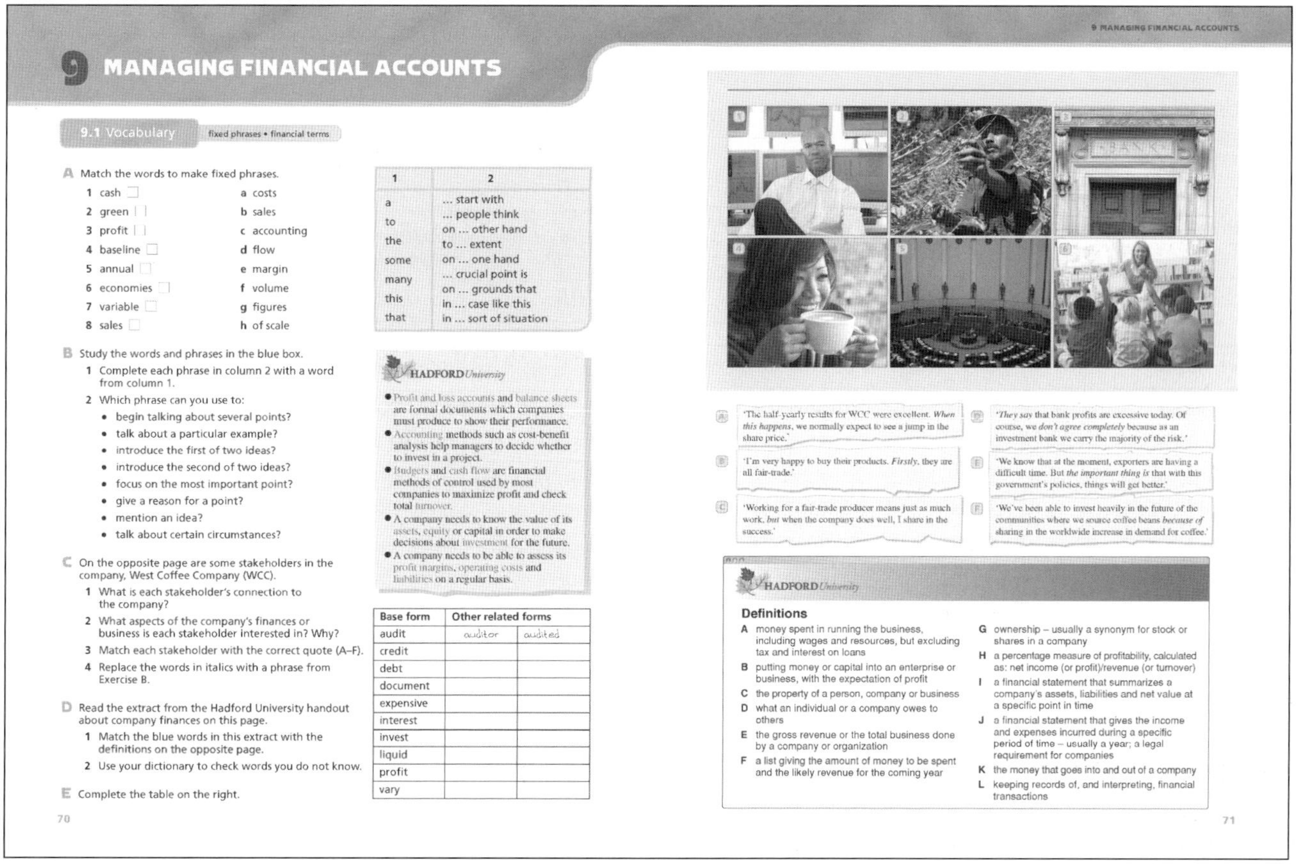

9 MANAGING FINANCIAL ACCOUNTS

9.1 Vocabulary fixed phrases • financial terms

A Match the words to make fixed phrases.

1 cash	a costs
2 green	b sales
3 profit	c accounting
4 baseline	d flow
5 annual	e margin
6 economies	f volume
7 variable	g figures
8 sales	h of scale

B Study the words and phrases in the blue box.

1 Complete each phrase in column 2 with a word from column 1.

2 Which phrase can you use to:

- begin talking about several points?
- talk about a particular example?
- introduce the first of two ideas?
- introduce the second of two ideas?
- focus on the most important point?
- give a reason for a point?
- mention an idea?
- talk about certain circumstances?

C On the opposite page are some stakeholders in the company, West Coffee Company (WCC).

1 What is each stakeholder's connection to the company?

2 What aspects of the company's finances or business is each stakeholder interested in? Why?

3 Match each stakeholder with the correct quote (A–F).

4 Replace the words in italics with a phrase from Exercise B.

D Read the extract from the Hadford University handout about company finances on this page.

1 Match the blue words in this extract with the definitions on the opposite page.

2 Use your dictionary to check words you do not know.

E Complete the table on the right.

1	2
a	... start with
to	... people think
the	on ... other hand
some	to ... extent
many	on ... one hand
this	... crucial point is
that	on ... grounds that
	in ... case like this
	in ... sort of situation

HADFORD *University*

- Profit and loss accounts and balance sheets are formal documents which companies must produce to show their performance.
- Accounting methods such as cost-benefit analysis help managers to decide whether to invest in a project.
- Budgets and cash flow are financial methods of control used by most companies to maximize profit and check total turnover.
- A company needs to know the value of its assets, equity or capital in order to make decisions about investment for the future.
- A company needs to be able to assess its profit margins, operating costs and liabilities on a regular basis.

Base form	Other related forms	
audit	auditor	audited
credit		
debt		
document		
expensive		
interest		
invest		
liquid		
profit		
vary		

70

A 'The half-yearly results for WCC were excellent. *When this happens*, we normally expect to see a jump in the share price.'

B 'I'm very happy to buy their products. *Firstly*, they are all fair-trade.'

C 'Working for a fair-trade producer means just as much work, *but* when the company does well, I share in the success.'

D '*They say* that bank profits are excessive today. Of course, we *don't agree completely* because as an investment bank we carry the majority of the risk.'

E 'We know that at the moment, exporters are having a difficult time. But *the important thing is* that with this government's policies, things will get better.'

F 'We've been able to invest heavily in the future of the communities where we source coffee beans *because of* sharing in the worldwide increase in demand for coffee.'

HADFORD *University*

Definitions

A money spent in running the business, including wages and resources, but excluding tax and interest on loans

B putting money or capital into an enterprise or business, with the expectation of profit

C the property of a person, company or business

D what an individual or a company owes to others

E the gross revenue or the total business done by a company or organization

F a list giving the amount of money to be spent and the likely revenue for the coming year

G ownership – usually a synonym for stock or shares in a company

H a percentage measure of profitability, calculated as: net income (or profit)/revenue (or turnover)

I a financial statement that summarizes a company's assets, liabilities and net value at a specific point in time

J a financial statement that gives the income and expenses incurred during a specific period of time – usually a year; a legal requirement for companies

K the money that goes into and out of a company

L keeping records of, and interpreting, financial transactions

71

General note

Read the *Vocabulary bank* at the end of the Course Book unit. Decide when, if at all, to refer your students to it. The best time is probably at the very end of the lesson or the beginning of the next lesson, as a summary/revision.

Lesson aims

- understand and use some fixed phrases/compound nouns from agribusiness
- understand and use some fixed phrases from academic English

Introduction

Introduce the topic. Ask students to say what accounting is and arrive at a definition on the board, such as:

> *Accounting is the 'recording, classifying, summarizing and interpreting' of financial transactions and events*
>
> (Adapted from www.investordictionary.com/definition/accounting.aspx)

Exercise A

This gives revision of some compound noun phrases (noun + noun, adjective + noun) connected with finance from previous units.

Set for individual work or pairwork. Check that students remember the meanings and that they can pronounce the compounds with the main stress on the correct word. Accept any reasonable alternatives which apply to the topic of finance. Ask students to make sentences with the compounds.

Answers

Model answers:

1 'cash	d flow	n + n
2 green	c 'accounting	adj + n
3 'profit	e margin	n + n
4 'baseline	g figures	n + n
5 annual	b 'sales, a costs	adj + n
6 economies	h of 'scale	
7 'variable	a costs*	adj + n
8 'sales	f volume	n + n

* when contrasted with *fixed*, the adjective not the noun takes the primary stress

Exercise B

Set for individual work and pairwork checking. Point out that some of the words in column 1 must be used more than once. Feed back with the whole class.

Answers

Model answers:

to start with	to begin talking about several points
many/some people think	to mention an idea
on the other hand	to introduce the second of two ideas
on the one hand	to introduce the first of two ideas
the crucial point is	to focus on the most important point
on the grounds that	to give a reason for a point
in a case like this	to talk about a particular example
in this/that sort of situation	to talk about certain circumstances

Exercise C

1/2 Set for pairwork discussion.

3 Set for individual work and pairwork checking.

4 Set for individual work. Check with the whole class, asking students to read out the quotation with the alternative phrase inserted in place of the original words in italics.

Answers

Model answers:

1/2

1 stocktraders – likely to be interested in profits, share price.
2 plantation workforce – manual worker in producer country interested in wages, job prospects and conditions.
3 banks – interested in good management, good year-on-year performance, capacity to pay back loans.
4 customer – interested in product quality and price.
5 government – politicians interested in jobs creation and securing supplies to meet consumer demands.
6 future generations – interested in long-term investment in the communities for education and health and natural environment.

3

1 (stocktraders): quote A
2 (plantation workers): quote C
3 (bank): quote D
4 (customers): quote B
5 (government): quote E
6 (future generations): quote F

4

A *When this happens*	*In this sort of situation,* we normally expect to see a jump in the share price.
B *Firstly*	*To start with,* they are all fair-trade.
C *but*	… *on the other hand,* when the company does well, I share in the success.
D *They say*	*Many/some people think* that bank profits are excessive today.
E *the important thing is*	But *the crucial point is* that with this government's policies, things will get better.
F *because of*	We've been able to invest heavily in the future of the communities where we source coffee beans *as a result of* sharing in the worldwide increase in demand for coffee.

Exercise D

This exercise gives some key business financial terms – and so is rather technical.

Set students to read the handout extract first and ask them to discuss in pairs which of the blue words they know and which are new for them. Feed back with the whole class, to establish how much is known. Where students give correct explanations, tell them they are right, and where they are wrong, also tell them, but do not give the right answer at this point.

Set questions 1 and 2 for individual work and pairwork checking. Feed back with the whole class, checking the meaning of other possibly unknown words. In particular, make sure students know *gross* and *net*.

The words will be used throughout the unit, so don't worry too much about practice at this point. However, for extra practice at this point if you wish, set students to work in pairs. One student should shut the book. The other student should say one of the words for the first student to explain. Then change over.

Answers

profit and loss accounts	J	a financial statement that gives the income, costs and expenses incurred during a specific period of time – usually a year; a legal requirement for companies
balance sheets	I	a financial statement that summarizes a company's assets, liabilities and net value at a specific point in time
accounting	L	keeping records of, and interpreting, financial transactions
budgets	F	a list giving the amount of money to be spent and the likely revenue for the coming year
cash flow	K	the money that goes into and out of a company
turnover	E	the gross revenue or the total business done by a company or organization
assets	C	the property of a person, company or business
equity	G	ownership – usually a synonym for stock or shares in a company
investment	B	putting money or capital into an enterprise or business, with the expectation of profit
profit margins	H	a percentage measure of profitability, calculated as: net income (or profit) revenue (or turnover)
operating costs	A	money spent in running the business, including wages and resources, but excluding tax and interest on loans
liabilities	D	what an individual or a company owes to others

Language note

There are many differences between financial terms used in the US and in Europe, which is sometimes confusing. For example, *revenue* is used in the US and *turnover* in Europe for the money that a company receives for its products or services. *Net income* is used in the US, while *net profit* is used in Europe for the actual income that a company receives after deducting the costs of doing business (i.e., costs of resources and raw materials, salaries, tax and interest payments).

Exercise E

Set for individual work. Tell students to use their dictionaries to check on the meanings and grammatical categories of the words if they are not sure. Explain that many of the base words only have one other related form.

Feed back with the whole class, pointing out that most of the words have a particular use in the area of finance (Those that can have more general use are shaded in the table in the Answers section). Check students can pronounce all the words correctly, particularly those where the word stress shifts.

Answers

Model answers:

Base form	Other related forms	
audit (n C, v)	auditor (n C)	audited (v, adj)
credit (n C/U, v)	creditor (n C)	
debt (n C/U)	debtor (n C)	indebted (adj)
document (n C, v)	documentation (n U)	
expense (n C/U)	expenditure (n U)	expensive
interest (n C/U, v)	interesting	
invest (v)	investment/s (n C/U)	investor(s)
liquid (n C/U, adj)	liquidity (n U)	
profit (n C/U, v)	profitability (n U)	profitable
vary (v)	variety (n C/U)	variable

Language note

With a good class, you might explore the issue of whether each noun is used as countable or uncountable or both, i.e., can the word be made plural, and if so, does that change the meaning?

Closure

It is important that students are familiar with the finance terminology from this lesson. On the board, write some terms from the lesson and ask students to give a definition; choose items from Exercises A and D. Or read out a definition and ask students to tell you the appropriate word or phrase. Check the pronunciation. This exercise can also be done as a dictation.

Alternatively, write the words and definitions on different cards and give a card to each student. The student then reads out the word or the definition and the rest of the class must produce the correct answer.

9.2 Listening

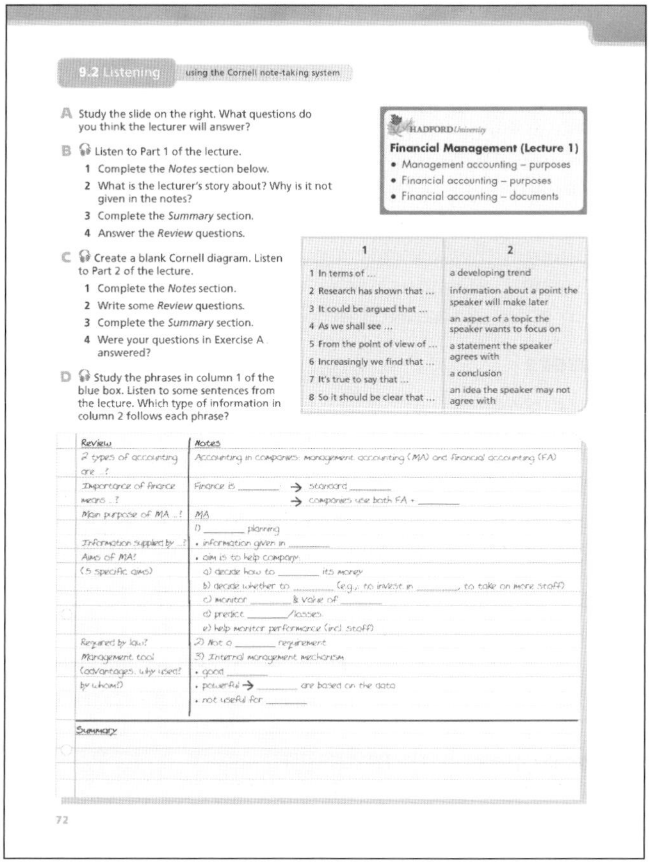

9.2 Listening using the Cornell note-taking system

A Study the slide on the right. What questions do you think the lecturer will answer?

B Listen to Part 1 of the lecture.

1 Complete the *Notes* section below.
2 What is the lecturer's story about? Why is it not given in the notes?
3 Complete the *Summary* section.
4 Answer the *Review* questions.

C Create a blank Cornell diagram. Listen to Part 2 of the lecture.

1 Complete the *Notes* section.
2 Write some *Review* questions.
3 Complete the *Summary* section.
4 Were your questions in Exercise A answered?

D Study the phrases in column 1 of the blue box. Listen to some sentences from the lecture. Which type of information in column 2 follows each phrase?

1	2
1 In terms of ...	a developing trend
2 Research has shown that ...	information about a point the speaker will make later
3 It could be argued that ...	an aspect of a topic the speaker wants to focus on
4 As we shall see ...	a statement the speaker agrees with
5 From the point of view of ...	a conclusion
6 Increasingly we find that ...	an idea the speaker may not agree with
7 It's true to say that ...	
8 So it should be clear that ...	

Review	Notes
2 types of accounting are ...?	Accounting in companies: management accounting (MA) and financial accounting (FA)
Importance of finance means ...?	Finance is ______ → standard ______
	→ companies use both FA + ______
Main purpose of MA ...?	MA
	1) ______ planning
Information supplied by ...?	• information given in ______
Aims of MA?	• aim is to help company:
(5 specific aims)	a) decide how to ______ its money
	b) decide whether to ______ (e.g., to invest in ______, to take on more staff)
	c) monitor ______ & value of ______
	d) predict ______/losses
	e) help monitor performance (incl. staff)
Required by law?	2) Not a ______ requirement
Management tool	3) Internal management mechanism
(advantages, why used?	• good ______
by whom?)	• powerful → ______ are based on the data
	• not useful for ______

Summary

72

General note

Read the *Skills bank – Using the Cornell note-taking system* at the end of the Course Book unit. Decide when, if at all, to refer students to it. The best time is probably at the very end of the lesson or the beginning of the next lesson, as a summary/revision.

Lesson aims

- use the Cornell note-taking system

Further practice in:

- listening for an established purpose
- understanding fractured text
- recognition of fixed phrases and what type of information comes next
- using abbreviations and symbols in note-taking

Introduction

1 Review key vocabulary from this unit by writing a selection of words from Lesson 9.1 on the board and asking students to put the words in groups, giving reasons for their decisions.

For example:

budgets, *forecasts* = looking into the future

balance sheet, *profit and loss accounts*, *cash flow forecasts* = documents

assets, *equity*, *capital*, *property* = things a company owns

liabilities, *debt* = things a company must pay

2 Revise note-taking symbols and abbreviations by using extra activity 3 at the end of this unit.

3 Introduce the elements of the Cornell note-taking system. Try to elicit some of the R words. Ask students to try to think of five words beginning with *re-* with six or seven letters that are good strategies to use when studying and taking notes. Write the words as follows on the board:

RE _ _ _ _ = *record*
RE _ _ _ _ = *reduce*
RE _ _ _ _ = *recite*
RE _ _ _ _ _ = *reflect*
RE _ _ _ _ = *review*

Discuss with the class what each word might mean when taking notes. Try to elicit the following, helping where needed.

record	Take notes during the lecture.
reduce	After the lecture, turn the notes into one- or two-word questions or 'cues' which help you remember the key information.
recite	Say the questions and answers aloud.
reflect	Decide on the best way to summarize the key information in the lecture.
review	Look again at the key words and the summary (and do this regularly).

Tell students that in this lesson they will be introduced to this system of note-taking – which can be used both for lectures, and also for reading and for revision for exams later. Do not say much more at this point; they will see how the system works as the lesson progresses.

Subject note

The Cornell system was developed by Walter Pauk at Cornell University, USA. (Pauk, W. and Owens, R. (2010). *How to Study in College* (10th edition). Boston: Houghton Mifflin). Pauk advised students to use a large, loose-leaf notebook, with holes punched for filing. This is better than a bound notebook, because you can organize the notes in a file binder. You can also take out notes and rewrite them. Pauk's method, which is now called the Cornell system, is based on a specific page layout.

Pauk told students to divide up the page into three areas. The first area is a column 5 cm wide on the left side of the page. This is the cue area. The main part of the page is the note-taking area. At the bottom of the page is a row 8 cm high, which is the summary area. The basic grid with information on what each section should contain is reproduced in the additional resources section (Resource 9B).

The note-taking and learning process involves the *Five Rs* in the order listed in the introduction to this lesson (and in the *Skills bank*). There are many references on the Internet for this system. Two useful ones at the time of writing are:

http://chomabiology.weekly.com/uploads/1/4/5/7/1457871/c-notes.pdf

http://lsc.cornell.edu//LSC_Resources/cornellsystem.pdf

Exercise A

Set for pairwork discussion. Refer students to the lecture slide. Tell them to look at the title and bullet points, and for each bullet point to devise questions which they expect the lecturer to answer. Do not explain any words from the slide, or allow students to check in their dictionaries at this point, as the meanings of these words will be dealt with in the lecture.

Feed back with the whole class, asking several students to read out their questions. Write some of the questions on the board if you wish.

Exercise B

1/2/3 Refer students to the notes at the bottom of the page. Tell them that this student has used the Cornell system to take notes but has not managed to complete everything and so has left some gaps. (Note that this is quite a normal occurrence in note-taking – details may need to be filled in later, for example by checking with other people.)

Allow students time to read the gapped notes. Also make sure they read question 2 and are ready to listen out for a story.

Play Part 1, pausing after each major point if you wish.

Tell students to work in pairs to compare their answers to questions 1 and 2, and to complete the summary in 3. Feed back with the whole class, using an OHT or other visual display of the answers if you wish. The completed notes are reproduced in Resource 9C, in the additional resources section, to facilitate this.

4 Now focus on the *recite* element of the Cornell system. Point out that here the student has completed the *Review* section. Tell students to cover up the *Notes* section of the answer and ask them if they can say anything about the first and second questions in the *Review*. Then put students in pairs to test each other on the remaining notes.

Answers

Model answers:

1/3/4 See below and Resource 9C.

Review	Notes
2 types of accounting are …?	Accounting in companies: management accounting (MA) and financial accounting (FA)
Importance of finance means …?	Finance is *important*: → standard *tools/ methods* → companies use both FA + *MA*
	MA
Main purpose of MA …?	1) *Forward* planning
Information supplied by …?	• information given in *annual budgets*
Aims of MA? (5 specific aims)	• aim is to help company: (a) decide how to *spend* its money (b) decide whether to *borrow funds* (e.g., to invest in *new machinery*, to take on more staff) (c) monitor *cash flows* & value of *investments* (d) predict *profits*/losses (e) help monitor performance (incl. staff)
Required by law?	2) Not a *legal* requirement
Management tool (advantages: why used? by whom?)	3) Internal management mechanism • good *practice* • powerful → *decisions* are based on the data • not useful for *outsiders*

Summary

There are two types of accounting: management and financial. Management accounting is a useful management tool, used internally within a company.
It's not a legal requirement, unlike financial accounting.

2 The lecturer talks about his/her own experience of bad financial management. It is not in the notes because it is a digression – that is, as a personal experience it is not essential information for the subject.

Transcript 2.9

Part 1

Good morning, everyone. I'm going to talk to you this morning in some more detail about two key aspects of agribusiness financial management: financial accounting and management accounting. OK, first of all, it is important to point out that in the EU, for example, about 95 per cent of agribusinesses are micro-businesses. As we've seen, a micro-agribusiness tends to have a financial manager who combines all the finance functions, and in farm businesses, the farm manager performs the role of financial manager. We're going to look first at what we mean by management accountancy, then what we mean by financial accountancy, in the context of a medium or large agribusiness. I'll also touch briefly on the treasury function and on green accountancy and then, before I finish, I'll give you a summary of the differences between financial and management accountancy.

Before I begin, I have a little story to tell you. As a young trainee manager, I worked for an agribusiness producing speciality ice-creams – Home Farm Ices, or HFI for short. The boss had been marketing manager for a large agribusiness, and decided to move to the country and start his own enterprise. HFI was very successful initially, with good sales and rapid income growth. Unfortunately, the boss wasn't as good at finances as he was at marketing. He didn't have proper financial systems in place and so he didn't realize that if he took overheads and finance costs into account, he was actually making a loss on over 50 per cent of his lines. After three years, HFI didn't have enough money to pay its taxes or its raw materials supplier, and its bank was threatening to call in its overdraft – effectively, to close the company.

The point of the story? Well, it's not just bad businesses that end up in trouble. A lot of potentially very good businesses do too, and the problems often start with the financial management ... or lack of it. Right ... to get back to the main part of my lecture.

Now to start with, as we have already noticed, it's pretty clear that in order to be successful, companies need to pay particular attention to finance and financial matters. In doing this, they make use of some fairly standard financial tools. In fact, as we will see, companies adopt a number of methods to help them to manage and control their finances. But I suppose the first point to note is that a major difference can be made between management accounting, or MA, and financial accounting, or FA. It's the first of these that I'm going to focus on now, although it's worth pointing out here that, in terms of finance and financial planning and monitoring, companies use both kinds of accounting to help them measure and plan their activities.

To illustrate the topic, I want to look at a larger enterprise. West Coffee Company (WCC) was established in the 1980s to trade in coffee – buying and selling on the international market. Today, they have stakes in agribusinesses producing raw materials in Kenya and Brazil. They retain their core business, but over time they have diversified downstream, and by downstream we mean new products created from their *core* product. So they've produced a range of instant coffees, for example, which they sell either under licence in supermarkets across Europe or under their brand name of 'Supra' coffee. Now, before we look at WCC and the way they organize their finance and accounting department, it's important to say that the company accountant today is a key member of the management team and will be expected to have good communication and interpersonal skills.

OK, so to start with, let's take a few moments to consider the role of management accounting. What is management accounting supposed to do? Research has shown that there are three important aspects of management accounting. It could be argued that these three aspects also exist for financial accounting. But, as we shall see here, these tend to have a different emphasis – I'll come back shortly and look at similarities and differences.

The first important point to note here is that management accounting deals with the forward planning aspect of business; that is to say, with how a company is *going* to perform, rather than concentrating on how well or how badly it has performed in the past. So, from the point of view of the senior management team at WCC, it's fair to say that they depend on their management accountant to interpret the data from the financial accountants. In addition, they need him or her to provide a Management Information System (MIS), which keeps the focus of decisions on profits. The MIS at West Coffee Company uses annual budgets – monthly reports enabling managers to decide how to spend money, and to know whether WCC will need to borrow money in the future if they want to invest in new machinery or pay for expansion in the short term, for example. The management accountant also helps monitor cash flows and the value of investments, and predicts future profits or losses. For a company like WCC, where changes in the external environment such as an increase in the price of coffee globally can be critical, increasingly we find that they use this predicting – this future-oriented – role of management accounting as an internal management tool. It helps to monitor the performance of divisions or separate units of the company.

Exercise C

Refer students to the lecture slide in Lesson 9.2. Ask them what they know already about financial accounting documents. What would they like to know?

Tell them to prepare a page to take notes using the Cornell system. Remind them that they may not get all the information. If they miss something, they should leave a space. They can fill it in after the lecture.

Let them read the questions through and tell them to listen out for the answers to questions 2, 3 and 4.

1 Play Part 3 straight through. Students should complete the *Notes* section.

2–4 Set for pairwork. Feed back with the whole class. Ask for suggestions for phrases to use to find out about the importance of digressions, e.g., *Why did the lecturer start talking about …? I didn't understand the bit about … Is it important?* and so on (see *Skills bank*). Set question 3 for pairwork.

5/6 Set for pairwork. Students compare their notes, complete any blank spaces and then write the *Review* and *Summary* sections.

Feed back with the whole class, building a set of notes on the board.

Answers

Possible answers:

1 See notes below.

2 The Cornell note-taking system.

3 It's important to know how to take good notes.

4 To find out about management accounting documents.

5/6 See notes. Note: The summary here is concerned with the section on financial accounting only – not the lecturer's overall summary of the differences between management accounting and financial accounting.

Transcript 2.13

Part 3

OK, so moving on to think about financial documents. OK, it's important to remember here that the three major distinctions that we described between management accounting and financial accounting are reflected in the sort of documentation which is provided for accounting purposes. Today, I'm going to look at the three types of document which are typically associated with financial accounting: balance sheets, profit and loss accounts, and cash flow statements.

Let's start with the baseline document: the balance sheet. OK, so we already know that companies are required by law to produce this document – that is,

Review	Notes
FA documents are …?	<u>Financial accounting documentation</u>
Balance sheet gives …?	1 <u>Balance sheet</u> (NB legal requirement)
(3 things)	• assets = what co. owns, e.g., factories, vehicles
	• liabilities = what co. owes, e.g., bank loan
	• shareholders' equity (also shareholders' capital) = money raised through shares (also = assets/liabilities)
Profit & loss acct. shows …?	2 <u>Profit & loss account</u> (shows net profit or loss) e.g., sales income
(give example)	• less cost of sales = gross profit
	• less other expenditure (e.g., salaries, rent) = net profit
US term for profit & loss acct …?	• US: profit & loss acct = 'income statement'
Who must have cash flow statements?	3 <u>Cash flow statement</u> (NB req. only for plc)
CFS gives information on …?	= info. on finance for activities, i.e., where the money came from (e.g., inside or outside the company)
Summary FA documents are: balance sheet, profit & loss account, cash flow statement. These documents give a good picture of company's finances.	

it is a statutory requirement. So let's look in a bit more detail at the balance sheet and the sections you will find in it. If you look at any balance sheet, you will see that it is normally divided into three main sections, and the three main pieces of information you find about the company are as follows.

Number one is their assets. That is, what they own and how much it's worth; assets include their factories and vehicles. Number two is a summary of the liabilities of a company, which if you like is the opposite of the assets. Liabilities, as we have already seen, are what the company owes to others. The company has a bank loan which attracts interest, and it also owes money to suppliers. Thirdly, and this is the final main piece of information on the balance sheet, we find the owner's equity or shareholders' equity. 'Shareholders' equity' is defined as the amount of money the company has raised through shares. In other words, this is the money which is helping to finance the company. For some companies, the amount of share equity is not very large, relatively, but with other companies, it can be enormous: For example, before BP started to have problems as a result of the Louisiana oil spill, its stockholder equity exceeded $100 billion. As Baker points out in *Principles of Business Finance,* one of your core texts – the 2nd edition was published in 2002 – we can think of shareholders' equity as total assets minus total liabilities; but you don't need to worry about that at this stage.

OK, so now we can see that these three sections or parts of the balance sheet – the assets, the liabilities and the shareholders' equity – together can provide shareholders and others with a good, accurate picture of the company. To quote Baker, 'At a specific point in time; these three parts of a balance sheet can give investors an idea about what the company has – that is, what it owns – as well as describing its debts or what it owes.' It also describes, as we have seen, the amount of money which has been invested by the shareholders, and which, therefore, the company is able to use.

By the way, I see that a number of you are using the Cornell note-taking system. That's excellent. Do you all know about Cornell? No? Right, well, I suggest you look at *How to study in college* by Walter Pauk, the 10th edition, published in 2010. It's very helpful, and it should be in the university library. I'm sure that you all know the importance of taking good notes – and this system is particularly useful.

So to get back to the main topic ... looking at the diagram on the slide, we can see that some other documentation is required for financial accounting. The second element is the profit and loss account, or P&L account, which can tell us if the company is financially healthy, and perhaps also whether it is being managed well. One definition of the P and L account given by *thefreedictionary.com* on the web is: 'An account compiled at the end of an accounting period to show gross and net profit or loss.' Typically, this account shows, for example, the cost of sales, which if it's taken away from the sales income, gives us the gross profit. You then deduct other items of expenditure, such as salaries, or rent, and you are left with a net profit, or perhaps a loss! By the way, in the US, a profit and loss account is known as an 'income statement'.

Now, the third element of financial accounting – that's the cash flow statement – is slightly different because it is not always necessary. To be more precise, this third element is only required when a company is a 'plc' – that's a public limited company, whose shares can be traded or bought and sold on the stock exchange. So, to summarize briefly: This particular part of the accounting process, the cash flow statement, answers questions about how the company was able to find the cash to finance its recent activities; for example, whether this money came from inside or from outside the company.

Now, I think that's all I'm going to say for the moment on the basic parts of financial accounting documentation. Are there any questions so far? No, good. Now, when I see you in tutorials, we'll look in more detail at management accounting. In the meantime, I'm going to set you a research task. Right, now listen carefully ... your task is to find out about the different documents that are used in management accounting. I'd like you to work in groups of four. Each group should find out about the various management accounting documents that are used and report back on your findings.

Exercise D

Tell students that lecturers will often give references while they talk and it is important to note down any references. The kinds of information may differ – they may just be names of books or articles, they may be an exact quotation (a 'direct quote') or they may be a paraphrase (sometimes called an 'indirect quotation'). Refer students to the table and check that they know what each row represents.

Play each extract and allow students time to complete the sections of the table. Check with the whole class.

Answers

	Extract 1	Extract 2	Extract 3	Extract 4
Name of writer	Baker	Walter Pauk		Baker
Title and date of source	*Principles of Business Finance* 2nd edition 2002	*How to Study in College* 10th edition 2010	thefreedictionary.com	as in 1
Location	core text	university library	web	as in 1
Type of reference	indirect quotation/paraphrase	name of book	direct quotation	direct quotation
Relevant to …?	shareholder equity; balance sheet	Cornell note-taking	profit and loss account	the (three sections of the) balance sheet
Introducing phrase	As Baker points out in …	… I suggest you look at …	One definition of … given by …	To quote Baker, …

Transcript 2.14

Extract 1

As Baker points out in *Principles of Business Finance*, one of your core texts – the 2nd edition was published in 2002 – we can think of shareholders' equity as total assets minus total liabilities; but you don't need to worry about that at this stage.

Extract 2

By the way, I see that a number of you are using the Cornell note-taking system. That's excellent. Do you all know about Cornell? No? Right, well, I suggest you look at *How to study in college* by Walter Pauk, the 10th edition, published in 2010. It's very helpful, and it should be in the university library.

Extract 3

The second element is the profit and loss account, or P and L account, which can tell us if the company is financially healthy, and perhaps also whether it is being managed well. One definition of the P and L account given by *thefreedictionary.com* on the web is: 'An account compiled at the end of an accounting period to show gross and net profit or loss.'

Extract 4

OK, so now we can see that these three sections or parts of the balance sheet – the assets, the liabilities and the shareholders' equity – together can provide shareholders and others with a good, accurate picture of the company. To quote Baker, 'At a specific point in time; these three parts of a balance sheet can give investors an idea about what the company has – that is, what it owns – as well as describing its debts or what it owes.'

Language note

A 'core text' is the main text for the course. Students are usually told they should buy a copy of the core text for their course.

Exercise E

Set for individual work – possibly homework – or else a pair/small group writing task. If the latter, tell students to put their writing on an OHT or other visual medium so that the whole class can see and comment on what has been written. You can correct language errors on the OHT.

Exercise F

Tell students to work in groups of three or four. Either give each group a topic (i.e., a type of documentation) or allow them to choose. Make sure that each topic is covered by at least one, and preferably two groups.

Feed back on questions 1–3 with the whole class. Tell them that each student should now carry out research into the group's topic. They should each look at a different source and so will need to decide who is going to look at which one. You will also need to arrange the date for the feedback and discussion of the information – this is the focus of Exercise D in Lesson 9.4. Tell students that in Lesson 9.4 they will take part in a seminar on this topic.

Answers

Possible answers:

1 Information to find: What are these documents? What are the purposes and uses of each type of document: Who? When? What for? Are there different categories within each type of document?

3 Use subject course books, the library and the Internet to find out the necessary information. Some example websites at the time of writing are:

www.investopedia.com

www.thefreedictionary.com

Alternatively (or in addition) – depending on your teaching situation and access to the sources of information – you can refer students to the information on page 107 of the Course Book.

Closure

Resource 9D, in the additional resources section, presents a (simplified) balance sheet for a large US company. Ask students to go to *Google define* and find out the meanings of these words if they don't know them. Then tell them to fill the spaces in the balance sheet.

cash
inventories
fixed assets
net receivables
accounts payable
long-term debt
shares outstanding

Answers

Model answers:

cash – assets held as money in bank accounts

inventories – stock or products

fixed assets – something tangible that is owned by the business such as equipment, buildings, etc.

net receivables – money that is owed to the business

accounts payable – money that is owed by the business to others (e.g., suppliers)

long-term debt – money owed to be paid back over a long period

shares outstanding – all the shares a company has that are held by investors

	Jan 06	Jan 05	Jan 04
Assets			
Cash	6,414	5,488	5,199
Net receivables	2,662	1,715	1,254
Inventories	32,191	29,447	26,612
Fixed assets	79,290	68,567	58,530
Other current assets	2,557	1,841	1,356
Total assets	123,114	107,058	92,951
Liabilities			
Accounts payable	40,178	35,107	31,051
Short-term debt	8,648	7,781	6,367
Long-term debt	30,171	23,669	20,099
Other liabilities	6,019	4,270	3,772
Total liabilities	85,016	70,827	61,289
Shareholders' equity			
Shares outstanding	4,165	4,234	4,296

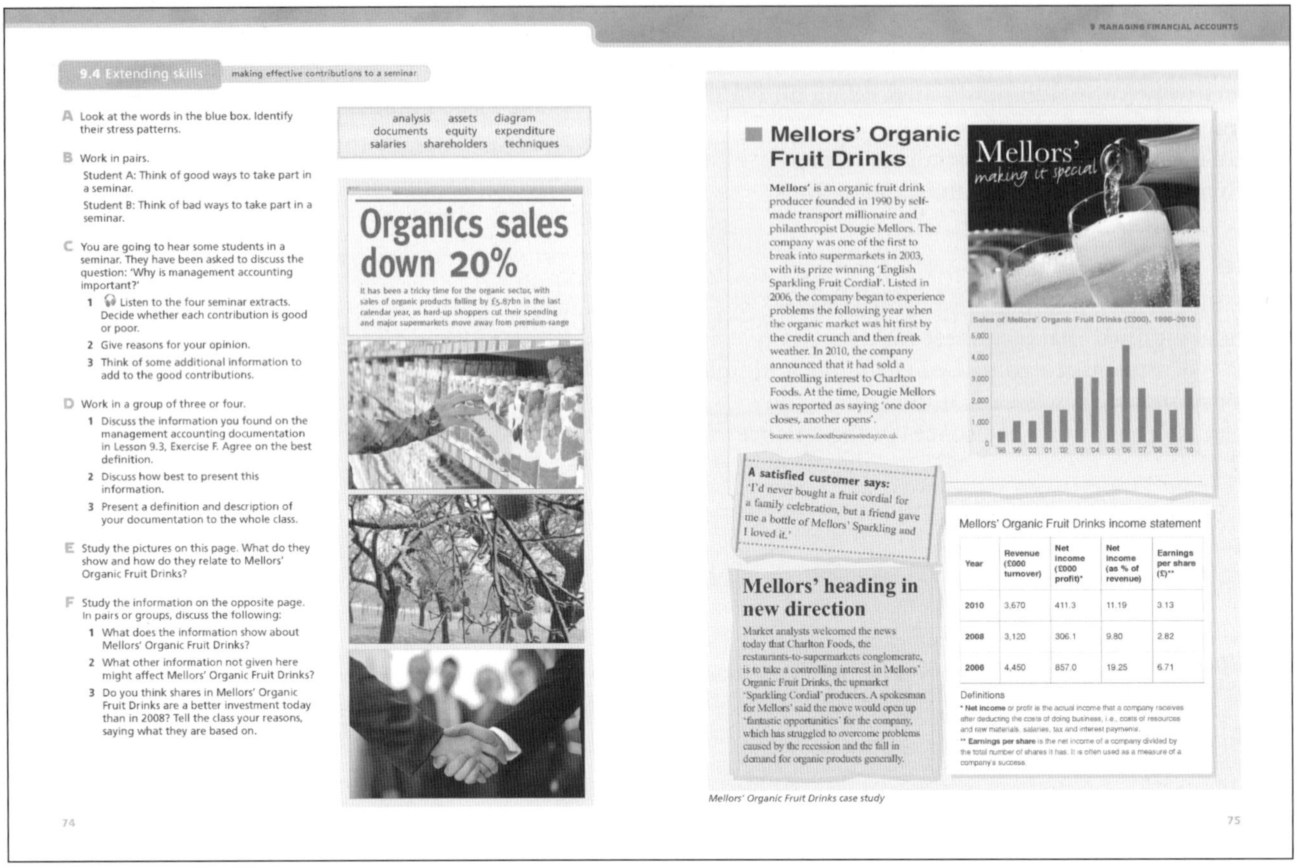

9.4 Extending skills | making effective contributions to a seminar

A Look at the words in the blue box. Identify their stress patterns.

analysis assets diagram
documents equity expenditure
salaries shareholders techniques

B Work in pairs.

Student A: Think of good ways to take part in a seminar.

Student B: Think of bad ways to take part in a seminar.

C You are going to hear some students in a seminar. They have been asked to discuss the question: 'Why is management accounting important?'

1 Listen to the four seminar extracts. Decide whether each contribution is good or poor.
2 Give reasons for your opinion.
3 Think of some additional information to add to the good contributions.

D Work in a group of three or four.

1 Discuss the information you found on the management accounting documentation in Lesson 9.3, Exercise F. Agree on the best definition.
2 Discuss how best to present this information.
3 Present a definition and description of your documentation to the whole class.

E Study the pictures on this page. What do they show and how do they relate to Mellors' Organic Fruit Drinks?

F Study the information on the opposite page. In pairs or groups, discuss the following:

1 What does the information show about Mellors' Organic Fruit Drinks?
2 What other information not given here might affect Mellors' Organic Fruit Drinks?
3 Do you think shares in Mellors' Organic Fruit Drinks are a better investment today than in 2008? Tell the class your reasons, saying what they are based on.

Organics sales down 20%

It has been a tricky time for the organic sector, with sales of organic products falling by £5.87bn in the last calendar year, as hard-up shoppers cut their spending and major supermarkets move away from premium-range

Mellors' Organic Fruit Drinks

Mellors' is an organic fruit drink producer founded in 1990 by self-made transport millionaire and philanthropist Dougie Mellors. The company was one of the first to break into supermarkets in 2003, with its prize winning 'English Sparkling Fruit Cordial'. Listed in 2006, the company began to experience problems the following year when the organic market was hit first by the credit crunch and then freak weather. In 2010, the company announced that it had sold a controlling interest to Charlton Foods. At the time, Dougie Mellors was reported as saying 'one door closes, another opens'.

Source: www.foodbusinesstoday.co.uk

Sales of Mellors' Organic Fruit Drinks (£000), 1998–2010

A satisfied customer says:
'I'd never bought a fruit cordial for a family celebration, but a friend gave me a bottle of Mellors' Sparkling and I loved it.'

Mellors' heading in new direction

Market analysts welcomed the news today that Charlton Foods, the restaurants-to-supermarkets conglomerate, is to take a controlling interest in Mellors' Organic Fruit Drinks, the upmarket 'Sparkling Cordial' producers. A spokesman for Mellors' said the move would open up 'fantastic opportunities' for the company, which has struggled to overcome problems caused by the recession and the fall in demand for organic products generally.

Mellors' Organic Fruit Drinks income statement

Year	Revenue (£000 turnover)	Net income (£000 profit)*	Net income (as % of revenue)	Earnings per share (£)**
2010	3,670	411.3	11.19	3.13
2008	3,120	306.1	9.80	2.82
2006	4,450	857.0	19.25	6.71

Definitions

* **Net income** or profit is the actual income that a company receives after deducting the costs of doing business, i.e., costs of resources and raw materials, salaries, tax and interest payments.

** **Earnings per share** is the net income of a company divided by the total number of shares it has. It is often used as a measure of a company's success.

Mellors' Organic Fruit Drinks case study

74 75

General note

Students will need the information they researched for Lesson 9.3, Exercise F.

Decide how you want students to present their information, e.g.,

- short talk with/without PowerPoint, OHT, or other visual medium
- to the whole class or to another group

Make sure that students understand the options for the presentation types.

Lesson aims

- make effective contributions to a seminar

Further practice in:

- stress within words

Introduction

Use a few of the review cues from the Cornell notes in Lesson 9.3 for students to try to recall the ideas on financial accounting in the lecture. If students appear to be having difficulty remembering, ask them to look again at their own notes from Exercise C in Lesson 9.3. Check that students are clear about the difference between management and financial accounting. Ask them to give the three main differences:

- financial accounting looks at the past and management accounting at the present and future
- financial accounting is a legal requirement, whereas management accounting is just good management practice
- management accounting is only available to managers inside the company, while the results of financial accounting are available to outsiders

Exercise A

Set for individual work and pairwork checking.

Answers

Oo	assets
oO	techniques
Ooo	diagram, documents, equity, salaries, shareholders
oOoo	analysis, expenditure

Exercise B

This is revision from Unit 5. Set for individual work and pairwork checking. Feed back with the whole class. Give a time limit and see which pair can think of the most Do's and Don'ts in the time. Refer to Unit 5 Lesson 4 for suggestions if you need to.

Answers

Possible answers:

Do	Don't
prepare the topic beforehand	
ask politely for information	demand information from other students
try to use correct language	
speak clearly	mumble, whisper or shout
say when you agree with someone	get angry if someone disagrees with you
link correctly with previous speakers	
build on points made by other speakers	
make a contribution, even if you are not sure if it is new or relevant	stay silent, waiting for 'the perfect moment'
be constructive	be negative
give specific examples to help explain a point	be vague
listen carefully to what others say	start a side conversation
allow others to speak	dominate the discussion
paraphrase to check understanding	
use clear visuals	

Exercise C

Check that students understand the topic for the seminar discussion. Ask them what they might expect to hear. Work through these extracts one at a time. Complete questions 1–3 for each extract before moving on to the next.

1 Set for individual work.

2 First check that students have understood the extract as well as possible. Then ask for opinions from the whole class on the contribution.

3 Once everyone has a clear notion of whether the contribution is a good one, ask for suggestions for additional points. Alternatively, set this part for pairwork after you have completed questions 1 and 2.

Answers

Model answers:

See table below.

	✓		Reasons	Possible additional information
Extract 1	✓		speaks clearly explains the point clearly answers correctly uses good fixed phrases	Any of these: Company can 1. decide how to spend its money 2. decide whether/ how to borrow funds, e.g., to invest in a product, to buy new equipment, to employ more staff 3. keep a check on cash flows & value of investments 4. predict profits/losses
Extract 2			doesn't speak clearly doesn't answer the question is talking about FA vs. MA poor use of visuals	
Extract 3			speaks clearly, but doesn't answer the question the points are not relevant to the question – is talking about accountants, not MA	
Extract 4	✓		speaks clearly explains the point clearly answers correctly uses good fixed phrases; has prepared well has a good visual	any of the other management accounting documents that students have researched into: break-even chart; contribution statements; budgets; investment appraisal

Transcript 🎧 2.15

Extract 1

It seems quite clear that management accounting is a very useful way of controlling what happens in a company. Let's look at three basic advantages: Firstly, management accountants can show where the strengths and weaknesses are in a company; secondly, it's possible to look at a wide range of different types of performance in different sections or units of a company; and thirdly, pretty important this, …

Extract 2

… erm, I think one big difference is management accounting. This is very important. It's possible … er … we can see how this is very important. So let's look at the chart and … oh, sorry, that's the wrong chart, just a minute … right, so here is some difference between financial and management … er … you can see I think, this difference … do you have any questions about this chart? …

Extract 3

We could ask the question: How much does it cost to use an accountant? Usually, this is very expensive, but it is necessary. If you use an accountant, you can get a good balance sheet. Then you can show the customers how well the company is performing. On the other hand, a good accountant is also very important for the public. In many countries, including the UK, you have to have a balance sheet for the public, and this balance sheet is the basic sheet for the whole company. Here, we must look also at the profit and loss account because that is …

Extract 4

So, this main difference – the fact that unlike financial accounting, which is public, management accounting is an internal mechanism. This difference shows us one of the main functions of management accounting. In fact, we could say that this summarizes the main importance of this type of accounting, and it also explains why more and more companies are using management accounting methods to improve their efficiency. If we look at the chart I've prepared here, we can see the main areas that management accounting will be looking at. For example, if we examine in a bit more detail the section called Cash Flow Forecasting, we can see that this can be a very effective method for a company to examine its profitability. Another area is …

Exercise D

Students should work in the same groups as their research groups from Lesson 9.3, Exercise F. They will need to have with them the research they have done individually on the group's chosen topic.

1 Tell each group to discuss the information that they have found and agree on the best definition and description of the type of documentation they have researched.

2 In discussing this question, students will need to decide who is going to speak when and say what. Encourage them to practise presenting to each other before talking to the whole class.

3 Allow each group a maximum of five minutes for the presentation. Then allow some time for questions. If more than one group have done the same topic, encourage disagreement and critical analysis. Remind the groups when discussing to use all the *good* techniques and phrases they have learnt.

Exercise E

This is a case study of a fictitious organic fruit drinks business which can be used to give more practice in discussion both of the main topic but also of business strategy, takeovers, markets and advertising. To set the context, you can ask students, prior to using the unit, to explore the market for, and marketing of, fruit drinks of all sorts using their local supermarket. Otherwise, or in addition, refer students to the picture of the orchard. With the whole class, elicit the words needed to discuss the different stages of fruit drinks production.

Discuss the background, including the linguistic culture of fruit drinks in the UK. You can introduce words and expressions such as 'perry', 'fruit cordial', 'fruit cup', 'beer' (as in ginger, i.e., non-alcoholic, beer) and 'syrup'. Ask for examples from students' own cultures. Use examples from, e.g., weekend magazines.

Without impinging on Ex F, you can discuss the concept of size in business: 'Is small beautiful?' Consider what sort of changes the acquisition of a company are likely to bring about for both parties. Elicit examples of businesses being taken over and discuss the possible or actual advantages/disadvantages.

Exercise F

Put students in pairs (or threes). For question 3, each pair can join another pair and agree a decision.

1 In their pairs, students can divide the information between them and then summarize the information for their partner. If there are words that they do not understand, they will need to check meanings in dictionaries or online.

2 Students will need to think about Mellors' Organic Fruit Drinks in PEST terms. For example, are the premium prices on organic products sustainable long term?

3 In coming to a conclusion on this, students can access information from the web, and think laterally and inventively in terms of how the company might go forward in a new partnership.

Closure

Use the *Vocabulary bank* at the end of the Course Book unit to check that the group can remember the meaning, spelling and pronunciation of the financial vocabulary. You may also want to refer students to the information on financial documentation on page 107 of the Course Book.

Extra activities

This unit contains a lot of technical and semi-technical vocabulary. Students will need plenty of practice with the vocabulary.

1 Work through the *Vocabulary bank* and *Skills bank* if you have not already done so, or as revision of previous study.

2 Use the *Activity bank* (Teacher's Book additional resources section, Resource 9A).

A Set the crossword for individual work (including homework) or pairwork.

Answers:

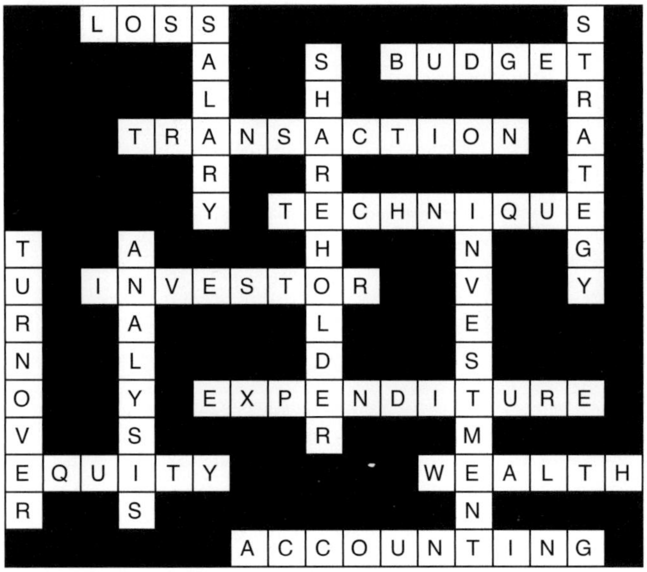

B Ask students to look at the nouns in the table: are plural forms of the nouns possible? Tell students to use an English–English dictionary or online definitions to help them find out the answers to the following questions.

1 Which forms are countable and which uncountable?

2 Do the countable and uncountable forms have different meanings?

Answers:

Noun	Countable or uncountable?	Notes
assets	C	
cash	U	
cost	C/U	
economy	C/U	*The economy* usually refers to that part of a country's affairs
liabilities	C/U	
loss	C/U	profit and loss account
margin	C/U	
profit	C/U	
sale	C/U	
share	C/U	
taxation	U	

3 Revise note-taking symbols – see the list at the back of the Course Book. Check back to Unit 5 if necessary. Give the meanings and ask students to write down the symbol (or do it the other way round). Then ask students to think about and discuss which ones they actually use. Are there any other ones that they have come across that they think are useful?

Alternatively, write the meanings on a set of cards. Put students in groups of about six with two teams in each group. Give each group a pile of cards. A student from each team picks a card and, without showing the members of his or her team, draws the appropriate symbol. The members of his or her team must say what the symbol stands for. If the student writes the correct symbol and the team gets the meaning right, the team gets a point. If the student writes the wrong symbol and/or the team gets it wrong, the team loses a point. The teams take it in turns to pick a card.

4 Identify finance and company abbreviations which will be useful for your students. Use the same procedure as in Activity 3. If they are studying in a UK or European-based context, the following may be useful:

Abbreviation	Meaning
FTSE100	Financial Times top 100 companies
AIM	alternative investment market
NASDAQ	National Association of Securities Dealers Automated Quotations (US electronic stock exchange)
HBOS	Halifax Bank of Scotland (now owned by Lloyds Banking Group and known only by the acronym)
RBS	Royal Bank of Scotland
AGM	annual general meeting
GAAP	generally accepted accounting practice
ROCE	return on capital employed
AMEX	American Express
Co	company
plc	public limited company
NGO	non-governmental organization
CEO	chief executive officer
SME	small to medium-sized enterprise
FedEx	Federal Express
VW	Volkswagen
GE	General Electric

Some useful glossaries are at:

www.spanish-translator-services.com/english-abbreviations/financial/index.htm

www.lse.co.uk/FinanceGlossary.asp

www2.hemscott.com/equities/GLOSS.HTM#OFEX

or for USA:

www.nysfaaa.org/docs/student_family/terms_acronyms.html

Answers

Model answers:

Word/phrase	Rating	Words which show less than 100% confidence
probably caused	**	probably
may have contributed to	*	may contributed (i.e., there were other reasons)
was possibly one of the factors which contributed to	*	possibly one of the factors (i.e., there were several factors) contributed
could have been a factor which led to	*	could a factor (i.e., there were other factors)
caused	***	–
seems to have caused	**	seems

Exercise B

Remind students that surveying the text means scanning and skim-reading to get an approximate idea of the text contents. They should:

- look at the title
- look at the first few lines and the final few lines of the text
- look at the first sentence of each paragraph

Note that this is in order to get a very approximate idea of the contents of the text. This will enable students to formulate questions about the text for which they might try to find answers. Students should be discouraged from reading deeply at this point, as they will be able to do this later.

Set for individual work and pairwork discussion. Each pair should agree on three questions. Feed back with the whole class. Write some research questions on the board.

Exercise C

Set for individual work followed by pairwork discussion. Feed back with the whole class. Ask whether the questions you have put on the board have been answered in the text.

Exercise D

These questions sometimes require students to 'infer' information – that is, understand what is not directly stated.

Set for individual work and pairwork checking. Feed back with the whole class, making sure that students understand the answers.

Answers

Model answers:

1 They can use consultants, the Internet or books.
2 Smaller businesses tend to use the Internet because they can't afford to employ consultants.
3 The key consideration is customer confidence.
4 Politicians tended to take the view that things were 'OK' in the past.
5 No, on the contrary, newspapers are typically very interested.
6 Not if they are rigorous and vigilant.

Exercise E

Set for individual work and pairwork checking. Feed back with the whole class. Point out that these phrases are very important in academic writing and will help to determine whether something is a fact or an opinion – an important aspect of reading comprehension. They are also used by writers in developing their arguments for or against a particular point of view.

Answers

Most experts agree that a proactive approach to food safety is essential when you are setting up a food-related agribusiness.	T
Obviously, when customers buy a food product, they need to know that it will not harm them.	C
In the past, politicians sometimes took the view that food in the UK was safe, and that they were doing all that was needed to protect consumers.	T
According to all available information, when consumers suffer food poisoning as a result of eating a particular product, the impact can be immediate and very damaging.	C
And according to Knowles (op. cit.), there remains significant evidence of 'the increasing emergence of rare serotypes of food-borne pathogens' in Europe.	T
Recent surveys carried out by the FSA and reported on their website suggest that things are getting better.	T
Given their careful approach, there is no evidence that Downland Foods need to be worried about problems arising from food safety.	C

Exercise F

Set for pairwork. Feed back with whole class. Discuss any differences in students' answers, and whether neutral equivalents are hard to find for some of the words.

Answers

Possible answers:

1 The key is, never be complacent. The UK in recent times has regularly suffered devastating outbreaks of food-related diseases such as salmonella and foot and mouth. And according to Knowles (*op. cit.*) there remains significant evidence of 'the increasing emergence of rare serotypes of food-borne pathogens' in Europe. In other words, a growth in the varieties of specific germs that cause diseases such as bovine spongiform encephalopathy (BSE). A BSE outbreak in the 1990s caused appalling and widespread suffering to animals as well as farmers, and did immeasurable damage to the UK rural economy. The number of cattle slaughtered soared to over 4 million, while experts estimated that the cost to the UK economy rocketed and 'severely affected an industry that was estimated to be worth £3.2bn a year' (Atkinson, 1999). As a result of that outbreak, confidence in the British food industry plummeted, taking years to recover.

2 The choice of words emphasizes the terrible impact of outbreaks of diseases such as these and conveys the message that the writer is emotionally involved.

3

Marked word	Neutral alternative
devastating	bad
appalling	serious
widespread	general
immeasurable	significant
slaughtered	killed
soared	rose
rocketed	climbed
plummeted	fell

Exercise G

Draw the table from the Answers section on the board. Ask students to look at the example sentence and say which box each part of the sentence should go in. Complete the table for the example sentence as shown. Point out how each of the noun phrases is made up of several words. In each case, elicit which words are the core of the noun phrases (shown in bold in the table in the Answers section). Do the same with the verb phrases. Ask students to suggest how the sentence can be rewritten in several short, very simple sentences in which noun phrases and verb phrases are reduced to the core meaning as far as possible. Demonstrate with examples if necessary.

Set questions 1–4 (relating to sentences A and B) for individual work and pairwork checking. Feed back with the whole class.

Answers

Model answers:

1/2 A A BSE outbreak in the 1990s | caused | appalling and widespread suffering to animals as well as farmers, | and | did immeasurable damage | to the UK rural economy.

B The number of cattle | slaughtered soared | to over 4 million, | while | experts estimated that | the cost to the UK economy rocketed | and |'severely affected | an industry | that was estimated to be worth £3.2bn a year' (Atkinson, 1999).

3 See table on next page.

	Subject noun phrases	Verb phrases	Object/complement noun phrases	Adverbial phrases	Notes
Example	The UK	**has suffered**	devastating **outbreaks** of food-related diseases such as salmonella and foot and mouth	in recent times regularly	*such as* introduces examples of the noun phrase items
A	A BSE **outbreak**	**caused**	appalling and widespread **suffering** to animals as well as farmers	in the 1990s	*as well as* is used here as *and*
	(and) (**it**)	**did**	immeasurable **damage** to the rural UK economy		*it* is assumed
B	the number of **cattle**	(which **were**) **slaughtered**			this is not a sentence – it is the subject – see next line; there is an embedded clause in the subject; we can omit the relative pronoun *which* and the auxiliary (*were*) in this case
	(**it**)	**soared**	**to over 4 million**		*soared* being intransitive *to over 4 million* is the indirect object
	(while) **experts**	**estimated** (that)	the **cost** to the UK economy		*while* is used here as a conjunction = *and*
		rocketed			
		(and) **affected**	an **industry** that was estimated to be worth £3.2bn a year'	severely	

4 Possible sentences:

A BSE is an example of recent public health problems.

BSE caused harm and suffering to animals and people.

BSE seriously damaged the economy of country areas in the UK.

B Action involved killing more than 4 million animals.

Experts assessed the cost of the BSE outbreak.

Experts believe the cost to the UK grew dramatically and very quickly.

Experts believe that BSE did serious damage to an important strategic industry.

Language note

1 Subjects and objects will always be nouns, with or without modifying adjectives.

Complements can be

- nouns: *He is a doctor.*
- adjectives: *He is French.*
- adverbs: *He arrived late.*

2 There are several types of conjunction in English.

Coordinating conjunctions such as *and, or, but* link elements of equal grammatical status.

Correlative conjunctions have two items: *either … or …; both … and ….*

Subordinating conjunctions relate clauses to each other using single words (e.g., *that* with verbs of saying, thinking, etc., *after, as, before, if, although, while*) or phrases (e.g., *as soon as, in order to, provided that …*).

See a good grammar reference book for full explanations.

3 Adverbial phrases add information about the actions or processes described by the verb phrase.

Closure

Here is some advertising language from the website of a company that sells food products. Ask students to identify any marked vocabulary items in each sentence and to suggest more neutral words. Feed back, comparing answers and discussing any differences of opinion.

Brilliant ingredients for pies 'just like your mother made'!

Give your friends delicious chilli-flavoured chocolates this Christmas!

Taste the difference with our super-fine flour!

Use this token and slash your food bills!

Buy one and get another for next-to-nothing!

Probably the best cakes in the world!

Possible changes:

(*Brilliant*) Good ingredients for pies ('*just like your mother made*') that taste home-made.

Give your friends (*delicious*) tasty chilli-flavoured chocolates this Christmas!

Taste the difference with our (*super-fine*) high-quality flour!

Use this token and (*slash*) reduce your food bills!

Buy one and get another for (*next-to-nothing*) a reduced price!

(*Probably the best cakes in the world!*) Very good cakes!

10.3 Extending skills

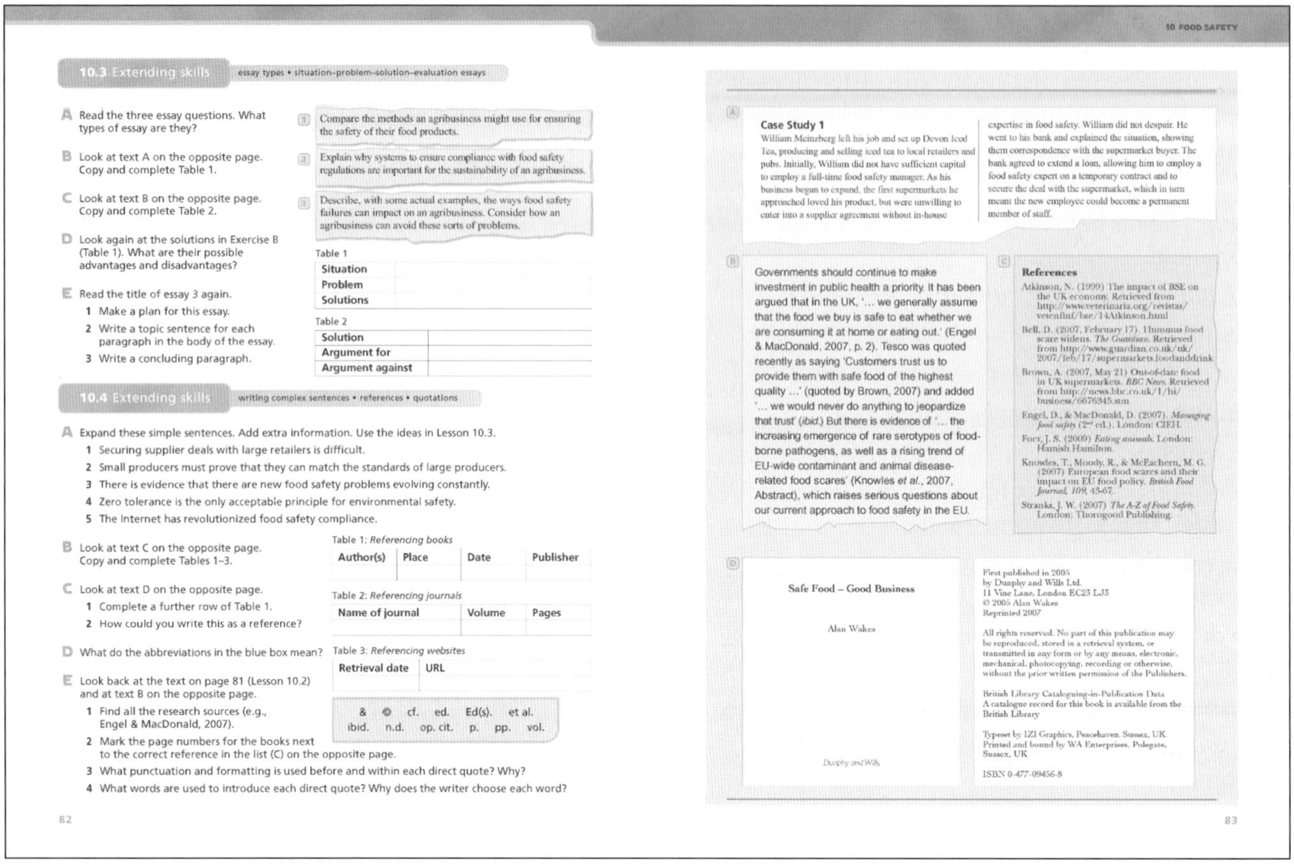

10.3 Extending skills

essay types • situation–problem–solution–evaluation essays

A Read the three essay questions. What types of essay are they?

B Look at text A on the opposite page. Copy and complete Table 1.

C Look at text B on the opposite page. Copy and complete Table 2.

D Look again at the solutions in Exercise B (Table 1). What are their possible advantages and disadvantages?

E Read the title of essay 3 again.

1 Make a plan for this essay.
2 Write a topic sentence for each paragraph in the body of the essay.
3 Write a concluding paragraph.

1 Compare the methods an agribusiness might use for ensuring the safety of their food products.

2 Explain why systems to ensure compliance with food safety regulations are important for the sustainability of an agribusiness.

3 Describe, with some actual examples, the ways food safety failures can impact on an agribusiness. Consider how an agribusiness can avoid these sorts of problems.

Table 1

Situation	
Problem	
Solutions	

Table 2

Solution	
Argument for	
Argument against	

10.4 Extending skills

writing complex sentences • references • quotations

A Expand these simple sentences. Add extra information. Use the ideas in Lesson 10.3.

1 Securing supplier deals with large retailers is difficult.
2 Small producers must prove that they can match the standards of large producers.
3 There is evidence that there are new food safety problems evolving constantly.
4 Zero tolerance is the only acceptable principle for environmental safety.
5 The Internet has revolutionized food safety compliance.

B Look at text C on the opposite page. Copy and complete Tables 1–3.

C Look at text D on the opposite page.

1 Complete a further row of Table 1.
2 How could you write this as a reference?

D What do the abbreviations in the blue box mean?

E Look back at the text on page 81 (Lesson 10.2) and at text B on the opposite page.

1 Find all the research sources (e.g., Engel & MacDonald, 2007).
2 Mark the page numbers for the books next to the correct reference in the list (C) on the opposite page.
3 What punctuation and formatting is used before and within each direct quote? Why?
4 What words are used to introduce each direct quote? Why does the writer choose each word?

Table 1: *Referencing books*

Author(s)	Place	Date	Publisher

Table 2: *Referencing journals*

Name of journal	Volume	Pages

Table 3: *Referencing websites*

Retrieval date	URL

& © cf. ed. Ed(s). et al.
ibid. n.d. op. cit. p. pp. vol.

A

Case Study 1

William Meinzberg left his job and set up Devon Iced Tea, producing and selling iced tea to local retailers and pubs. Initially, William did not have sufficient capital to employ a full-time food safety manager. As his business began to expand, the first supermarkets he approached loved his product, but were unwilling to enter into a supplier agreement without in-house expertise in food safety. William did not despair. He went to his bank and explained the situation, showing them correspondence with the supermarket buyer. The bank agreed to extend a loan, allowing him to employ a food safety expert on a temporary contract and to secure the deal with the supermarket, which in turn meant the new employee could become a permanent member of staff.

B

Governments should continue to make investment in public health a priority. It has been argued that in the UK, '… we generally assume that the food we buy is safe to eat whether we are consuming it at home or eating out.' (Engel & MacDonald, 2007, p. 2). Tesco was quoted recently as saying 'Customers trust us to provide them with safe food of the highest quality …' (quoted by Brown, 2007) and added '… we would never do anything to jeopardize that trust' (*ibid.*) But there is evidence of '… the increasing emergence of rare serotypes of food-borne pathogens, as well as a rising trend of EU-wide contaminant and animal disease-related food scares' (Knowles *et al.*, 2007, Abstract), which raises serious questions about our current approach to food safety in the EU.

C

References

Atkinson, N. (1999) The impact of BSE on the UK economy. Retrieved from http://www.veterinaria.org/revistas/vetenfinf/bse/14Atkinson.html

Bell, D. (2007, February 17). Hummus food scare widens. *The Guardian*. Retrieved from http://www.guardian.co.uk/uk/2007/feb/17/supermarkets.foodanddrink

Brown, A. (2007, May 21) Out-of-date food in UK supermarkets. *BBC News*. Retrieved from http://news.bbc.co.uk/1/hi/business/6676345.stm

Engel, D., & MacDonald, D. (2007). *Managing food safety* (2nd ed.). London: CIEH.

Foer, J. S. (2009) *Eating animals*. London: Hamish Hamilton.

Knowles, T., Moody, R., & McEachern, M. G. (2007) European food scares and their impact on EU food policy. *British Food Journal, 109*, 43-67.

Stranks, J. W. (2007) *The A-Z of Food Safety*. London: Thorogood Publishing.

D

Safe Food – Good Business

Alan Wakes

Dunphy and Wills

First published in 2005
by Dunphy and Wills Ltd.
11 Vine Lane, London EC23 LJ3
© 2005 Alan Wakes
Reprinted 2007

All rights reserved. No part of this publication may be reproduced, stored in a retrieval system, or transmitted in any form or by any means, electronic, mechanical, photocopying, recording or otherwise, without the prior written permission of the Publishers.

British Library Cataloguing-in-Publication Data
A catalogue record for this book is available from the British Library

Typeset by IZI Graphics, Peacehaven, Sussex, UK
Printed and bound by WA Enterprises, Polegate, Sussex, UK

ISBN 0-477-09456-8

Lesson aims

- understand situation–problem–solution–evaluation structure in essays
- understand the use of information in this type of essay structure to:

  describe
  give cause and effect
  compare
  evaluate
  argue for

Further practice in:

- identifying required essay types
- producing an outline
- writing key sentences – which can be expanded in the next lesson into longer sentences

Introduction

Revise the different types of essay that were examined in Unit 8. Say or write on the board some key words or phrases from essay titles such as the following:

State …
Outline …
Describe …
Compare …
Evaluate …
Discuss …
Why …?
How …?
To what extent …?
How far …?

Ask students to say

- what type of essay is required
- what type of organizational structure should be used

If students find this difficult, refer them to the *Skills bank* for Unit 8.

Exercise A

Set for individual work and pairwork checking.

Feed back with the whole class. Point out that in real life, essays given by lecturers often involve several types of writing in one essay. This is the case with essay 3. Tell students that in fact a possible structure for essay 3 would be the following, which is commonly found in many types of writing (including newspapers and academic writing).

Situation: Description of a state of affairs, often giving some reasons and background information.	description
Problem(s): The problems which are caused by the state of affairs; plus the effects of these problems.	description (cause and effect)
Solution(s): Ways of dealing with the problems (i) which may have been tried in the past or are being tried now; (ii) which will, may or could be tried in the future; suggestions for further solutions.	description (+ possibly suggestion)
Evaluation of solution(s): Comparison of solutions; opinion on how successful the solutions are or could be + justification; an opinion on which is the best option + justification.	comparison and argument

Tell students they will plan (and possibly write) this essay.

Answers

Model answers:

1 Comparison, plus some evaluation.

2 Analysis.

3 Description, then comparison and evaluation/argument/opinion, plus support (see table above).

Exercise B

Set for individual work and pairwork checking. Feed back with the whole class.

Answers

Model answers:

Situation	William Meinzberg left his job and set up an iced tea business, producing and selling iced tea to local retailers and pubs.
Problem	As his business began to expand, the first supermarkets he approached would not enter into a supplier agreement without in-house expertise in food safety.
Solutions	Went to his bank and secured a loan. Employed a temporary food safety expert.

Exercise C

Set for individual work and pairwork checking. Feed back with the whole class.

Answers

Model answers:

Solution	investment in public health
Argument for	evidence of new forms of food-borne diseases
Argument against	companies like Tesco can provide guarantees of safe food

Exercise D

Set for pairwork discussion. Feed back with the whole class. Accept any reasonable suggestions. Common-sense answers are also suitable here.

Answers

Possible answers:

	Advantages	Disadvantages
went to his bank and secured a loan	meets the need	risk: incurs debt; risk of failure to meet repayments; possible requirement to provide security (e.g., a house or existing business)
employed a temporary food safety expert	no long-term commitment; opportunity to test solution; relatively low-cost solution	cost implications

Exercise E

1 Set for pairwork discussion. Remind students to refer back to the text in Lesson 10.2 for ideas and information, as well as the texts they have discussed in this lesson. Remind students about the basic structure of an essay (introduction – main body – conclusion).

If you wish, you can give students the first two columns of the table in the Answers section, with the third column empty for them to complete. The table is reproduced in the additional resources section (Resource 10B) for this purpose.

Feed back with the whole class. Build the plan on the board, using the ideas in the Answers section to help.

2 Ask students to write some topic sentences for the four body paragraphs, using the information in the plan. Remind students that topic sentences need to be very general. Set for individual work.

Feed back with the whole class, writing some examples on the board.

3 Set for pairwork, then discussion, with the whole class. Or if you prefer, set for individual homework. The ideas should be those of the students. Remind them to introduce their ideas with suitable phrases.

Note: Students will need their essay plans again in Lesson 10.4.

Answers

1 Possible essay plan:

Introduction	**Examples of ideas**	
introduce the topic area give the outline of the essay	agribusinesses involved in food = risk *In this essay, I will discuss the issue of …* *I will illustrate/describe* … (examples) *I will consider* … (solutions) *Finally, I will suggest* … (best strategy)	
Body	**Para 1:** situation/problems (general)	agribusiness food safety and environmental health a special issue ∵ 1. impacts livestock, business, community (workers, suppliers, etc.); local/regional/national economy; closure/suspension of businesses (evidence: research data) 2. impact is direct (local authority action/court action/penalties) and indirect (media reports = damaged reputation) 3. an agribusiness doesn't have to be at fault to suffer
	Para 2: problems (specific examples)	examples of cases: use the Internet and search, e.g., 'food safety stories', BSE
	Para 3: solutions	1. invest in expert staff and good systems (secure finance understanding bank) 2. train all new staff in food hygiene and safety 3. secure good information, e.g., using reputable websites 4. establish strict safety procedures 5. adapt product/manufacturing strategy avoid imported products / high-risk ingredients (e.g., pork) use local/organic/trusted suppliers
	Para 4: evaluations of solutions	1. essential for most businesses if securing funds, then be realistic and honest with yourself/funders and assume 'worst case' when planning, e.g., repayments 2. creates good image with suppliers and customers but expensive and time consuming 3. excellent ROI (return on investment); cost/benefit 4. even small failures can = disaster 5. common sense, e.g., shell fish pie generally = higher risk than potato pie
Conclusion	*In my view/As I see it, the best strategy is … because …* *Firstly …* *Secondly …* *Thirdly …*	

2 Possible topic sentences:

Para 1	Agribusinesses specializing in food processing or food product manufacturing face special problems.
Para 2	There are numerous examples of the sort of problems that can affect even the largest and best-established food businesses.
Para 3	There are a range of solutions to the problem, but the important thing to remember is that they are all interconnected.
Para 4	In conclusion, there is overwhelming evidence that successful, sustainable agribusinesses have one thing in common: an absolute commitment to food safety.

3 Students' own concluding paragraphs.

Language note

Although 'situation–problem–solution–evaluation of solution' is often said to be an organizing principle in writing, in practice it is sometimes difficult to distinguish between the situation and the problem: they may sometimes seem to be the same thing. The important thing is to be clear about the main *focus* of the essay – that is, the answer to the question *What am I writing about?* – and to structure the essay around this.

Closure

Set up a *Dragon's Den*-style role-play. *Dragon's Den* is a TV series in which entrepreneurs try to persuade a panel of investors to invest in their business.

For this activity, the investors are retail buyers, successful and experienced businessmen and women. The entrepreneurs are people who have been running a new food-producing agribusiness for three years with some success, and who are now looking to establish a new supplier relationship. The buyers have to be confident that the entrepreneurs have established a sustainable business and, in particular, have quality systems – including food safety systems – that meet stringent requirements.

Put students in groups of four. Within each group, students work in pairs. Each pair should think of a product/product range. They should prepare a short (one-minute) presentation. They should provide a realistic back story including current market/supplier deals and key policies: organic/non-organic, ingredient sourcing, traceability (ensuring that sources of ingredients are established and meet required standards), training, quality management and justification for expansion into regional/national markets.

When both pairs are ready, they should take it in turns to be the entrepreneurs and the buyers. The buyers should make food safety/quality systems/compliance a priority.

10.4 Extending skills

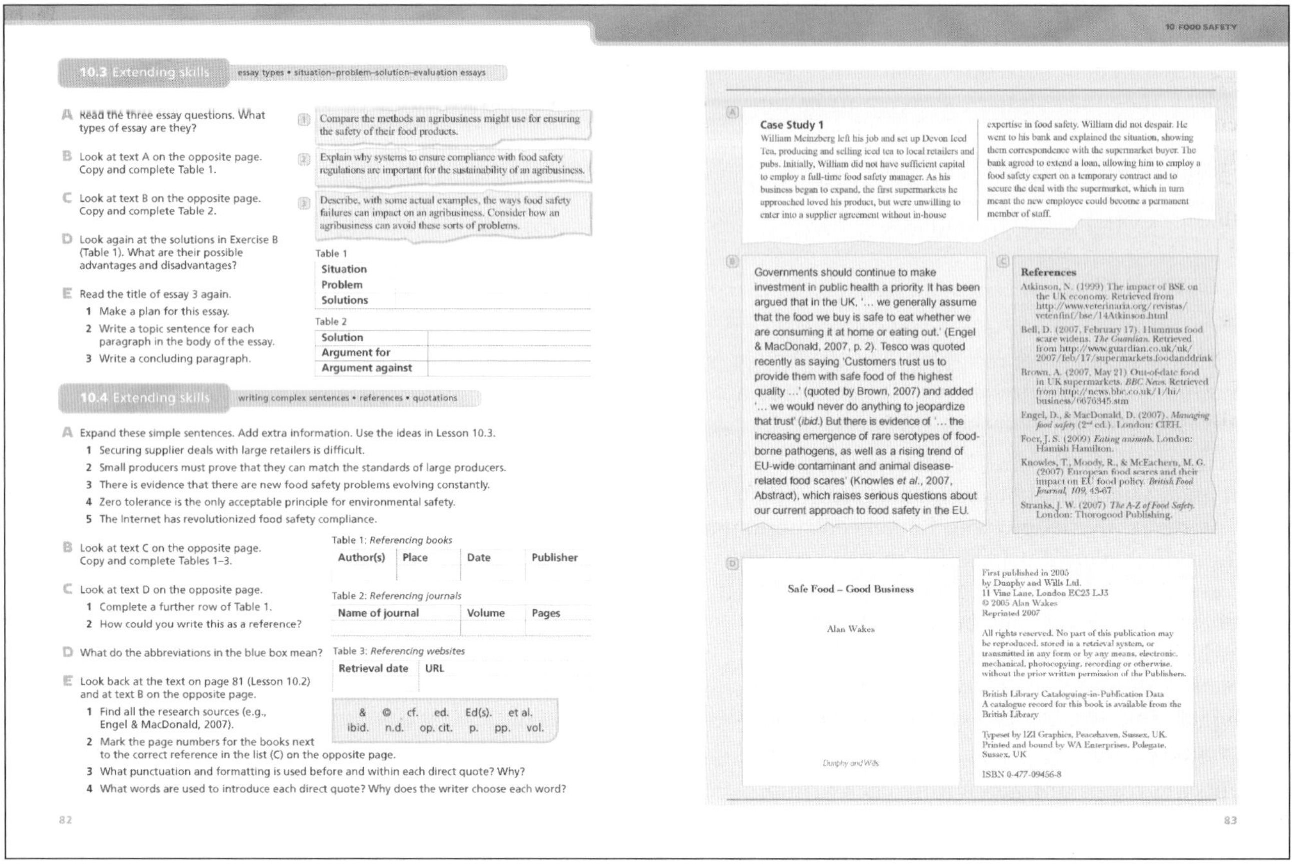

10.3 Extending skills essay types • situation–problem–solution–evaluation essays

A Read the three essay questions. What types of essay are they?

B Look at text A on the opposite page. Copy and complete Table 1.

C Look at text B on the opposite page. Copy and complete Table 2.

D Look again at the solutions in Exercise B (Table 1). What are their possible advantages and disadvantages?

E Read the title of essay 3 again.
1 Make a plan for this essay.
2 Write a topic sentence for each paragraph in the body of the essay.
3 Write a concluding paragraph.

1 Compare the methods an agribusiness might use for ensuring the safety of their food products.

2 Explain why systems to ensure compliance with food safety regulations are important for the sustainability of an agribusiness.

3 Describe, with some actual examples, the ways food safety failures can impact on an agribusiness. Consider how an agribusiness can avoid these sorts of problems.

Table 1

Situation	
Problem	
Solutions	

Table 2

Solution	
Argument for	
Argument against	

10.4 Extending skills writing complex sentences • references • quotations

A Expand these simple sentences. Add extra information. Use the ideas in Lesson 10.3.
1 Securing supplier deals with large retailers is difficult.
2 Small producers must prove that they can match the standards of large producers.
3 There is evidence that there are new food safety problems evolving constantly.
4 Zero tolerance is the only acceptable principle for environmental safety.
5 The Internet has revolutionized food safety compliance.

B Look at text C on the opposite page. Copy and complete Tables 1–3.

C Look at text D on the opposite page.
1 Complete a further row of Table 1.
2 How could you write this as a reference?

D What do the abbreviations in the blue box mean?

E Look back at the text on page 81 (Lesson 10.2) and at text B on the opposite page.
1 Find all the research sources (e.g., Engel & MacDonald, 2007).
2 Mark the page numbers for the books next to the correct reference in the list (C) on the opposite page.
3 What punctuation and formatting is used before and within each direct quote? Why?
4 What words are used to introduce each direct quote? Why does the writer choose each word?

Table 1: *Referencing books*

Author(s)	Place	Date	Publisher

Table 2: *Referencing journals*

Name of journal	Volume	Pages

Table 3: *Referencing websites*

Retrieval date	URL

& © cf. ed. Ed(s). et al.
ibid. n.d. op. cit. p. pp. vol.

A

Case Study 1
William Meinzberg left his job and set up Devon Iced Tea, producing and selling iced tea to local retailers and pubs. Initially, William did not have sufficient capital to employ a full-time food safety manager. As his business began to expand, the first supermarkets he approached loved his product, but were unwilling to enter into a supplier agreement without in-house expertise in food safety. William did not despair. He went to his bank and explained the situation, showing them correspondence with the supermarket buyer. The bank agreed to extend a loan, allowing him to employ a food safety expert on a temporary contract and to secure the deal with the supermarket, which in turn meant the new employee could become a permanent member of staff.

B

Governments should continue to make investment in public health a priority. It has been argued that in the UK, '... we generally assume that the food we buy is safe to eat whether we are consuming it at home or eating out.' (Engel & MacDonald, 2007, p. 2). Tesco was quoted recently as saying 'Customers trust us to provide them with safe food of the highest quality ...' (quoted by Brown, 2007) and added '... we would never do anything to jeopardize that trust' (*ibid.*) But there is evidence of '... the increasing emergence of rare serotypes of food-borne pathogens, as well as a rising trend of EU-wide contaminant and animal disease-related food scares' (Knowles *et al.*, 2007, Abstract), which raises serious questions about our current approach to food safety in the EU.

C

References

Atkinson, N. (1999) The impact of BSE on the UK economy. Retrieved from http://www.veterinaria.org/revistas/vetenfinf/bse/14Atkinson.html

Bell, D. (2007, February 17). Hummus food scare widens. *The Guardian*. Retrieved from http://www.guardian.co.uk/uk/2007/feb/17/supermarkets.foodanddrink

Brown, A. (2007, May 21) Out-of-date food in UK supermarkets. *BBC News*. Retrieved from http://news.bbc.co.uk/1/hi/business/6676345.stm

Engel, D., & MacDonald, D. (2007). *Managing food safety* (2nd ed.). London: CIEH.

Foer, J. S. (2009) *Eating animals*. London: Hamish Hamilton.

Knowles, T., Moody, R., & McEachern, M. G. (2007) European food scares and their impact on EU food policy. *British Food Journal, 109*, 43-67.

Stranks, J. W. (2007) *The A-Z of Food Safety*. London: Thorogood Publishing.

D

Safe Food – Good Business

Alan Wakes

Dunphy and Wills

First published in 2005
by Dunphy and Wills Ltd.
11 Vine Lane, London EC23 LJ3
© 2005 Alan Wakes
Reprinted 2007

All rights reserved. No part of this publication may be reproduced, stored in a retrieval system, or transmitted in any form or by any means, electronic, mechanical, photocopying, recording or otherwise, without the prior written permission of the Publishers.

British Library Cataloguing-in-Publication Data
A catalogue record for this book is available from the British Library

Typeset by IZI Graphics, Peacehaven, Sussex, UK.
Printed and bound by WA Enterprises, Polegate, Sussex, UK

ISBN 0-477-09456-8

General note

This lesson focuses on writing references for a bibliography according to the APA (American Psychological Society) system. Before the lesson, it would be useful to familiarize yourself with this system. See the *Skills bank*, and for more detailed information, websites such as http://owl.english.purdue.edu/owl/resource/560/01/

Lesson aims

- use quotations with appropriate punctuation and abbreviations such as ibid.
- write a reference list (APA system)

Further practice in:

- the reverse activity to Lesson 10.2, i.e., putting extra information into simple sentences in an appropriate way

Introduction

Introduce the idea of using sources in writing. Look back at the text in Lesson 10.2 and ask students to find all the places where a reference to a source is mentioned.

Ask them to find a quotation and a paraphrase. What are the main differences?

Exercise A

Do the first sentence with the whole class as an example on the board. Students should feel free to add words as appropriate to make a coherent sentence; they can also paraphrase.

Set the remaining sentences for individual work.

Answers

Possible answers:

1 There is substantial anecdotal and other evidence to indicate that agribusiness initially will have significant difficulty in securing supplier deals with major retailers.
2 Small producers in particular have reported difficulty in demonstrating that they have met the quality and service standards shown by large producers.
3 There is significant data to support the view that new-variant food-related diseases are on the increase.
4 The evidence suggests that agribusinesses which adopt a zero tolerance approach to food safety and other forms of risk are the enterprises that will build sustained success in the marketplace.

5 Judging not only by the increase in traffic on sites such as the FSA's website, it is clear that the information revolution has had a major impact in areas such as food safety and compliance.

Exercise B

Tell students that this is a list of references from the text in Lesson 10.2. Note that it is called 'References' because it lists all the references actually given (it is not a list of all the references the author might have consulted but not referred to – this is a bibliography).

Set for individual work and pairwork checking. Note that these tables are intended to help students identify some key information. For a full set of categories to include in a reference list, see the *Skills bank*. Tell students that when writing a reference list, they will need to pay close attention to the detail of the layout which is in the APA style (the American Psychological Association). See the *Skills bank* for a website which (at the time of writing) gives further details. In particular, students should note and will need to practise:

- putting the names of writers and multiple writers in the correct alphabetical order according to family name, with the right spacing and punctuation
- writing all numbers correctly, including dates and page references
- using punctuation including the role and placing of full stops, commas and colons
- laying out the references in the correct style with the correct positions (e.g., of indents and tabs)
- using standard APA style features such as italic and brackets

Answers

Table 1:

Author(s)	Place of publication	Date of publication	Publisher
Engel, D., & MacDonald, D.	London:	2007.	CIEH.

Table 2:

Name of journal	Volume	Pages
British Food Journal	*109*	43–67.

Table 3:

Retrieval date	URL
May 21, 2007	http://news.bbc.co.uk/1/hi/business/6676345.stm

Language and subject note

In the case of journals, there is an increasing tendency to refer to the volume number only in reference lists, omitting the issue number. Thus, for example, *English for Specific Purposes, 16*(1), 47–60 might become *English for Specific Purposes, 16*, 47–60.

Exercise C

Set for individual work and pairwork checking.

Answers

1

Author(s)	Place of publication	Date of publication	Publisher
Wakes, A.	London:	2005.	Dunphy and Wills Ltd.

2 Wakes, A. (2005). *Safe food – good business.* London: Dunphy and Wills Ltd.

Language and subject note

In the APA system, titles of books (but not titles of articles) are in italics, sentence case – that is, initial capital letter only, unless the title contains a proper noun. If the title contains a colon, the first word after the colon is also capitalized.

Journal *titles* are in italics and 'headline' or 'title' style – that is, all key words are capitalized but not conjunctions and prepositions. For example: *Journal of Small Business Management.*

Journal *articles* are sentence case, no italics.

Exercise D

Many of these were covered in Unit 5, so ask students to check back if they are not sure, or they can refer to the list at the back of their books; they can also check online at the APA site and/or the other site given in the *Skills bank*.

Set for individual work and pairwork checking.

Answers

&	and
©	copyright
cf.	compare
ed.	edition
Ed(s).	editor(s)
et al.	and other authors
ibid.	same place in a work already referred to
n.d.	no date (used in a reference list if there is no date – as is often the case with web articles)
op. cit.	the work already referred to
p.	page
pp.	pages
vol.	volume

Exercise E

Remind students (if you have not done so already) of the two main ways in which students can use sources (i.e., references to other writers' work) in their writing:

- by giving the exact words used by another writer
- by paraphrasing another writer's ideas, i.e., rewriting the ideas using their own, different words but retaining the meaning

The first method is referred to as quotation or direct quotation. Short direct quotations should be in quotation marks, and incorporated into the paragraph. Quotations of more than one sentence should be 'display quote' style, i.e., on a new line, and indented.

The second method is referred to as paraphrase, summary or indirect quotation. Note that around 90% of the paraphrase should be new words.

1/2 Set for individual work. Tell students to identify the research sources. They should then locate the source in the reference list on page 83 of the Course Book. This may seem a mechanical exercise, but it is useful for students to get into the habit of doing this. It will enable them to find an original source book, refer to the relevant part of the book, and read more about the subject.

3/4 Students should identify the punctuation and introducing phrases used. Feed back with the whole class.

Language and subject note

If the quotation is a full sentence, it begins with a capital letter inside the opening quotation mark and ends with a full stop inside the closing quotation mark.

If there are some words missing from the original quotation that were at the start of the original sentence, the quotation does not begin with a capital letter.

Original spelling/punctuation conventions are retained in a quotation.

Page references are normally given with direct quotations.

Answers

Model answers:

See table on next page.

Quote	Source	Punctuation/formatting before/within each direct quote	Introducing phrase + reason for choice
'not responding is a response – we are equally responsible for what we don't do.'	Foer, J. S. (2009) *Eating Animals*. London: Hamish Hamilton.	'xxx'.	asserts that reason: the quote highlights the importance of a proactive active food safety strategy
'... rising trend' ...	Knowles, T., Moody, R., & McEachern, M. G. (2007) European food scares and their impact on EU food policy. *British Food Journal*, *109*(1) pp. 43–67	... 'xxx' ...	research has continued to indicate reason: it is reasonable to assume that what follows, is true
'the increasing emergence of rare serotypes of food-borne pathogens'. ...	See above.	'xxx.'	there remains significant evidence of reason: what follows is evidence
'... supermarkets across the country emptied their shelves of hummus yesterday after salmonella was found in dips from one of the UK's main suppliers' ...	Bell, D. (2007, February 17). Hummus food scare widens. *The Guardian*. Retrieved from http://www.guardian.co.uk/uk/2007/feb/17/supermarkets.foodanddrink	'... xxx.'	Headlines ... proclaiming that reason: the story illustrates the impact of bad publicity
... 'severely affected an industry that was estimated to be worth £3.2bn a year?'	Atkinson, N. (1999) The impact of BSE on the UK economy. Retrieved from http://www.veterinaria.org/revistas/vetenfinf/bse/14Atkinson.html	... 'xxx'.	experts estimated ... economy rocketed and reason: what follows is an informed but not exact statistic
'... we generally assume that the food we buy is safe to eat whether we are consuming it at home or eating out.'	Engel. D., & MacDonald, D., (2007). *Managing food safety* (*2nd ed.*) London: CIEH	'... xxx.'	it has been argued that reason: what follows is an argument
'Customers trust us to provide them with safe food of the highest quality ...'	Brown, A. (2007, May 21). Out-of-date food in UK Supermarkets. *BBC News*. Retrieved from http://news.bbc.co.uk/1/hi/business/6676345.stm	'Xxx ...'	Tesco was quoted recently as saying reason: Tesco is a market leader in food supply
'... we would never do anything to jeopardize that trust.'	See above.	'... xxx.'	See above.

Language and subject note

An ampersand (&) is used with multiple authors, preceded by a comma.

The full stop at the end of the reference is omitted in the case of URLs.

Dates are (for example) April 7, not April 7th.

Closure

Refer students to the *Skills bank* for a summary of writing references. Study how the following are used:

- names (order)
- punctuation (capital letters, full stops, commas, colons)
- layout (indentation, spacing)
- style features (italics, brackets)

For further practice, use Resource 10C from the additional resources section. Ask students to check the references on a library database or on the Internet (discuss which sources are likely to be the most accurate and give them all the information they need – often the best way to check bibliographical details is to use a university library catalogue, as information found on the Internet is frequently inaccurate or incomplete). They should also make any necessary changes to ensure the references fit the APA models used in this unit. If possible, they should use the online website references (see *Skills bank*) to help them. Remind students that they will also need to put the references in the right alphabetical order.

Correct versions are:

Adams, C. R., & Early, M. P. (2004). *Principles of horticulture* (4th ed.). Oxford: Elsevier/Butterworth-Heinemann

Baker, H. G. (1974). The evolution of weeds. *Annual Review of Ecology & Systematics, 5*, 1–24.

Batish, D., Kohli, K., & Singh, H. P. (2006). *Handbook of sustainable weed management.* Birmingham: Harworth Press.

Bond, W., & Turner, R. J. (2005). A review of mechanical weed control. HDRA. Retrieved September 14, 2007, from http://www.gardenorganic.org.uk/organicweeds/downloads/mechanical.pdf

Cousens, R., & Mortimer, M. (1995). *Dynamics of weed populations.* Cambridge: Cambridge University Press.

Morrison, J. R., & Williams, J. B. (1987). *ADAS colour atlas of weed seedlings.* Netherlands: Wolfe Publishing Ltd.

Extra activities

1 Work through the *Skills bank* and *Vocabulary bank* if you have not already done so, or as revision of previous study.

2 Use the *Activity bank* (Teacher's Book additional resources section, Resource 10A).

A Set the wordsearch for individual work (including homework) or pairwork.

Answers

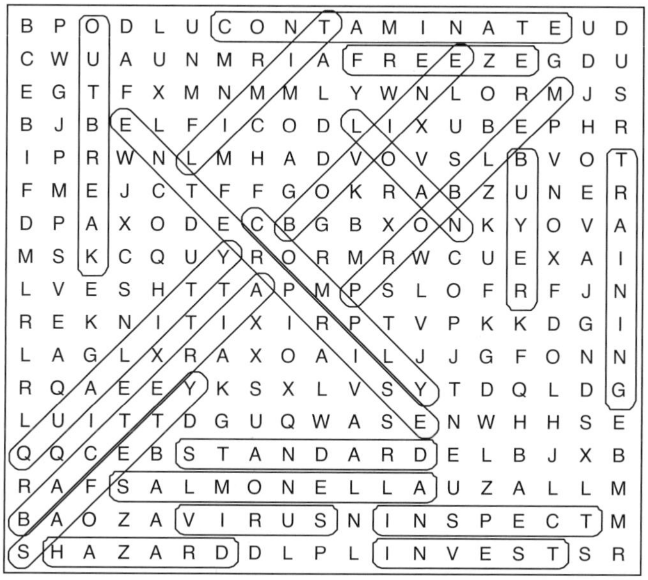

B Set for individual work (including homework) or pairwork. Accept all reasonable answers. Students should be able to explain the meaning.

Answers

Possible answers:

cross	contaminate
environmental	health
food	hygiene/safety/poisoning/-borne
hazard	analysis/control
hygiene	training
imported	foods
local	authority
notifiable	cases
smoking	ban
spoilable	foodstuffs

3 Ask students to choose one of the other essays in Lesson 10.3 and make a plan. They can also write topic sentences for each paragraph in the essay.

11 THE FUTURE OF FARMING

This unit looks in more detail at the future of agribusinesses, with a focus on political, economic and environmental factors, and some brief mention of other factors. The concept of PEST was introduced in Unit 1 and technological factors were examined in Unit 4.

Skills focus

Listening

- recognizing the speaker's stance
- writing up notes in full

Speaking

- building an argument in a seminar
- agreeing/disagreeing

Vocabulary focus

- words/phrases used to link ideas (*moreover*, *as a result*, etc.)
- stress patterns in noun phrases and compounds
- fixed phrases from academic English
- words/phrases related to environmental issues

Key vocabulary

See also the list of fixed phrases from academic English in the *Vocabulary bank* (Course Book page 92).

animal welfare
assure
award scheme
barren battery
biodiversity
campaign (v and n)
carbon credit scheme
carbon dioxide
carbon storage
catastrophic
charitable sector
climatologist
compassion
consultation exercise
credit crunch
decline (v and n)
donate
drastically
drive up
drought
emerging economies
endorsement
fossil fuels
fuel price inflation
given (n)
Global Positioning System (GPS)
habitat
habitat destruction
multifunctional
precision farming
pressure group
radical
raise awareness
reduction
relax
sceptic
shortfall
slowdown (n)
speculation
stabilize
stalk
stewardship
temporary
till (v)
tree felling
trend
win-win situation

11.1 Vocabulary

11 THE FUTURE OF FARMING

11.1 Vocabulary linking ideas

A Look at the diagram on the opposite page.
1. Name the five factors.
2. Discuss how the example(s) of each factor might influence farming.
3. Give more examples of each factor.

B Study the linking words and phrases in box a.
1. Put them into two groups for:
   a discussing reasons and results
   b building an argument
2. Is each linking word used to join ideas:
   a within a sentence?
   b between sentences?
3. Can you think of similar linking words?
4. Put the linking words in question 1b in a suitable order to list points in support of an argument.

C Study the words in box b.
1. Sort the words into two groups according to whether they are concerned with *climate and weather* or with *agriculture*.
2. In pairs, explain your decisions.
3. Are the words nouns, verbs or adjectives? What is the stress pattern of each word?
4. What other words or phrases have the same meaning?

D Read the text on the right.
1. Complete each space with a word or phrase from box a or box b. Change the form if necessary.
2. Can you think of other words or phrases with the same meaning as the blue words?
3. Find all the words or phrases in the text clearly connected with *agriculture* and all the words connected with *climate and weather.*
4. Match the phrases in the table on the page opposite with a later phrase that refers back to them. One has been done for you.

E Do the general knowledge quiz on the opposite page.

a
another point is as a result because finally firstly for example in addition moreover one result of this secondly since so

b
access arable difficulties drought flooding forage grow labourer livestock machinery moisture output yields

According to a recent scientific report, farms in the UK need to start adapting to cope with reduced water due to higher temperatures and summers with lower rainfall. ________, there is a clear risk of more frequent droughts as well as ________, both of which could lead to falling ________. The information suggests that an upland ________ farmer might need to consider bringing animals indoors earlier in the season because of the rainier autumns in this more unpredictable environment. ________, farmers need to look at the consequences for their livestock. The report suggests that there may be a longer ________ period for ________ crops. ________ could be that yields decrease after the early season, meaning farms may need to preserve more forage for winter feeding. Not only this, but there is also a warning that for ________ farms in certain elevated or vulnerable areas, if there are increased temperatures and a drier summer, ________, there could be problems establishing crops because of a lack of ________. ________ that lower precipitation in the summer could mean wetter autumns, which could in turn mean ________ for farmers accessing their land. The report advises that this might have a knock-on effect firstly on farm labourers, and ________ on machinery use. What is more, it could encourage more spring sowing.

Source: Maxwell, J. (2009). Dealing with the impact of reduced rainfall on farming in the UK. *Hadford University Scientific Journal*, 7, 34–38.

86

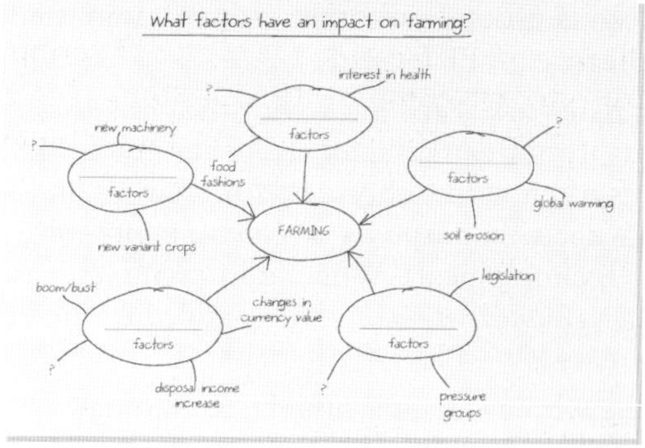

First phrase	Second phrase
a recent scientific report	the information/the report
higher temperatures and summers with lower rainfall	
farms ... need to start adapting	
upland	

HADFORD *University*

General Knowledge Quiz

1 What do these letters stand for?
- EU
- CAP
- WWF
- FAWC
- GM
- PETA
- FAO
- BBC

2 Who or what are these?
- Friends of the Earth
- James E Hansen
- Good Egg Awards
- Facebook
- conservation tillage
- Twitter
- battery
- credit crunch
- United Nations

87

General note

Read the *Vocabulary bank* at the end of the Course Book unit. Decide when, if at all, to refer your students to it. The best time is probably at the very end of the lesson or the beginning of the next lesson, as a summary/revision.

Lesson aims

- use rhetorical markers: to add points in an argument; to signal cause and effect (between- and within-sentence linking)
- further understand lexical cohesion: the use of superordinates/synonyms to refer back to something already mentioned; building lexical chains

Further practice in:

- synonyms, antonyms and word sets from the discipline
- abbreviations and acronyms

Introduction

1. Revise some vocabulary from previous units. Give students some key words from the previous unit (in italics below) and ask them to think of terms connected with these words (for example, some key phrases from food safety):

   food-borne, contaminated *food*, *food* safety, *food* poisoning
   health scare, *health* and safety.

2. Introduce the topic: before asking students to open their books, ask them what kinds of thing businesses need to pay attention to outside their own organizations that may affect their future. Accept any reasonable suggestions. Tell students that they first met some of these concepts in Unit 1 and later Unit 4, but here they are going to consider these topics in more detail.

Exercise A

Ask students to open their books and look at the diagram on page 87.

Check the meaning of the words in the diagram. If necessary, give some examples of pressure groups such as animal rights groups and other groups committed to animal welfare, e.g., PETA, etc.

1 Set for pairwork. Feed back with the class. Ask students if they can remember the acronym: PEST – which is often used in business for analysis and strategy decision purposes (Political, Economic, Social, Technological). What additional factor is here? (*Environment* – sometimes added to the PEST acronym as EV – PEST EV.)

2/3 With the whole class, discuss the technology section. What can they remember about technology in Unit 4? Ask students to explain how exactly computers affect agribusinesses. Discuss with the class other possible future technological developments and how they might affect agribusinesses: For example, what effect might the development of efficient and cheap non-carbon fuels have on farms? Answer: Changing the way their machinery is powered would initially be a major expense, but might save money ultimately.

Next look at the social factors. Discuss with the class the likely effects of changing age structure and improved education on agribusiness (note that this topic is considered further in the text in this lesson). Ask the class to suggest more social factors which might affect agribusinesses. Accept any reasonable suggestions.

Set the remaining factors (political, economic and environmental) for pairwork discussion. It would be a good idea to ask each pair to think about a different factor. Ask a few pairs to feed back to the class. Accept any reasonable suggestions.

Answers

Possible answers:

1 See diagram.

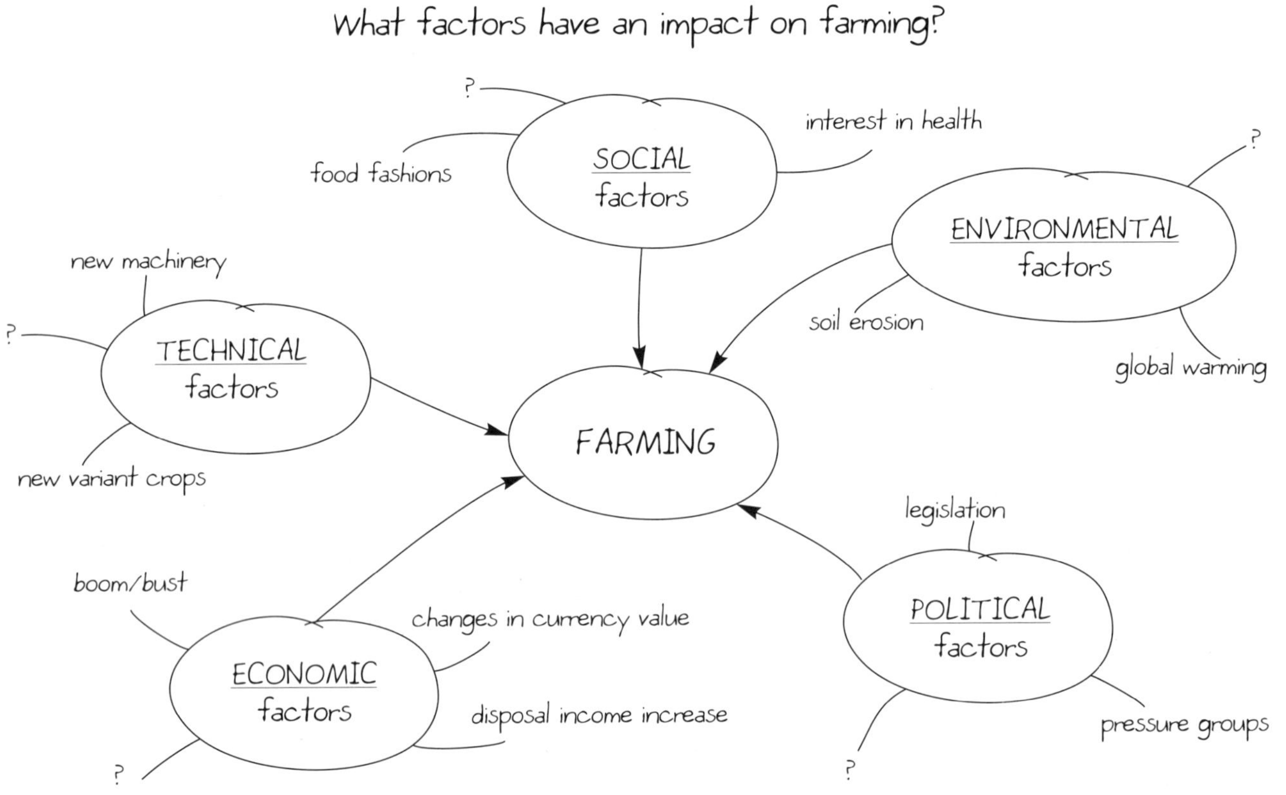

2 Answers depend on the students.

3 For the question marks, some more examples of factors are:

Technological factors

- innovations in, e.g., pest management

Social factors

- attitudes to animal cruelty
- attitudes to fair trading
- population growth

Political factors

- the impact of government
- global agreements
- anti-competition legislation
- subsidies
- employment law
- health and safety
- animal regulations
- farming/countryside lobby
- price capping
- environmental legislation

Economic factors

- economic/business cycle
- exchange rates
- emerging economies
  - speculation
  - inflation
  - wage inflation
  - credit crunch

Environmental factors

Some of these are covered in detail in Lesson 11.4. Here students should be thinking about:

- climate change
- water shortages
- extreme weather
- positive/negative impact of new technologies

Add some of these to the diagram, as appropriate and as students suggest them. It would be a good idea to make a large poster-sized copy of the diagram or put it on an OHT or other visual medium to which you can add more examples as the unit progresses.

Exercise B

1 Set for individual work and pairwork checking. Feed back with the whole class, building the table in the Answers section.

2 Explain what is meant by 'within' and 'between' sentences: 'Within-sentence' linking words or phrases join clauses in a sentence;

'between-sentence' linking words or phrases connect two sentences. Demonstrate with the following:

Within-sentence linking words:

Farmers need to understand the issues because the answers will impact on the performance of a farming business in the UK, positively or negatively.

Make sure students can see that within-sentence linking words precede dependent clauses.

Between-sentence linking words:

The answers will impact on the performance of a farming business in the UK, positively or negatively. As a result, farmers need to understand the issues.

Point out that with between-sentence linking words there is usually a comma before the rest of the sentence.

Ask students to say which of the other words in box a are 'between' and which are 'within'.

3 Ask for suggestions for synonyms and add to the table.

4 First, make sure that students understand the basic principle of an argument, which is:

Statement
+
one or more support(s) for statement (= more facts, reasons, results, examples, evidence, etc.)

Constructing a complex argument will usually entail a statement plus several supports.

With the whole class, elicit suggestions for how to use the linking words/phrases when constructing an argument. Build the table in the Answers section on the board.

Answers

Possible answers:

1/2/3

Linking words/phrases	Use for	Within or between sentence	Other similar words/phrases
another point is	building an argument	between	and another thing
as a result	reasons and results	between	consequently
because	reasons and results	within	as
finally	building an argument	between	lastly
firstly	building an argument	between	to begin with/to start with for one thing
for example	building an argument	between	for instance
in addition	building an argument	between	also
moreover	building an argument	between	furthermore
one result of this	reasons and results	between	one consequence of this is because of this
secondly	building an argument	between	next then
since	reasons and results	within	as
so	reasons and results	between	therefore thus hence

4 A typical argument is constructed like this:

firstly	making the first major support point
for example	supporting the point with a specific example
in addition	adding another related point in support
secondly	making the second major support point
another point is	adding another related point in support
moreover	adding more information to the point above
finally	making the last point

Language note

1 Note that within-sentence linking words may be placed at the beginning of the sentence with a comma after the first clause, as in:

__Because__ the answers will impact on the performance of a farming business in the UK, positively or negatively, farmers need to understand the issues.

2 Although the between-sentence linking words are described above as joining two sentences, they can of course link two independent clauses joined by coordinating linking words *and* or *but*, as in:

The answers will impact on the performance of a farming business in the UK, positively or negatively, and, __as a result__, farmers need to understand the issues.

Exercise C

1 Set for individual work. Note that students should try to put each word into one of the two categories, even if it is not immediately clear how it could be relevant. If they are not sure which category to use, they should try to think of a phrase containing the word and imagine how it could be relevant to one of the categories.

2 Ask students to compare their answers and to justify their choices. Feed back with the whole class, discussing the words for which students feel the category is not obvious. If no decision can be reached, say you will come back to the words a little later.

3/4 Set for pairwork. Feed back with the whole class if you wish.

Answers

Model answers:

Word	Suggested categories	Part of speech	Other words/phrases
'access	agriculture	v (T) n (U)	reach entrance, way in
'arable	agriculture	adj	non-livestock, crop farming
'difficulties	agriculture	n (C)	problems
'drought	climate and weather	n (C)	acute water shortage
'flooding	climate and weather	n (U)	excess precipitation
'forage	agriculture	n (U)	crop (planted for animals in field)
'grow	agriculture	v (I/T)	cultivate
'labourer	agriculture	n (C)	agricultural worker
'livestock	agriculture	n (U)	(farm) animals
ma'chinery	agriculture	n (U)	equipment
'moisture	climate and weather	n (U)	slight amount of water, dampness
'output	agriculture	n (C)	production, yield
precipi'tation	climate and weather	n (U)	rain
'yields	agriculture	n (C usually plural)	productivity

Exercise D

Note: Students may need to use dictionaries in question 2.

Students should first read through the text to get an idea of the topic.

1/2 Set for individual work and pairwork checking. Feed back with the whole class.

Point out that having done these questions, it should now be possible to say whether the words in box b can be listed as 'climate and weather' or 'agriculture' or both. The point here is that the context will make clear what the meaning of a word should be. This is important when it comes to making a guess at the meaning of a word you are not sure of initially.

3 Set for individual work and pairwork checking. Feed back with the whole class, if you wish, using an OHT and two coloured pens, or other visual medium. Point out how the two themes of 'climate and weather' and 'agriculture' run through the text. Tell students that a particular topic will have groups of words which are connected to or associated with it – known as 'lexical chains'. These lexical chains show us the themes that run through the text and which help 'glue' the ideas together to make a cohesive piece of text. It is a good idea, therefore, to learn vocabulary according to topic areas.

4 It is also common to use synonymous words and phrases to refer back to something already mentioned. Ideally, use an OHT, or other visual medium, of the text (see Resource 11B in the additional resources section), and with a coloured pen draw a line to show how *a recent scientific report* is referred to later in the text as *the information* and *the report*. Set for individual work and pairwork checking. Feed back with the class (linking the phrases with coloured pens if using Resource 11B).

Answers

Model answers:

1 According to a recent scientific report, farms in the UK need to start adapting to cope with reduced water due to higher temperatures and summers with lower rainfall. Moreover, there is a clear risk of more frequent droughts as well as flooding, both of which could lead to falling outputs. The information suggests that an upland livestock farmer might need to consider bringing animals indoors earlier in the season because of the rainier autumns in this more unpredictable environment. In addition, farmers need to look at the consequences for their livestock. The report suggests that there may be a longer growing period for forage crops. One result of this could be that yields decrease after the early season, meaning farms may need to preserve more forage for winter feeding. Not only this, but there is also a warning that for arable farms in certain elevated or vulnerable areas, if there are increased temperatures and a drier summer, as a result, there could be problems establishing crops because of a lack of moisture. Another point is that lower precipitation in the summer could mean wetter autumns, which could in turn mean difficulties for farmers accessing

their land. The report advises that this might have a knock-on effect firstly on farm labourers, and secondly on machinery use. What is more, it could encourage more spring sowing.

2

Word	Synonym
scientific	science-based
adapting	changing
reduced	less
clear	apparent, evident
suggests	proposes
preserve	keep, retain
vulnerable	exposed
establishing	creating, founding
accessing	getting to, reaching
sowing	planting

3

Agriculture	Climate and weather
farms	reduced water
outputs	higher temperatures
upland livestock farmer	lower rainfall
animals	more frequent droughts
livestock	flooding
growing period	rainier autumns
forage crops	unpredictable environment
yields	increased temperatures
season	drier summer
winter feeding	lack of moisture
arable farms	lower precipitation
establishing crops	wetter autumns
accessing … land	
farm labourers	
machinery	
spring sowing	

4

First phrase	Second phrase
a recent scientific report	the information/the report
higher temperatures and summers with lower rainfall	this more unpredictable environment/lower precipitation in the summer/increased temperatures and a drier summer
farms … need to start adapting	farmer … might need to consider … /farmers need to look at …
upland	elevated

Exercise E

Time this general knowledge quiz if you wish. Alternatively, set it for homework for students to research the answers. Make sure that students understand they need to say what the initials stand for and also what the organization/body does.

In order to be able to follow the lecture in Lesson 11.2, students will need to be familiar with these organizations and bodies.

Answers

Model answers:

1 EU – European Union
CAP – Common Agricultural Policy
WWF – World Wildlife Fund
FAWC – Farm Animal Welfare Committee
GM – genetically modified
PETA – People for the Ethical Treatment of Animals
FAO – Food and Agriculture Organization of the United Nations
BBC – British Broadcasting Corporation

2 Friends of the Earth – environmental action campaigning group
James E Hansen – American academic best known for work on climate change
Good Egg Awards – a Europe-wide scheme to encourage food retailers to support more humane egg production
Facebook – a social networking site that connects individuals with their friends and social groups
conservation tillage – a form of arable farming that does not involve ploughing/which minimizes CO_2 emissions
Twitter – an online social networking and microblogging site that enables its users to send and read text-based posts of up to 140 characters
battery – a form of chicken/egg farming that uses small wire cages which deprive the chicken of its natural environment
credit crunch – alternative name given to the global financial crisis of the late 2000s
United Nations – international organization of 193 states, founded in 1945

Closure

Ask students to review the lesson and list ten of the factors which affect farming businesses. Then divide the class into groups and tell the groups to discuss the factors and try to rank them in order of importance. They will need to think of specific countries if possible and the effects the factors may have on these countries. They should give reasons for their ranking.

11.2 Listening

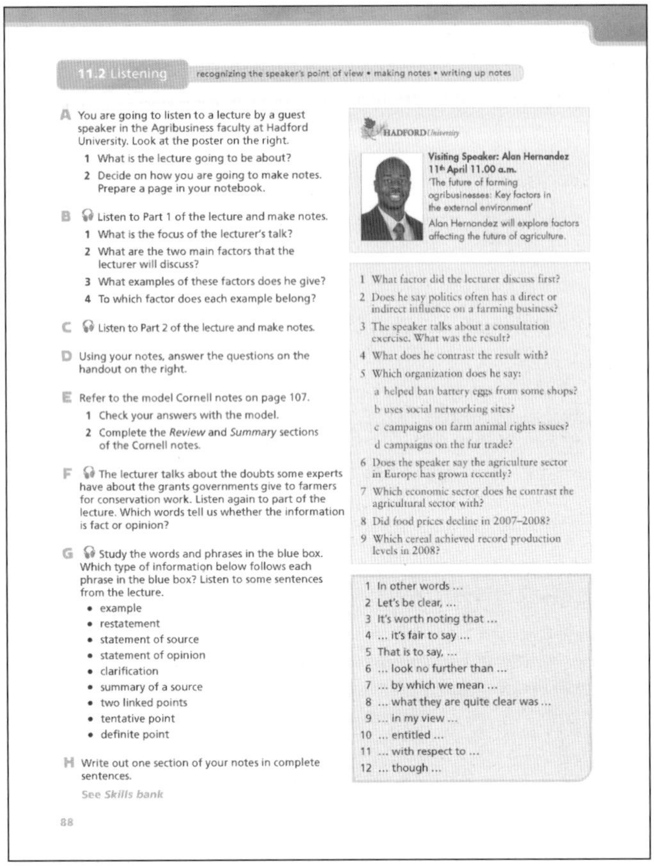

11.2 Listening recognizing the speaker's point of view • making notes • writing up notes

A You are going to listen to a lecture by a guest speaker in the Agribusiness faculty at Hadford University. Look at the poster on the right.
1 What is the lecture going to be about?
2 Decide on how you are going to make notes. Prepare a page in your notebook.

B Listen to Part 1 of the lecture and make notes.
1 What is the focus of the lecturer's talk?
2 What are the two main factors that the lecturer will discuss?
3 What examples of these factors does he give?
4 To which factor does each example belong?

C Listen to Part 2 of the lecture and make notes.

D Using your notes, answer the questions on the handout on the right.

E Refer to the model Cornell notes on page 107.
1 Check your answers with the model.
2 Complete the *Review* and *Summary* sections of the Cornell notes.

F The lecturer talks about the doubts some experts have about the grants governments give to farmers for conservation work. Listen again to part of the lecture. Which words tell us whether the information is fact or opinion?

G Study the words and phrases in the blue box. Which type of information below follows each phrase in the blue box? Listen to some sentences from the lecture.
- example
- restatement
- statement of source
- statement of opinion
- clarification
- summary of a source
- two linked points
- tentative point
- definite point

H Write out one section of your notes in complete sentences.

See *Skills bank*

HADFORD University

Visiting Speaker: Alan Hernandez
11th April 11.00 a.m.
'The future of farming agribusinesses: Key factors in the external environment'
Alan Hernandez will explore factors affecting the future of agriculture.

1 What factor did the lecturer discuss first?
2 Does he say politics often has a direct or indirect influence on a farming business?
3 The speaker talks about a consultation exercise. What was the result?
4 What does he contrast the result with?
5 Which organization does he say:
a helped ban battery eggs from some shops?
b uses social networking sites?
c campaigns on farm animal rights issues?
d campaigns on the fur trade?
6 Does the speaker say the agriculture sector in Europe has grown recently?
7 Which economic sector does he contrast the agricultural sector with?
8 Did food prices decline in 2007–2008?
9 Which cereal achieved record production levels in 2008?

1 In other words ...
2 Let's be clear, ...
3 It's worth noting that ...
4 ... it's fair to say ...
5 That is to say, ...
6 ... look no further than ...
7 ... by which we mean ...
8 ... what they are quite clear was ...
9 ... in my view ...
10 ... entitled ...
11 ... with respect to ...
12 ... though ...

88

General note

Read the *Skills bank – Writing out notes in full* at the end of the Course Book unit. Decide when, if at all, to refer your students to it. The best time, as before, is probably at the very end of the lesson or the beginning of the next lesson, as a summary/revision.

Lesson aims

- recognize and understand phrases that identify the speaker's point of view
- use background knowledge in listening comprehension
- convert notes into full sentences and paragraphs

Further practice in:

- making notes (use of heading systems and abbreviations)
- referring to sources
- general academic words and phrases

Introduction

1 Review phrases indicating a speaker's view of the truth value of a statement. Write a sentence such as the following on the board (i.e., a 'fact' about which there may be differences of opinion): *Agricultural productivity has increased in inverse proportion to the number of people employed in farming, in Europe.*

Ask students to say whether they think this is true or not. Elicit phrases they can use before the sentence to show how certain they are about their opinion.

Dictate or write on the board the following phrases. Ask students to say what the effect of each phrase is when put before the sentence on the board. In each case, does it make the writer sound confident or tentative?

The research shows that ...
A survey found that ...
The evidence does not support the idea that ...
It appears to be the case that ...
The evidence suggests that ...
The evidence shows that ...
It is clear that ...
It is possible that ...

2 Revise the Cornell note-taking system. Elicit the R words. Ask students to describe how to divide up the page (refer to Unit 9). Revise the other ways to take notes (see Units 1 and 3).

3 Revise note-taking symbols and abbreviations (see Units 5 and 9, and Unit 9 extra activity 3).

Exercise A

Refer students to the Hadford University lecture announcement. Tell them to look at the title and the summary of the talk.

Set the exercises for pairwork discussion. Feed back with the whole class.

Answers

Possible answers:

1 Accept any reasonable suggestions.

2 The lecturer is clearly going to list causes and effects. This suggests that possibly a flowchart or spidergram might be a suitable form of notes (as in Unit 1), as well as the Cornell system (which is used here) or the more conventional numbered points system.

Exercise B

Play Part 1 once through *without* allowing students time to focus on the questions.

Put students in pairs to answer the questions by referring to their notes. Feed back with the whole class, building a set of notes on the board if you wish. Note that 4 is a general knowledge question – which students should be able to answer from their (background) general knowledge.

Add the examples to your spidergram from Lesson 11.1 if you have not already done so.

Ask students which method they are going to use to make notes, now that they have listened to the introduction. They should make any adjustments necessary to the page they have prepared in their notebooks.

Answers

Model answers:

1 External influences on farming and farming businesses.
2 Political and economic factors.
3 Inflation, interest rates, impact of new competitors, changes in legislation, changes in public opinion.
4 Political: changes in legislation, changes in public opinion; economic: inflation, interest rates, impact of new competitors

Transcript 2.16

Part 1

Good morning. My name is Alan Hernandez and I'm a specialist in agricultural business strategy. It's a pleasure to be here today. I'm going to examine some of the external factors that will affect the future of agriculture. In other words, I'll look at some of the outside influences that affect farming agribusinesses in a country like the UK today.

Let's be clear, I don't want to suggest that there aren't a lot of *internal* issues that affect a farming agribusiness. We know that companies need to be concerned with the way they manage their operations and finance. But there are issues at a national and international level which have a significant impact on business decisions. To some extent, each farming business will be affected in a different way, but it is clear that all farming agribusinesses will have to be aware of changes in inflation and interest rates, and the impact on their markets of new competitors in developing countries. Not only that, but they also need to be aware of changes in legislation and public opinion in the areas where they operate. So to try and make sure we keep the discussion of *external* pressures on agricultural business reasonably simple, I'm going to summarize some of the key points to do with first political, then economic, factors.

Exercise C

Play the whole of the rest of lecture through *once* without stopping. Students should make notes as they listen.

Answers

See *Notes* section of the table in the Answers section for Exercise E.

Transcript 2.17

Part 2

To start with, then: political influences on farming. Whether it is decisions by the World Trade Organization or national election results, politics has a key and often indirect – as well as direct – influence on a farming business. If we take the European Union, for example, then changes to the Common Agricultural Policy, or CAP, will affect a UK farmer – but how? Will changes in the CAP encourage the growth of small farm businesses? Will changes in the CAP favour sustainable farming or industrial farming in the European Union, or both? The answers will impact on the performance of a farming business in the UK, positively or negatively. Farmers need to understand the issues.

So, let's stay with the example of the EU and the CAP. The CAP recently underwent a consultation exercise which produced some radical thinking. The aims of the CAP, however, remained pretty much unchanged. It has a number of priorities. Firstly, its aim is to increase productivity. Secondly, it ensures a fair standard of living for the agricultural community. Thirdly, it works to stabilize markets and assure food supply. And finally, very importantly, it is responsible for ensuring reasonable prices for the consumer. The results were: not much change. In the same period, it's worth noting that there's been a major and global *credit crunch*. And it's fair to say the result has been a 20 per cent fall in the sale of organic products in supermarkets in the UK.

Are economic pressures more powerful then? Well, no, they're just different, perhaps.

And staying with political pressures on a farm business, another example is the pressure group. That is to say, those organizations whose job it is to influence government policy. Let's be clear, no government or enterprise in the agribusiness sector today can ignore the power of the pressure group. As an example, we need look no further than egg production. Because until quite recently, the vast majority of the eggs we bought were still produced in 'batteries', by which we mean huge sheds where chickens are packed in small wire cages to produce the maximum number of eggs at the minimum unit cost. Compassion in World Farming and other pressure groups mounted a highly effective campaign, including the Good Egg Awards scheme, raising awareness of what they are quite clear was an unnecessary and cruel method of farming. The resulting public backlash led to the EU banning barren batteries from 2012. It also led directly to companies like McDonald's and Sainsbury's supermarket refusing to use or sell eggs produced using these methods. Not as significant as the impact of global recession, but significant nonetheless, I'm sure you'll agree.

There are many other pressure groups dedicated to improving the treatment of farm animals. Most of these are charitable organizations. People for the Ethical Treatment of Animals – or PETA as it's known – claims to be the largest animal rights group in the world. It tends to concentrate on specific issues such as factory farming and the fur trade. However, pressure groups aren't all in the charitable sector. The Farm Animal Welfare Committee (FAWC), a case in point, is an expert committee funded by the UK government to provide advice on the welfare of farmed animals – another sort of pressure group.

Friends of the Earth, which was founded in 1969, is a high-profile organization – 90 per cent funded by individual donations – working on global campaigns such as 'Fix the Food Chain'. Amongst other things, this highlights the threat caused by clearing rainforests to make room for soy production. And the Royal Society for the Protection of Birds (RSPB) has long been instrumental in highlighting issues such as the effects of modern farming methods on biodiversity, helping persuade governments to give farmers a financial incentive to protect bird habitats. Because, biodiversity, as we know, has an impact on agriculture. Interestingly, the major global pressure groups such as the World Wildlife Fund (WWF) are now more and more harnessing the enormous power of social networking sites. The name WWF is associated with protecting wild animals, but with campaigns like their One Planet Future and using sites like Facebook and Twitter, they are increasingly active in the mainstream sustainability debate.

It's important to be aware that farms aren't simply machines, producing the raw materials we eat. They rely on things like rainfall, sunshine and good-quality soil. And as stewards of our landscape and natural heritage, they are held to account by governments and consumers. What's more, concerns about the effect of modern farming practices on birdlife and bees, for example, are important to the people who pay taxes, as well as the people who donate to pressure groups like the RSPB. So, they influence governments who allocate taxpayers' money to support conservation and pro environment actions.

Not everyone, of course, supports these actions. Some experts say that such funding does not, in fact, represent a good investment of taxpayers' money. I have to say, in my view, they may have a point. An excellent recent BBC investigative programme entitled *The Price of Britain's Disappearing Wildlife* raised some very interesting issues. In the programme, with respect to effectiveness and value for the taxpayer's money, it was explained that the government invests heavily in biodiversity grants for farmers. For example, they gave £546 million in biodiversity grants in 2009 to farmers to encourage wildlife. Actually, though some farms have seen an increase in birdlife, overall the farmland bird population has experienced a 50 per cent *fall* in the last 30 years despite this high level of investment by the taxpayer.

Be that as it may, we can see clearly that the effect of politics on the sector is particularly important, given the globalization of markets.

Now let's move on from politics and look at some of the economic influences on the sector. It is well known that in Europe, there has been a huge reduction in the agriculture sector compared with other sectors of the economy such as the service sector, over the last century. Productivity, in contrast, has increased dramatically. As a result, while food prices in Europe have been volatile over the last 30 years, overall, in general, there has been a fall in the cost of food. In 2007, however, this trend was broken. Prices increased suddenly, reaching exceptional levels in early 2008. The impact on economies globally has been significant. The increase in prices resulted from a number of economic factors. There is a correlation between global population growth, rising incomes, consumption in emerging economies, and the development of new markets and the increase in demand. And, at the same time, global supply has been under pressure for some time due to a slowdown in the growth of food crop yields and fuel price inflation, which impacted on agricultural commodity prices. In 2007, these underlying problems were exacerbated by temporary problems, such as shortfalls caused by bad weather, the imposition of trade restrictions, exchange rates and speculation in the markets, as well as the loss of agricultural land to non-food crops including biofuels, driving up agricultural commodity prices.

After 2008, however, the economic environment changed again. Agricultural commodity prices fell, with wheat production, for example, reaching record levels in 2008, leading to a decline in wheat prices. Weather conditions were also more favourable: energy prices came down, the export restrictions imposed earlier were lifted, and the EU itself relaxed rules on production, while at the same time providing developing countries with help to alleviate the worst impact of shortages.

So, if we assume that economic pressures can have a devastating effect on the sector, can we reduce the worst impacts on the most vulnerable, at least? Well, looking to the future, there are initiatives that suggest this may be so. Charities like Farm Africa, for example, who focus on training and support for poor communities in East Africa, help families work themselves out of poverty through improved ways to manage their crops, livestock, forests and access to water. Charities like this offer a pragmatic, low-cost approach, based on finding what works, and applying this knowledge.

Exercise D

Put students in pairs to answer the questions by referring to their notes. Feed back with the class to see how much they have been able to get from the lecture. If they can't answer something, do not supply the answer at this stage.

Answers

Model answers:

1 political
2 He says it has both a direct and indirect influence.
3 The consultation exercise produced some radical thinking, but CAP aims were unchanged.
4 He contrasts it with the effects of the credit crunch.
5 a Compassion in World Farming.
  b WWF
  c The Farm Animal Welfare Committee
  d PETA
6 No, he says it has declined over the last century
7 service industries
8 No, they rose dramatically
9 wheat

Exercise E

1 Set for individual work.
2 Set for individual work and pairwork checking. Feed back with the whole class.

Answers

Possible answers: See table on next page.

Review	Notes
Types of political influence are …?	<u>1 Political factors</u>: e.g., decisions by the World Trade Organization, national election results
Example of a political influence on farming …?	(a) European policy, e.g., CAP/recent consultation exercise → principles unchanged → increase productivity → stable markets → fair standard of living → food supply → fair prices Much influence?
Economic influence on farming …? (1)	BUT NB global credit crunch → 20% ↓ in sales of organic products bought in supermarkets!
Examples of pressure groups …?	(b) pressure groups, e.g., (i) Compassion in World Farming: egg production → ban battery hens → 'Good Egg Awards' → EU bans barren batteries (ii) Friends of the Earth: 'Fix the Food Chain' → end deforestation (iii) WWF/PETA → worldwide charitable organizations campaign on animal rights issues, e.g., factory farming (iv) RSPB → highlights farming impact on biodiversity → e.g., birdlife
Criticism of Government grants …?	BUT e.g., BBC programme: questions effectiveness and value for money biodiversity grants 2009 = £546m → farmland bird population falls 50% in last 30 years
Economic influence on farming …? (2)	<u>2 Economic factors</u>: Given → last 30 years → ++ reduction in agriculture sector: e.g., service sector BUT
Impact of changes: (underlying) …?	• productivity ++ increase: → food price volatility → food price falls BUT
Impact of changes: (exceptional) …?	• 2007/2008: ++ increase in prices: → pop. growth → emerging economies → income/consumption growth = ++ demand → fall in supply → slow yields growth → fuel price inflation Also temp. problems – bad weather + trade restrictions ± exchange rate and speculation = ++ agri. commodity prices
Future …?	New initiatives, e.g., Farm-Africa → reduce neg. impact on vulnerable → focus on pragmatic → low cost → holistic development programmes

Summary

There are many political influences on agricultural businesses. They include government actions and pressure groups. Economic influences on farming tend to be more dramatic. Recently developed countries have seen a reduction in the agricultural sector, but increased productivity with growth of service industries. In 2007/2008, population growth and increased demand for agricultural commodities in new markets led to commodity shortages and massive price rises. There are indications of a global determination to protect the most vulnerable globally, while measures are taken to prevent a recurrence.

Exercise F

Discuss the question with the whole class. Ask them if they can remember any phrases which signal whether comments are fact or just opinion.

Play the extract. Ask students to tell you to stop the recording when they hear key phrases. Write the phrases on the board.

Remind students that it is important to recognize when someone is giving only their opinion, with which others might well disagree.

Answers

Model answers:

Some experts say that …	The speaker doesn't say 'it's true' or 'it's not true'; he quotes unnamed 'experts'.
… funding does not, in fact, represent a good investment …	A strong statement would be 'experts say this isn't good …'. Using 'in fact' and 'represent' makes the statement less positive.
I have to say, in my view, …	This is the lecturer's opinion, but 'I have to say' and 'in my view' serves the purpose of reducing the strength of 'they are right!'
they may have a point …	Again, the lecturer endorses the view, but use of 'may' qualifies that endorsement.
An *excellent* recent BBC investigative programme …	The effect of an unequivocal endorsement of a programme (by an identified, respected, high-profile organization) is to highlight the impression that the rest of paragraph comprises personal views that are unsubstantiated.
In the programme, *… it was explained that …*	The lecturer is quoting another source, *not* endorsing the views 'explained' directly.
Actually, though some farms …	The lecturer means 'contrary to what most people think'. Again, the lecturer is stating the case very strongly to persuade the listener.

Transcript 2.18

Not everyone, of course, supports these actions. Some experts say that such funding does not, in fact, represent a good investment of taxpayers' money. I have to say, in my view, they may have a point. An excellent recent BBC investigative programme entitled *The Price of Britain's Disappearing Wildlife* raised some very interesting issues. In the programme, with respect to effectiveness and value for the taxpayer's money, it was explained that the government invests heavily in biodiversity grants for farmers. For example, they gave £546 million in biodiversity grants in 2009 to farmers to encourage wildlife. Actually, though some farms have seen an increase in birdlife, overall the farmland bird population has experienced a 50 per cent *fall* in the last 30 years despite this high level of investment by the taxpayer.

Exercise G

Allow students time to read the phrases and the types of information, making sure that they understand any difficult words. Remind students that 'type' of information tells you what the speaker *intends to do* with the words. The words themselves are something different.

Ask students to try to match the phrases and types of information as far as they can. Note that it is not always possible to say what the function of a phrase is outside its context, so they may not be able to match all the phrases and information types before hearing the extracts. Note that some types of information are needed more than once.

When they have done as much as they can, play the extracts one at a time, allowing time for students to identify the type of information which follows. Check answers after each extract, making sure that students understand the information that actually follows the phrase. If possible, students should also give the actual words.

Answers

Model answers:

Fixed phrase	Type of information which follows the phrase
1 In other words …	restatement
2 Let's be clear …	clarification
3 It's worth noting that …	example
4 … it's fair to say …	tentative point
5 That is to say, …	restatement
6 … look no further than …	example
7 … by which we mean …	restatement
8 … what they are quite clear was …	definite point
9 … in my view …	statement of opinion
10 … entitled …	statement of source
11 … with respect to …	summary of a source
12 … though …	two linked points

Transcript 2.19

Extract 1

I'm going to examine some of the external factors that affect the future of agriculture. In other words, I'll look at some of the outside influences that affect farming agribusinesses in a country like the UK.

Extract 2

Let's be clear, I don't want to suggest that there aren't a lot of *internal* issues that affect a farming agribusiness ...

Extract 3

The results were: not much change. In the same period, by contrast, it's worth noting that there's been a major and global *credit crunch*.

Extract 4

And it's fair to say the result has been a 20 per cent fall in the sale of organic products in supermarkets in the UK.

Extract 5

And staying with political pressures on a farm business, another example is the pressure group. That is to say, those organizations whose job it is to influence government policy.

Extract 6

As an example, we need look no further than egg production.

Extract 7

Because until quite recently the vast majority of the eggs we bought were still produced in 'batteries', by which we mean huge sheds where chickens are packed in small wire cages to produce the maximum number of eggs at the minimum unit cost.

Extract 8

Compassion in World Farming and other pressure groups mounted a highly effective campaign including the Good Egg Awards scheme, raising awareness of what they are quite clear was an unnecessary and cruel method of farming.

Extract 9

I have to say, in my view, they may have a point.

Extract 10

An excellent recent BBC investigative programme entitled *The Price of Britain's Disappearing Wildlife* raised some very interesting issues.

Extract 11

In the programme, with respect to effectiveness and value for the taxpayer's money, it was explained that the government invests heavily in biodiversity grants for farmers.

Extract 12

Actually, though some farms have seen an increase in birdlife, overall the farmland bird population has experienced a 50 per cent *fall* in the last 30 years despite this high level of investment by the taxpayer.

Exercise H

Use this section from the Cornell notes to demonstrate what to do:

> **Notes**
>
> 1 Political factors: e.g., decisions by the World Trade Organization, national election results
>
> (a) European policy, e.g., CAP/recent consultation exercise
>
> principles unchanged
> increase productivity
> fair standard of living
> stable markets
> food supply
> fair prices
>
> Much influence?

Elicit from students suggestions on how to write up the notes in complete sentences. Write the suggestions on the board.

Ask students to say what they need to add in to the notes to make a good piece of writing, e.g.,

Grammar: relative pronouns, articles and determiners, prepositions, auxiliary verbs, linking words, 'there was/were' clauses (in italics in the model notes below).

Vocabulary: some vocabulary may need to be added, particularly where symbols are used in the notes, or where extra words are needed to make sense of the information or give a good sense of flow in the writing (in bold below).

Note that this of course works the other way: when making notes, these elements can be excluded from the notes.

Possible rewrite of the notes:

Examples *of* political factors *which* **impact** *on* **farming businesses include**: decisions by the World Trade Organization *and* national election results.

Taking *a* **specific example,** *the* **EU Common Agricultural Policy** *or* CAP, recent*ly* **carried out** *a* consultation exercise. One **outcome** of the **exercise was** a **decision** to **maintain** the principles **underpinning** *the* **CAP, including commitments** *to* increase productivity, **work for** *a* fair standard of living *for* **members, maintenance** *of* stable markets *and* food supplies *within a* fair pric*ing* **framework.**

This **begs** *the* **question** *as to* **whether** *the* **consultation exercise brought about significant change** *for the* **sector.**

Set another section for individual writing in class or for homework. Either ask students to refer to their own notes, or to the Cornell notes on page 107 of the Course Book.

Closure

1 Tell students to review and make a list of the main topics and arguments presented in this lesson. Then ask them to try and summarize the viewpoints, using some of the language they have practised.

2 They could also give a two- or three-sentence summary of anything that they themselves have read, e.g., *I read a useful article on X by Y. It said that …*

3 Ask students to do some research and to make a list of useful or interesting books/articles/websites on the topics in this lesson. They should draw up a list, including correct referencing, and share their sources with other students.

11.3 Extending skills

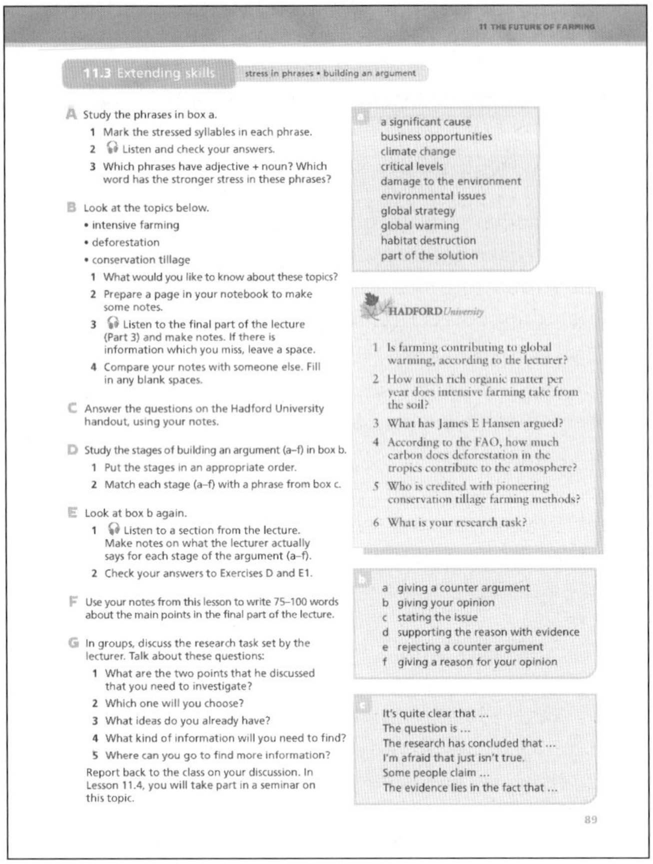

11 THE FUTURE OF FARMING

11.3 Extending skills — stress in phrases • building an argument

A Study the phrases in box a.

1 Mark the stressed syllables in each phrase.
2 Listen and check your answers.
3 Which phrases have adjective + noun? Which word has the stronger stress in these phrases?

B Look at the topics below.

- intensive farming
- deforestation
- conservation tillage

1 What would you like to know about these topics?
2 Prepare a page in your notebook to make some notes.
3 Listen to the final part of the lecture (Part 3) and make notes. If there is information which you miss, leave a space.
4 Compare your notes with someone else. Fill in any blank spaces.

C Answer the questions on the Hadford University handout, using your notes.

D Study the stages of building an argument (a–f) in box b.

1 Put the stages in an appropriate order.
2 Match each stage (a–f) with a phrase from box c.

E Look at box b again.

1 Listen to a section from the lecture. Make notes on what the lecturer actually says for each stage of the argument (a–f).
2 Check your answers to Exercises D and E1.

F Use your notes from this lesson to write 75–100 words about the main points in the final part of the lecture.

G In groups, discuss the research task set by the lecturer. Talk about these questions:

1 What are the two points that he discussed that you need to investigate?
2 Which one will you choose?
3 What ideas do you already have?
4 What kind of information will you need to find?
5 Where can you go to find more information?

Report back to the class on your discussion. In Lesson 11.4, you will take part in a seminar on this topic.

a
a significant cause
business opportunities
climate change
critical levels
damage to the environment
environmental issues
global strategy
global warming
habitat destruction
part of the solution

HADFORD University

1 Is farming contributing to global warming, according to the lecturer?
2 How much rich organic matter per year does intensive farming take from the soil?
3 What has James E Hansen argued?
4 According to the FAO, how much carbon does deforestation in the tropics contribute to the atmosphere?
5 Who is credited with pioneering conservation tillage farming methods?
6 What is your research task?

b
a giving a counter argument
b giving your opinion
c stating the issue
d supporting the reason with evidence
e rejecting a counter argument
f giving a reason for your opinion

c
It's quite clear that ...
The question is ...
The research has concluded that ...
I'm afraid that just isn't true.
Some people claim ...
The evidence lies in the fact that ...

89

Lesson aims

- recognize stress patterns in noun phrases
- understand how to develop an argument:
  stating the issue
  giving a counter argument
  rejecting a counter argument
  giving opinions
  supporting opinions
- understand more general academic words and phrases mainly used in speaking

Further practice in:

- expressing degrees of confidence/tentativeness
- reporting back

Introduction

1 Revise the lecture in Lesson 11.2. Ask students to use the model Cornell notes on page 107. They should cover up the *Notes* section and use the *Review* and *Summary* sections (which they completed in Lesson 11.2) to help recall the contents of the lecture. They could work in pairs to do this.

2 Revise phrases which express degrees of confidence in 'facts'. Dictate these phrases. Do they show that the speaker is certain or tentative?

There is no question that (= certain)
We have to accept the evidence (= certain)
Some people claim that (= tentative)
What seems obvious is that (= certain)
As everyone is aware (= certain)
To some degree (= tentative)
This means ultimately that (= certain)
It's quite clear that (= certain)
We could argue that (= tentative)

Exercise A

1/2 Set for individual work and pairwork checking. This is an exercise in perceiving rhythm. At this point, there is no need to distinguish between different levels of stress. Students can underline all the stressed syllables. They will also need to count all the syllables.

Feed back with the whole class, checking pronunciation of the phrases and meanings.

3 Discuss this with the class first. Demonstrate with *ˌglobal ˈwarming*, showing how if you say *ˈglobal ˌwarming*, it appears that a contrast is being made with another type of warming. Tell students that the usual pattern for the adjective + noun phrase is for a heavier stress to go on the noun. *Climate change* is, however, different: it is a compound made from a noun + noun, and the stress is: *ˈclimate ˌchange* – i.e., the heavier stress goes on the first noun. Set students to pick out the other adjective + noun patterns, writing each one on the board. Elicit the stress patterns and give students time to practise the phrases.

Answers

1/2 a sigˌnificant ˈcause
ˈbusiness opporˌtunities
ˈclimate ˌchange
ˌcritical ˈlevels
ˌdamage to the enˈvironment
environˌmental ˈissues
ˌglobal ˈstrategy
ˌglobal ˈwarming
ˈhabitat deˌstruction
ˌpart of the soˈlution

3 Adjective + noun (second word has stronger stress): critical levels, environmental issues, global strategy, global warming, (a) significant cause. Please note that the stress can move to the adjective where a contrast is being made, e.g., '*global strategy*, not '*local strategy*.

Transcript 2.20

a sig,nificant 'cause
'business oppor,tunities
'climate ,change
,critical 'levels
,damage to the en'vironment
environ,mental 'issues
,global 'strategy
,global 'warming
'habitat de,struction
,part of the so'lution

Exercise B

1 Look at the three topics. Discuss with the class what they know already about these topics and find out what opinions they may have. Put students in pairs and ask each pair to write down one question to which they would like an answer in the lecture.
2 Set for individual work.
3 Play Part 3 straight through; students make notes.
4 Put students in pairs to compare their notes and fill in any gaps they may have.

Transcript 2.21

Part 3

Looking again at the effect of environmental issues on agricultural businesses, of course there is no industry that is more directly affected by environmental change than farming. So how serious are the problems for the future of farming? Well, first of all, there is no doubt that we are seeing significant changes in the natural environment, most catastrophically in the developing world. But the real question is: Is farming contributing to these changes and can farming help? Some people in agriculture claim that it is nothing to do with them. I'm afraid that whether you look at global warming or habitat destruction or loss of biodiversity, that view is simply not credible. For example, research has suggested that intensive farming currently takes something like 20 tons of rich organic matter out of every hectare every year, leading to a release of carbon into the atmosphere where it contributes to climate change. The evidence is contained in the research of climatologists like James E Hansen, who has argued that human activity is now a greater cause of global warming than natural climatological events. The result of activities like burning fossil fuels and deforestation means that we have pushed carbon dioxide concentrations in the atmosphere to critical levels.

So, if agriculture is a significant cause of the environmental problems we face, then what is the future of farming and how can it become part of the solution? Well first of all, let's be clear that not everyone is convinced of the science. There are sceptics who believe the changes we are seeing to our climate, for example, are not the result of man's activities. However, they are in a minority.

If we accept agriculture is responsible for damage to the environment, we can identify a number of areas where we could change and improve things. Each year, somewhere in the region of 33 million acres of forest globally is cut down, according to the Food and Agriculture Organization, or FAO, of the United Nations. Tree felling in the tropics contributes 1.5 billion metric tons of carbon to the atmosphere annually. That's 20 per cent of man-made greenhouse gas emissions. It is clear that the future of farming must include a strategy to promote sustainable forestry and eliminate the clearing of forests, in particular the dwindling rain forests, to make way for non-sustainable forms of agriculture.

What else can farming do? Well, there's interest in the so-called conservation tillage farming methods pioneered by Edward Faulkner in the 1940s. This is a way of farming which involves leaving unharvested crop remains in the field – you don't plough or disturb them. This encourages carbon storage. It can lead to a short-term fall in yields. But it has a double benefit because it increases the amount of carbon retained in the soil and reduces fuel costs for the farmer by taking away the need to harvest the stalks with tractors. It won't solve global warming, but it might help.

Of course, I am looking here at issues to do with the environment from the perspective of the *threat* they pose to agriculture and agribusinesses. There is a view that, in the future, farming businesses could benefit from opportunities created by things like the move away from fossil fuels. Carbon credit schemes, for example, are opening up new business opportunities. And mainstream, as well as alternative, agricultural economists are arguing that moving away from farming animals to growing and eating more vegetables and fruit will not only deliver a beneficial impact on global warming, but lead to a 'win-win' situation with farming businesses improving long-term profitability. Not particularly radical ideas, but they may prove important, given the scale of the challenge.

OK. Now I'm going to set you a task involving some of the points I've raised. Firstly, I'd like you to investigate areas of farming and related agribusinesses that might be affected by a new global strategy designed to reduce agriculture's impact on the environment. I'd like you to start by looking at the sort of changes needed to make their processes more environmentally friendly. Secondly, I'd like you to think about whether these changes could benefit farming businesses in the future or damage them by reducing productivity and profitability.

Exercise C

Set for individual work and pairwork checking. Feed back with the class on question 6 to make sure that it is clear.

Answers

1 yes
2 20 tons per hectare
3 Human activity is now a greater cause of global warming than natural climatological events.
4 1.5 billion metric tons
5 Edward Faulkner.
6 Firstly, investigate areas of farming and related agribusinesses that might be affected by global strategies designed to reduce agriculture's impact on the environment. Secondly, think about whether these changes might benefit or damage farms.

Exercise D

1 Set for pairwork discussion. Point out that there is no, one 'correct' order; students should try to identify the most logical sequence for the argument. Explain that a 'counter argument' means an opinion which you do not agree with or think is wrong. 'Issue' means a question about which there is some debate.

2 Set for individual work and pairwork checking. Do not feed back with the class at this point, but move on to Exercise E where the answers will be given.

Exercise E

1 Play the extract. Tell students to stop you when they hear each item. Make sure students can say exactly what the words are in each case. Ask them also to paraphrase the words so that it is clear that they understand the meanings.

2 If necessary, play the extract again for students to check that they have the phrases and types of statement correct, as the lecturer's words may vary from the phrases given in the Course Book. Ask how many students had the stages of an argument (Exercise D, question 1) in the same order as the recording/model answers below. Discuss any alternative possibilities (see *Language note* opposite).

Answers

Answers for Exercises D and E: see table below.

Type of statement	Phrase	Lecturer's actual words
c stating the issue	The question is …	But the real question is: Is farming contributing to these changes …?
a giving a counter argument	Some people claim …	Some people in agriculture claim that it is nothing to do with them.
e rejecting a counter argument	I'm afraid that just isn't true.	I'm afraid that … that view is simply not credible.
b giving your opinion	It's clear that …	It is clear that the future of farming … forms of agriculture.
f giving a reason for your opinion	The research has concluded that …	… research has suggested that intensive farming currently takes …
d supporting the reason with evidence	The evidence lies in the fact that …	The evidence is contained in the work of climatologists like James E Hansen, …

Language note

A common way in which an argument can be built is to give a counter argument, then reject the counter argument with reasons and evidence. There are, of course, other ways to build an argument. For example, the counter arguments may be given after the writer/speaker's own opinion. Or all the arguments against may be given followed by all the arguments for an issue (or vice versa), concluding with the speaker/writer's own opinion.

Transcript 2.22

But the real question is: Is farming contributing to these changes and can farming help? Some people in agriculture claim that it is nothing to do with them. I'm afraid that whether you look at global warming or habitat destruction or loss of biodiversity, that view is simply not credible. For example, research has suggested that intensive farming currently takes something like 20 tons of rich organic matter out of every hectare every year, leading to a release of carbon into the atmosphere where it contributes to climate change. The evidence is contained in the research of climatologists like James E Hansen, who has argued that human activity is now a greater cause of global warming than natural climatological events. The result of activities like burning fossil fuels and deforestation means that we have pushed carbon dioxide concentrations in the atmosphere to critical levels.

Exercise F

Set for individual work – possibly homework – or else a pair/small group writing task. If the latter, tell students to put their writing on an OHT or other visual medium, so that the whole class can look and comment on what has been written. You can correct language errors on the OHT.

Exercise G

Set students to work in groups of three or four. Have each group carry out the task in relation to one of three topics: global warming, pollution or waste. Allow each group to choose their topic. Make sure that each topic is covered by at least one, preferably two groups. Ask one person from each group to present the results of the group's discussion.

Tell the class that they should carry out research into their group's topic. You will also need to arrange the date for the feedback and discussion of the information – this is the focus of Exercise G in Lesson 11.4.

Closure

Arguments, counter arguments and giving opinions

Ask students to think about the methods seen above to build an argument. As they do this, write the statements below on the board or display them on an OHT or other visual medium. Ask them to think about whether they agree with the statements. They should prepare a brief summary of their viewpoints on the topics; they should also try and use some of the phrases used in this lesson.

1. Intensive farming contributes to global warming – therefore its profits should be heavily taxed to offset the damage it does.
2. Space exploration is a waste of money which could be better spent on the environment.
3. Low prices are the best indicator of a good farming sector.
4. Politics has no place in agriculture.
5. Global warming is an invention of a small group of scientists who oppose progress.
6. Deforestation has been going on for centuries with no one worrying about climate change, so we shouldn't start worrying now.

11.4 Extending skills

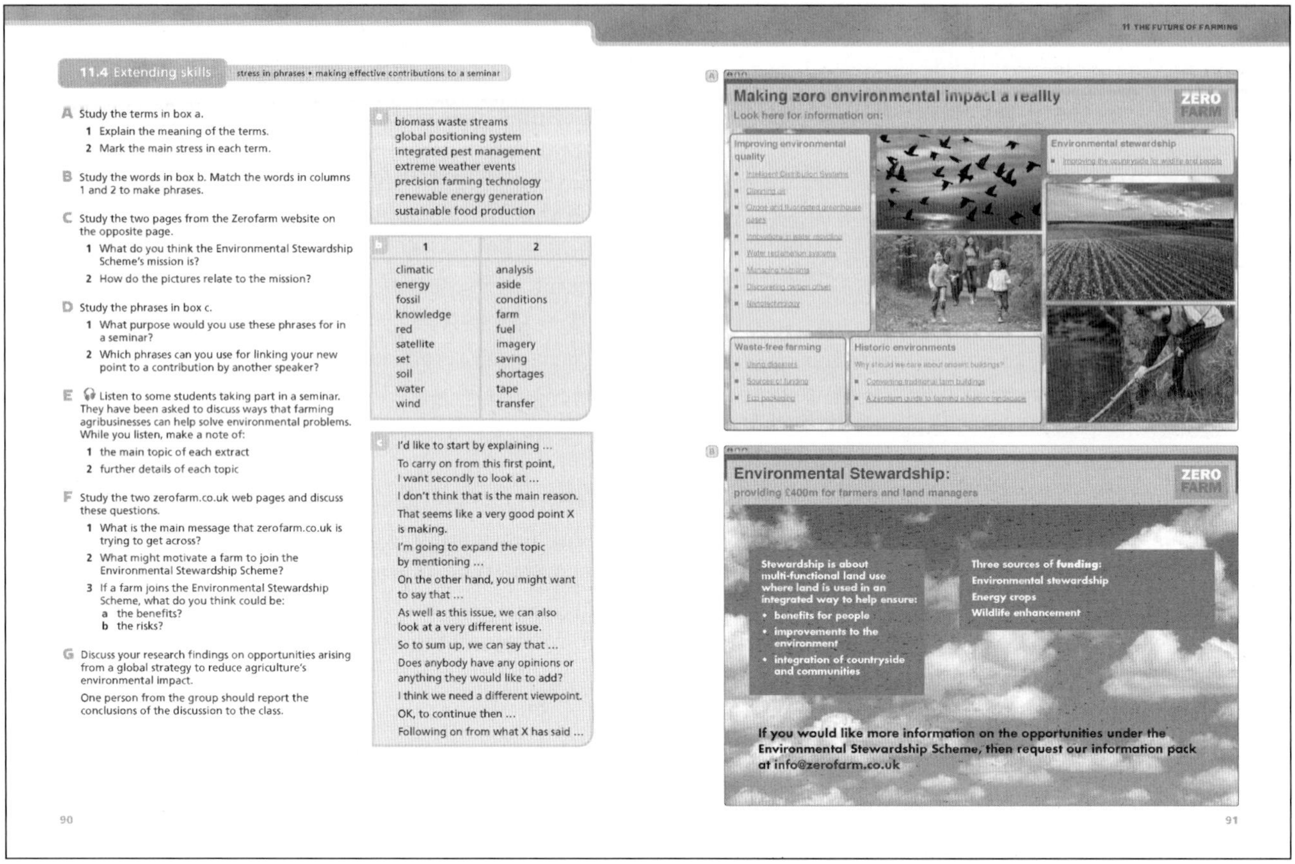
11 THE FUTURE OF FARMING

11.4 Extending skills — stress in phrases • making effective contributions to a seminar

A Study the terms in box a.
1 Explain the meaning of the terms.
2 Mark the main stress in each term.

B Study the words in box b. Match the words in columns 1 and 2 to make phrases.

C Study the two pages from the Zerofarm website on the opposite page.
1 What do you think the Environmental Stewardship Scheme's mission is?
2 How do the pictures relate to the mission?

D Study the phrases in box c.
1 What purpose would you use these phrases for in a seminar?
2 Which phrases can you use for linking your new point to a contribution by another speaker?

E Listen to some students taking part in a seminar. They have been asked to discuss ways that farming agribusinesses can help solve environmental problems. While you listen, make a note of:
1 the main topic of each extract
2 further details of each topic

F Study the two zerofarm.co.uk web pages and discuss these questions.
1 What is the main message that zerofarm.co.uk is trying to get across?
2 What might motivate a farm to join the Environmental Stewardship Scheme?
3 If a farm joins the Environmental Stewardship Scheme, what do you think could be:
a the benefits?
b the risks?

G Discuss your research findings on opportunities arising from a global strategy to reduce agriculture's environmental impact.
One person from the group should report the conclusions of the discussion to the class.

a
biomass waste streams
global positioning system
integrated pest management
extreme weather events
precision farming technology
renewable energy generation
sustainable food production

b

1	2
climatic	analysis
energy	aside
fossil	conditions
knowledge	farm
red	fuel
satellite	imagery
set	saving
soil	shortages
water	tape
wind	transfer

c
I'd like to start by explaining ...
To carry on from this first point, I want secondly to look at ...
I don't think that is the main reason.
That seems like a very good point X is making.
I'm going to expand the topic by mentioning ...
On the other hand, you might want to say that ...
As well as this issue, we can also look at a very different issue.
So to sum up, we can say that ...
Does anybody have any opinions or anything they would like to add?
I think we need a different viewpoint.
OK, to continue then ...
Following on from what X has said ...

A
Making zero environmental impact a reality
Look here for information on:
Improving environmental quality
Environmental stewardship
Waste-free farming
Historic environments

B
Environmental Stewardship:
providing £400m for farmers and land managers
Stewardship is about multi-functional land use where land is used in an integrated way to help ensure:
• benefits for people
• improvements to the environment
• integration of countryside and communities
Three sources of funding:
Environmental stewardship
Energy crops
Wildlife enhancement
If you would like more information on the opportunities under the Environmental Stewardship Scheme, then request our information pack at info@zerofarm.co.uk

90

91

Lesson aims

- recognize stress in compound phrases
- link contribution to previous contributions when speaking in a seminar
- understand vocabulary in the area of environmental issues

Further practice in:

- taking part in seminars:
  introducing, maintaining and concluding a contribution
  agreeing/disagreeing with other speakers

Introduction

1 Remind students that they are going to be presenting their research findings later in this lesson. Check that they can remember the main points from Lesson 11.3 lecture extracts; key phrases from the lecture could be used as prompts, e.g.,

But the real question is: Is farming contributing to these changes [in the environment and climate]? and can farming help?

I'm afraid that whether you look at global warming or habitat destruction or loss of biodiversity … [climate change is linked in part to modern farming methods]

What is the future of farming and how can it become part of the solution? [if agriculture is a significant cause of the environmental problems we face]

We can identify a number of areas where we could change and it would improve things [if we accept that agriculture is responsible for damaging the environment]

2 The following activity is a good way to check that students are familiar with the terminology and vocabulary from Lesson 11.3. Ask students to write down 5–10 words or expressions from the previous lesson relating to the future of farming and the current problems. Then use two or three students as 'secretaries'. Ask the class to dictate the words so that the secretaries can write the vocabulary on the board. Use this as a brainstorming session.

Exercise A

These are more complex noun phrases than in Lesson 11.3, since they are made up of three words. In some cases, the pattern is noun + noun + noun. In this case, there may be a compound made from the first two nouns, or the last two nouns. In other cases, the pattern is adjective + noun + noun, in which the second and third words make a compound. These patterns should become clear once the meaning is understood.

1 Discuss *global positioning system* with the class as an example. Elicit that it is a technology in general use – in cars and boats – which is being used in farming to reduce waste. Set the remaining phrases for individual work and pairwork discussion. Feed back with the whole class, writing each phrase on the board and underlining the words which make a compound noun.

2 Tell students to try to identify where the main stress should come in each phrase. The key to this is finding the two- or three-word compound which is at the base of the three-word phrase. The stress will normally fall in the same place as if this two-word compound was said without the third word. Demonstrate this with *global po'sitioning system*. The two-word compound here is not *global positioning*, but *positioning system*. This is a noun + noun compound, so the rules say this will normally be stressed on the first noun *po'sitioning system*. The main stress remains in its original place when the third word is added.

Tell students only to identify the syllable on which the heaviest stress in the phrase falls. (See also *Language note*.)

Answers

Model answers:

The basic compound is underlined in each case.

biomass 'waste streams	using material such as wood chips or forest thinnings and crop residues to provide renewable energy for power generation
global po'sitioning system	a system adapted to improve efficient use of resources such as watering in agriculture
integrated 'pest management	environmentally sensitive way of controlling pests on crops
extreme 'weather events	weather events such as heatwaves that occur rarely. The increase in incidence is associated with global warming.
precision 'farming technology	a sophisticated agricultural management technique that, e.g., identifies different types of soil within a field for different applications of water
renewable 'energy generation	the business of producing energy from renewable sources
sustainable 'food production	a method of producing food which meets 'sustainable' criteria, i.e., meets the needs of today without compromising our ability to meet the needs of tomorrow

Language note

Stress placement, especially in complex compound noun phrases, is notoriously unstable. There will be apparent exceptions to the rules explained in Exercise A. For example, in *precision 'farming technology* (noun + compound noun), the brain appears to see *farming* as an adjective rather than a noun.

Stress may often move, depending on the context: for example, *bad-'tempered* – but *'bad-tempered 'teacher*. It's also possible that some native speakers may not agree about some of the phrases above. The main point is to try to notice where the main stresses fall.

Exercise B

Set for individual work and pairwork checking. Tell students that although in some cases it will be possible to make a phrase with more than one option, they must use each word once, and they must use all the words.

Feed back with the whole class. Check that the meaning of all other phrases is understood. Check pronunciation.

Answers

1	2
climatic	conditions
energy	saving
fossil	fuel
knowledge	transfer
red	tape
satellite	imagery
set	aside
soil	analysis
water	shortages
wind	farm

Language note

Although in most noun–noun compounds the main stress comes on the first element, there are some compounds where this is not true. Definitive pronunciation of compounds can be found in a good pronunciation dictionary.

Exercise C

Refer students to web page A on page 91 of the Course Book. Set for pair or small group discussion. In both questions, encourage creative and lateral thinking (there is no 'right' answer).

Answers

1 environmental action, protecting the countryside, integrating people with the natural landscape
2 protecting wildlife/protecting landscape/protecting natural resources/promoting access

Exercise D

This is mainly revision. Set for individual work or pairwork discussion. Feed back with the whole class.

Answers

Model answers:

I'd like to start by explaining …
= beginning

To carry on from this first point, I want secondly to look at …
= maintaining/continuing a point

I don't think that is the main reason.
= disagreeing

That seems like a very good point X is making.
= confirming

I'm going to expand the topic by mentioning …
= adding a new point to someone else's previous contribution

On the other hand, you might want to say that …
= disagreeing

As well as this issue, we can also look at a very different issue.
= adding a new point to someone else's previous contribution

So to sum up, we can say that …
= summarizing/concluding

Does anybody have any opinions or anything they would like to add?
= concluding

I think we need a different viewpoint.
= disagreeing

OK, to continue then …
= maintaining/continuing a point

Following on from what X has said …
= adding a new point to someone else's previous contribution

Exercise E

Before students listen, tell them to look at the exercise and questions. Check that students understand the topic for the seminar discussion. Ask them what they might expect to hear.

Play each extract one at a time and ask students to identify the main topic and some further details. Feed back with the whole class.

Answers

Model answers: see table below.

	Main topic	Further details
Extract 1	species loss and solutions	impact on pollination of plants and potential cost/impact of disappearance of bees → new actions
Extract 2	land use for animals vs, e.g., biofuels	real cost of feeding animals; alternatives
Extract 3	urban agriculture	Seattle campaign; vertical farms
Extract 4	conventional strategies	renewable energy; hemp; vegetarianism

Transcript 2.23

Extract 1

MAJED: The lecturer we listened to last week introduced some interesting issues. In my part of the seminar, I'd like to build on what he said and talk about the way some farms have responded to the problem of species loss and the impact this could have on pollination of plants, for example. What researchers are discovering is that a lot of the bee population are in danger of disappearing in the UK. This could lead to a situation where farmers have to pollinate their crops by hand. It might sound like science fiction, but it's happening in parts of China, and apart from the problem of losing valuable species forever, the additional cost could be as much as £1.5 billion. I'd like to talk about some of the schemes where farms have been successful in providing habitat for birds and insects within their arable systems and the benefits they're having.

Extract 2

EVIE: OK, following on from what Majed has said, I'd like to bring in some ideas I've been reading about to do with encouraging people to eat more vegetables. I know it's fairly off the wall, but in the US for example, more than half the agricultural land is used for growing animal feed. And it takes something like seven pounds of animal feed to produce a pound of beef. For chickens, the figure is under one and a half pounds of feed. A lot of deforestation in recent years has been carried out to clear land for beef production. If we ate less meat, then we would see a reduction in the amount of land needed to fulfil our food needs. Perhaps this land could then be used instead for producing biofuels or for recreation. Farmers potentially don't need to lose out: In New Zealand, the government has been running something called a 'carbon trading scheme', where farmers are given credits which they can sell to businesses to offset their carbon emissions – it's a sort of tax on business, but the farmers gain. In 2008, forests grown for carbon credits totalled about 3,500 hectares, but that is expected to rise to 30,000. And this new industry is offsetting some of the problems caused by a decline in sheep prices, and it's benefitting the environment.

Extract 3

JACK: Right, that seems like a very good point Evie is making. I'm going to expand the topic by mentioning another fairly radical but important area to do with the future of farming, and that's urban agriculture. It's not clear how realistic this is, but Seattle, for example, has recently begun a campaign to encourage food production in parts of the city which were previously designated for housing or commercial use only. This isn't about encouraging farms to start up in the city – it's more about encouraging integration of food production. But there are also experiments going on in other places to see if 'vertical farms' are viable, that's farms which are constructed like multi-storey car parks, where the aim is to reduce energy and land use and increase productivity.

Extract 4

LEILA: Well, I'd like to start by explaining some more conventional strategies, where I think we're going to make the biggest impact. In the UK, a significant number of farms are now generating income from renewable energy – wind turbines and biomass crops, such as hemp. But it's important to make the point that actually, if you grow hemp using modern, intensive methods, you run the risk of causing the sort of soil erosion that we've identified as one of the key issues for the future of farming. To carry on from this first point, I want secondly to look at some more radical ideas. For example, Bruce McCarl is an American agricultural economist who has highlighted the potential benefits of vegetarianism, in terms of optimizing the use of our cultivated land and reducing farming's carbon footprint. It might be a drastic solution, but when it's linked to the move towards efficient, local supply chains, greater numbers of vegetarian consumers represent a sound commercial opportunity for a lot of farmers, particularly organic farmers.

Exercise F

1/2 Set all for pairwork. Tell students to study the zerofarm web pages on page 91 of the Course Book. Feed back with the whole class.

3 Set for small group discussion. Feed back with the whole class.

Answers

Possible answers:

1 The possibilities for zero environmental impact farms, the business opportunities for farmers and the environmental benefits for people and wildlife.

2 Income from grants to develop access for visitors, grow energy crops, or develop wildlife habitats, for example.

3 a grant income, capital improvements, job satisfaction, public relations

b commitment to provide access to visitors and potential for damage to conventional farming activities, additional cost above the grants for actions required by membership, forms and administration, external interference from, e.g., audits.

Exercise G

In their groups, students should now present their research findings on the business opportunities or threats to farms from a global strategy to reduce environmental damage to the rest of the group. Remind them that the task was to:

- investigate areas of farming and related agribusinesses that might be impacted by a new global strategy designed to reduce agriculture's impact on the environment;
- to think about whether these changes could benefit the prospects for farming businesses in the future or damage them by reducing productivity and profitability.

Encourage students to use the seminar language practised in this unit and earlier. In addition, students can, of course, make use of the information in Lesson 11.4.

They should be looking at, or at least mentioning, some or all of the following:

- explaining the issues: the overwhelming view that farming has a detrimental impact on the environment; political consequences of not taking action
- mentioning specifics: nature of impact, extent of impact
- saying what farming can do: range of solutions, potential impact on, e.g., global warming

- suggesting ways farm businesses could benefit:
  - grants
  - diversification
  - reduced costs
  - business sustainability
  - PR and profile of farming
  - job satisfaction

As a group, students should try to come to an overall conclusion. This conclusion should be presented to the rest of the class, together with supporting evidence from students' own research.

Closure

Ask students to imagine that they are 10–15 years in the future. What changes might we see in farm businesses? Ask them to think about the following:

- agriculture in the developed world
- agriculture in the developing world
- size of the agricultural sector compared with the services sector
- profitability of the sector
- the sort of issues being discussed
- the role played by agriculture in our countryside and community
- radical innovations

Extra activities

1 Work through the *Vocabulary bank* and *Skills bank* if you have not already done so, or as revision of previous study.

2 Use the *Activity bank* (Teacher's Book additional resources section, Resource 11A).

A Set the crossword for individual work (including homework) or pairwork.

Answers

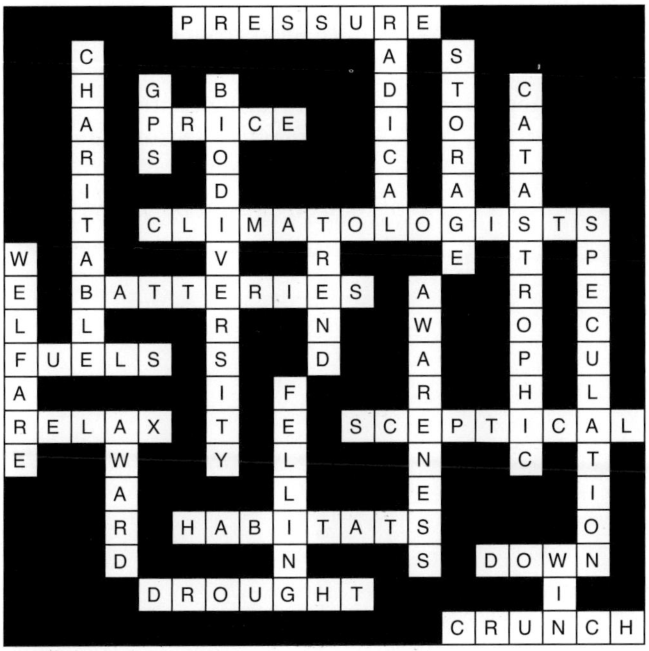

B Set for individual work or pairwork. Check students understand meanings.

Possible answers:

carbon dioxide

consultation exercise

drive up

economic downturn

environmentally friendly

global warming

greenhouse gases

habitat destruction

raise awareness

renewable energy

3 Tell students to add other words to each of the words below to make as many two-word phrases as possible. Elicit one or two examples, then set for individual work or pairwork.

- carbon
- environmental
- waste

Possible phrases:

carbon trading, carbon footprint, carbon offset(ting), carbon dioxide, carbon market

environmental issues, environmental concerns, environmental pressure group, environmental action

waste management, waste reduction, business waste, zero waste

4 Use an extended activity to allow students to practise some of the concepts they have studied in this unit. Tell students to work in groups. They are going *to do an* exercise designed to create a zero-emission farming business. Divide the activities into stages as follows:

A Identify the various stages involved in building such a business, e.g., writing a mission statement, drafting a business plan, researching techniques, getting planning permission, preparing the land, growing the crops, marketing the farm and its products, gaining accreditation for zero emissions.

B Create an outline business plan in some detail. The groups should make specific suggestions for:

- location, taking account of the need for distribution, etc.
- emissions policy
- product/service in detail
- farm layout
- plant and machinery
- sales and marketing
- HR

C Draw up a flow plan showing the various stages of the crop-growing/harvesting/distribution process – make sure that zero-emission factors are included at each stage. (Plans can be put onto A2 sheets, flipcharts or another visual medium and displayed for other groups to compare.)

12 THE GOOD AGRIBUSINESS

This unit provides an opportunity for revision of many of the concepts and vocabulary items used in the book. It features a case study of a soft fruit grower that is prospering as a result of undertaking a strategic review and business re-structuring; reasons for the decision and the benefits accruing are analyzed.

Skills focus

Reading

- understanding how ideas in a text are linked

Writing

- deciding whether to use direct quotation or paraphrase
- incorporating quotations
- writing research reports
- writing effective introductions/conclusions

Vocabulary focus

- verbs used to introduce ideas from other sources (*X contends/suggests/asserts that …*)
- linking words/phrases conveying contrast (*whereas*), result (*consequently*), reasons (*due to*), etc.
- words for quantities (*a significant minority*)

Key vocabulary

assert
biodiversity
carbon neutral
case study
commitment
concede
conservation
contend
continuous improvement process (CIP)
contract (n)
cost-sensitive
economic cycle
findings
grant funding
healthy profit
helpfulness
hurricane
hydroponic
incentive scheme
incremental steps
joint venture
leadership
lease (n and v)
legislation
majority
market share
minority
paraphrase
pesticide-free
questionnaire
random sample
respondent
soft fruit
substantial
supplier
survey
transparency
trends
under glass
value added tax (VAT)
veterinary services
water scarcity

12.1 Vocabulary

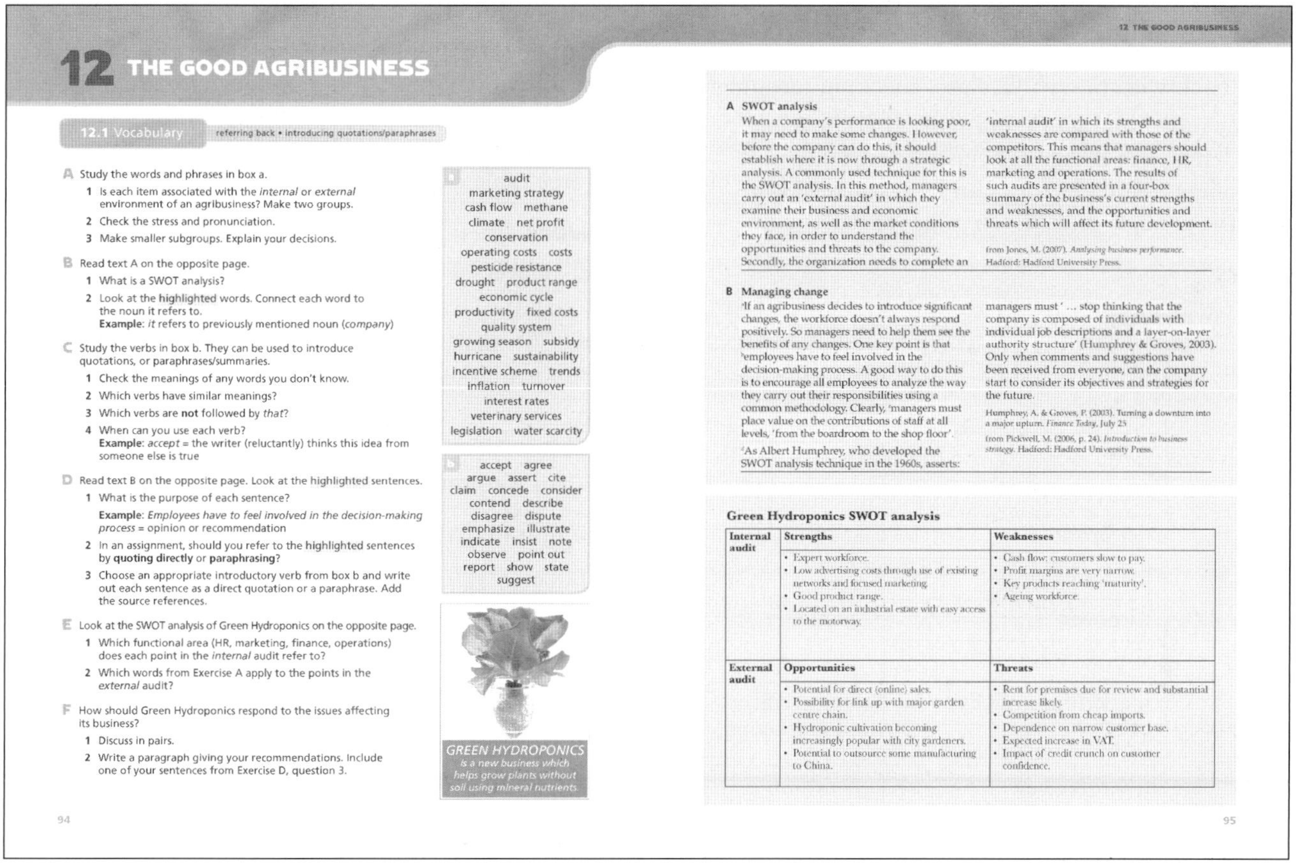

12 THE GOOD AGRIBUSINESS

12.1 Vocabulary — referring back • introducing quotations/paraphrases

A Study the words and phrases in box a.

1 Is each item associated with the *internal* or *external* environment of an agribusiness? Make two groups.
2 Check the stress and pronunciation.
3 Make smaller subgroups. Explain your decisions.

B Read text A on the opposite page.

1 What is a SWOT analysis?
2 Look at the **highlighted** words. Connect each word to the noun it refers to.
Example: *it* refers to previously mentioned noun (*company*)

C Study the verbs in box b. They can be used to introduce quotations, or paraphrases/summaries.

1 Check the meanings of any words you don't know.
2 Which verbs have similar meanings?
3 Which verbs are **not** followed by *that*?
4 When can you use each verb?
Example: *accept* = the writer (reluctantly) thinks this idea from someone else is true

D Read text B on the opposite page. Look at the highlighted sentences.

1 What is the purpose of each sentence?
Example: *Employees have to feel involved in the decision-making process* = opinion or recommendation
2 In an assignment, should you refer to the highlighted sentences by **quoting directly** or **paraphrasing**?
3 Choose an appropriate introductory verb from box b and write out each sentence as a direct quotation or a paraphrase. Add the source references.

E Look at the SWOT analysis of Green Hydroponics on the opposite page.

1 Which functional area (HR, marketing, finance, operations) does each point in the *internal* audit refer to?
2 Which words from Exercise A apply to the points in the *external* audit?

F How should Green Hydroponics respond to the issues affecting its business?

1 Discuss in pairs.
2 Write a paragraph giving your recommendations. Include one of your sentences from Exercise D, question 3.

a
audit
marketing strategy
cash flow methane
climate net profit
conservation
operating costs costs
pesticide resistance
drought product range
economic cycle
productivity fixed costs
quality system
growing season subsidy
hurricane sustainability
incentive scheme trends
inflation turnover
interest rates
veterinary services
legislation water scarcity

b
accept agree
argue assert cite
claim concede consider
contend describe
disagree dispute
emphasize illustrate
indicate insist note
observe point out
report show state
suggest

GREEN HYDROPONICS
is a new business which helps grow plants without soil using mineral nutrients

94

12 THE GOOD AGRIBUSINESS

A SWOT analysis

When a company's performance is looking poor, it may need to make some changes. However, before the company can do this, it should establish where it is now through a strategic analysis. A commonly used technique for this is the SWOT analysis. In this method, managers carry out an 'external audit' in which they examine their business and economic environment, as well as the market conditions they face, in order to understand the opportunities and threats to the company. Secondly, the organization needs to complete an 'internal audit' in which its strengths and weaknesses are compared with those of the competitors. This means that managers should look at all the functional areas: finance, HR, marketing and operations. The results of such audits are presented in a four-box summary of the business's current strengths and weaknesses, and the opportunities and threats which will affect its future development.

from Jones, M. (2007). *Analysing business performance*. Hadford: Hadford University Press.

B Managing change

¹If an agribusiness decides to introduce significant changes, the workforce doesn't always respond positively. So managers need to help them see the benefits of any changes. One key point is that ²employees have to feel involved in the decision-making process. A good way to do this is to encourage all employees to analyze the way they carry out their responsibilities using a common methodology. Clearly, ³managers must place value on the contributions of staff at all levels, 'from the boardroom to the shop floor'.

⁴As Albert Humphrey, who developed the SWOT analysis technique in the 1960s, asserts: managers must '... stop thinking that the company is composed of individuals with individual job descriptions and a layer-on-layer authority structure' (Humphrey & Groves, 2003). Only when comments and suggestions have been received from everyone, can the company start to consider its objectives and strategies for the future.

Humphrey, A. & Groves, P. (2003). Turning a downturn into a major upturn. *Finance Today*, July 25

from Pickwell, M. (2006, p. 24). *Introduction to business strategy*. Hadford: Hadford University Press.

Green Hydroponics SWOT analysis

Internal audit	Strengths	Weaknesses
	• Expert workforce. • Low advertising costs through use of existing networks and focused marketing. • Good product range. • Located on an industrial estate with easy access to the motorway.	• Cash flow: customers slow to pay. • Profit margins are very narrow. • Key products reaching 'maturity'. • Ageing workforce.
External audit	**Opportunities**	**Threats**
	• Potential for direct (online) sales. • Possibility for link up with major garden centre chain. • Hydroponic cultivation becoming increasingly popular with city gardeners. • Potential to outsource some manufacturing to China.	• Rent for premises due for review and substantial increase likely. • Competition from cheap imports. • Dependence on narrow customer base. • Expected increase in VAT. • Impact of credit crunch on customer confidence.

95

General note

Read the *Vocabulary bank* at the end of the Course Book unit. Decide when, if at all, to refer your students to it. The best time is probably at the very end of the lesson or the beginning of the next lesson, as a summary/revision.

Lesson aims

- understand deictic reference – pronouns and determiners
- refer to sources: the choice of introductory verb and stance of writer towards reference
- choose whether to quote or paraphrase

Further practice in:

- words and phrases from the discipline

Introduction

1 Revise the following words and phrases from the two previous units. Ask students to say which grammar class the words belong to and to provide definitions of them.

assure (v, T)
baseline (adj; n, C)
investment (n, C)
margin (n, C)
misleading (adj)
reduction (n, C)
shareholder (n, C)
shortfall (n, C)
speculation (n, U)
validate (v, T)
withdraw (v, T)

2 Introduce the topic of the unit: Write the words *the good agribusiness* on the board. Ask students: What does *good* mean when it is applied to an agribusiness? What factors do agribusinesses need to take into account when they plan their strategy? What processes will they need to go through in order to have a sustainable future? Have a class discussion on what they think makes a business *good* or *bad*. Check against profitable/unprofitable and successful/unsuccessful. Accept all suggestions. Do not elaborate, but tell students that this will be the topic of this unit.

Exercise A

1/2 Set for pairwork. Warn students that they may find that some words can be both internal and external. Ask students to mark the most strongly stressed syllable in compounds.

Feed back with the whole class, checking meanings.

3 Set for individual work and pairwork checking. Feed back with the whole class. Accept any groupings as long as they can be justified. Make sure the groupings below are also understood.

Answers

Model answers:

Internal	'audit	finance
	'cash flow	
	'costs	
	fixed 'costs	
	'net profit	
	'operating costs	
	'turnover	
	sustaina'bility	
	'marketing strategy	commercial and marketing
	'product range	
	in'centive scheme	operations
	produc'tivity	
	'quality system	
	veterinary 'services	
External	'climate	environment and nature
	conser'vation	
	'drought	
	'growing season	
	'hurricane	
	'methane	
	'pesticide resistance	
	'water scarcity	
	economic 'cycle	political and economic factors
	in'flation	
	'interest rates	
	legis'lation	
	'subsidy	
	'trends	

Exercise B

Introduce the idea of textual cohesion, created by referring back to words or ideas already mentioned with pronouns such as *it* and *this* (pronouns and determiners). Say that this an important way in which the sentences in a text are 'held together'. In reading and understanding, it is important to know what is being referred to by such words.

You can build up the answers to question 2 by copying Resource 12B in the additional resources section onto an OHT or other visual medium.

Set for individual work and pairwork checking. Feed back with the whole class, building the table below. Establish why a writer might use a particular referring word (see table on next page).

Answers

Model answers:

1 A technique which enables a company to evaluate itself by analyzing Strengths, Weaknesses, Opportunities and Threats.

2 SWOT analysis:

When a company's performance is looking poor, **it** may need to make some changes. However, before **the** company can do **this**, it should establish where it is now through a strategic analysis. A commonly used technique for **this** is the SWOT analysis. In **this method**, managers carry out an 'external audit' in which **they** examine **their** business and economic environment, as well as the market conditions **they** face, in order to understand the opportunities and threats to the company. Secondly, the organization needs to complete an 'internal audit' in which **its** strengths and weaknesses are compared with **those** of the competitors. **This** means that managers should look at all the functional areas: finance, HR, marketing and operations. The results of **such audits** are presented in a four-box summary of the business's current strengths and weaknesses, and the opportunities and threats which will affect **its** future development.

Word	Refers to	Comments
the + noun	a previously mentioned noun	one of several ways in which choice of article is governed
it, they	a noun	generally refers to the nearest suitable noun previously mentioned or the subject of the previous sentence
its, their	a previously mentioned noun, indicating possession	other possessive pronouns used in text for reference: *his, her, hers, theirs,* etc.
this	an idea in a phrase or a sentence	• often found at the beginning of a sentence or a paragraph; a common mistake is to use *it* for this purpose • also used with prepositions (e.g., *for this*)
this/these + noun	a previously mentioned noun/noun phrase	also used with prepositions (e.g., *in this method*).
those	a previously mentioned noun/noun phrase	• also used with prepositions • in this text 'those of the competitors' means 'the strengths and weaknesses of the competitors'; there is no need to repeat 'strengths and weaknesses'. *Those of* + noun is a useful construction to learn • *those* – not *these* – is used to show distance between the writer/speaker and the objects/concepts themselves
such + plural noun	a previously mentioned noun	meaning is: 'Xs like this'. Note that when referring to a singular noun, *such a X* is used (e.g., *in such a situation*)

Language note

Clearly, in this text, there are also relative pronouns which refer back to previously mentioned nouns in relative clauses. However, the grammar of relative pronouns is not covered here.

This is a complex area of written language. The reference words here are commonly found and arguably students should be able to use them in their writing. There are, of course, various other ways to refer back to a word or idea, such as when comparing: *the former ... the latter ...* ; *some ... others ...* .

For more information on this, see a good grammar reference book.

Exercise C

1–3 Set for individual work or pairwork. Feed back. Discuss any differences of opinion in question 2 and allow alternative groupings, with reasonable justifications. Establish that not all verbs have equivalents.

4 Discuss this with the whole class, building the table in the Answers section. Point out to students that the choice of introductory verb for a direct or indirect quote or a paraphrase or summary will reveal what they think about the sources. This is an important way in which, when writing essays, students can show a degree of criticality about their sources. Critically evaluating other writers' work is an important part of academic assignments, dissertations and theses.

Point out also that some verbs have a degree of markedness, that is, extra meaning or connotation.

Answers

Possible answers:

2 accept, agree, concede
argue, assert, claim, contend, insist
consider, note, observe, point out, state
disagree, dispute
illustrate, indicate, show

3/4 See table on next page.

		Used when the writer …
accept	that	(reluctantly) thinks this idea from someone else is true
agree	that	thinks this idea from someone else is true
argue	that	is giving an opinion that others may not agree with
assert	that	is giving an opinion that others may not agree with
cite	+ noun	is referring to someone else's ideas
claim	that	is giving an opinion that others may not agree with
concede	that	reluctantly thinks this idea from someone else is true
consider	that	is giving his/her opinion
contend	that	is giving an opinion that others may not agree with
describe	how; + noun	is giving a description
disagree	that; with + noun	thinks an idea is wrong
dispute	+ noun	thinks an idea is wrong
emphasize	that	is giving his/her opinion strongly
illustrate	how; + noun	is explaining, possibly with an example
indicate	that	is explaining, possibly with an example
insist	that	is giving an opinion that others may not agree with
note	that	is giving his/her opinion
observe	that	is giving his/her opinion
point out	that	is giving his/her opinion
report	that	is giving research findings
show	that	is explaining, possibly with an example
state	that	is giving his/her opinion
suggest	that; + gerund	is giving his/her opinion tentatively; *or* is giving his/her recommendation

Exercise D

Discuss with the students when it is better to paraphrase and when to quote directly. Refer to the *Skills bank* if necessary.

1/2 Set for individual work and pairwork checking. Feed back with the whole class.

3 Set for individual work. Remind students that if they want to quote another source but to omit some words, they can use three dots (…) to show some words are missing.

Answers

Possible answers:

See table on next page.

Language note

When deciding between quoting directly and paraphrasing, students need to decide whether the writer's original words are special in any way. If they are, then a direct quote is better – for example, with a definition, or if the writer has chosen some slightly unusual words to express an idea. If the writer is giving factual information or description, a paraphrase is better. Opinions also tend to be paraphrased.

A page reference should be given for a direct quotation but is not necessary for a paraphrase.

Answers

Possible answers:

1–3

Original sentence	The writer is ...	Direct quote or paraphrase?	Suggested sentence
a If an agribusiness decides to introduce significant changes, the workforce don't always respond positively.	making a statement of fact	paraphrase	Pickwell (2006) points out that an agribusiness workforce may not respond positively to the prospect of change.
One key point is that **b** employees have to feel involved in the decision-making process.	giving an opinion or recommendation	paraphrase	Pickwell (2006) argues that staff need to be consulted about any possible changes.
Clearly, **c** managers must place value on the contributions of staff at all levels, 'from the boardroom to the shop floor'.	giving a strong opinion	paraphrase with a direct quotation of the last seven words, which make a 'special' phrase	Pickwell (2006) emphasizes the need to take everyone's views seriously, 'from the boardroom to the shop floor' (p. 24).
d As Albert Humphrey, who developed the SWOT analysis technique in the 1960s, asserts: managers must '... stop thinking that the company is composed of individuals with individual job descriptions and a layer-on-layer authority structure' (Humphrey & Groves, 2003).	quoting from another writer; the other writer is making a strong statement	quote the other writer directly	Pickwell (2006) cites Humphrey & Groves (2003), who insist that managers must '... stop thinking that the company is composed of ... a layer-on-layer authority structure' (Humphrey & Groves, 2003, pp. 124–126).

Exercise E

This is an exercise in relating the vocabulary discussed in Exercise A to meanings in agribusiness. It will help prepare students for work later in the unit. Refer students to the picture and the SWOT analysis. Make sure they understand what Green Hydroponics is.

Set for pairwork discussion. Feed back with the whole class if necessary.

Answers

Possible answers:

1 Internal audit

Strengths	Functional area	Weaknesses	Functional area
Expert workforce.	HR	Cash flow: customers slow to pay.	finance
Low advertising costs through use of existing networks and focused marketing.	marketing	Profit margins are very narrow.	finance/marketing
Good product range.	marketing and operations	Key products reaching 'maturity'.	marketing and operations
Located on an industrial estate with easy access to the motorway.	operations	Ageing workforce.	HR

2 External audit

Opportunities	Words from box a	Threats	Words from Exercise A
Potential for direct (online) sales.	net profit	Rent for premises due for review and substantial increase likely.	economic cycle; fixed costs
Possibility for link up with major garden centre chain.	marketing strategy; product range	Competition from cheap imports.	productivity; net profit; quality system
Hydroponic cultivation becoming increasingly popular with city gardeners	marketing strategy; trends	Dependence on narrow customer base	cash flow; costs; economic cycle; product range; turnover
Potential to outsource some manufacturing to China	operating costs	Expected increase in VAT. Impact of credit crunch on customer confidence.	economic cycle; interest rates

Exercise F

1 Set for pairwork discussion, followed by class discussion.

2 Before setting the students to write, tell them they should refer to text B in their answer.

Answers

Possible answers:

1 Accept any reasonable suggestions.

2 It is clear that Green Hydroponics' business is heading for trouble. This is probably because they have not invested in new products, their profit margins are too narrow and they do not have young employees coming through. In addition, their fixed costs are likely to grow substantially as a result of a rent increase. However, there are a number of courses of action open to them: they can outsource some of their production to China, they have the chance to enter into a trading arrangement with a major retail chain, and they can diversify into direct sales and thereby improve their profit margins. However, Pickwell (2006) points out that an agribusiness workforce may not respond positively to the prospect of change.

Closure

Ask students to think about their local high street, food shops, supermarkets or restaurants. They should do the following (the exercise can be done in groups or pairs):

- List their favourites, the ones most used/visited, and so on; specify what they buy where, when they go, reasons for visiting, etc.
- Using this list, add what they like and don't like about shops or restaurants, plus any problems they have noticed or encountered with them.
- Use the points they have listed above to create a SWOT framework – allocating each of the points in their list to the appropriate category in the SWOT framework.
- Finally rank the shops/restaurants in some sort of order of excellence.

The object of the exercise is to look at SWOT from an informal, consumer viewpoint. It should also help to prime students for the questionnaire and the research report in Lessons 12.3 and 12.4.

12.2 Reading

12 THE GOOD AGRIBUSINESS

12.2 Reading linking ideas in a text • quoting and paraphrasing

A Discuss the following questions.

1 What do supermarkets expect from a soft fruit supplier?
2 How can soft fruit growers ensure that they are successful?

B Survey the text on the opposite page. What will the text be about? Write three questions to which you would like answers.

C Read the text. Does it answer your questions?

D Number the sentences on the right 1–8 to show the order in which they happened.

	Stefan Smith called in consultants.
	HSF secured grants for biodiversity.
	HSF achieved turnover of more than £6 million.
1	Stefan Smith sold his property business.
	Stefan Smith recognized the threat of dependence on a narrow customer base.
	HSF underwent re-engineering.
	HSF committed to becoming carbon neutral.
	HSF received a strategic marketing analysis report.

E For each paragraph:

1 Identify the topic sentence.
2 Think of a suitable title.

F Look at the underlined words in the text. What do they refer back to?

G Study the highlighted words and phrases.

1 What do they have in common?
2 What linking words or phrases can you use to show:
   - addition?
   - contrast?
   - concession?
   - result?
   - reason?
3 Write the sentences with the highlighted items again, using other linking words or phrases with similar meanings.

H Read the text on the right. A student has written about Stefan Smith and the sequence of events that led to major change of direction, but the quotations and paraphrases have not been correctly done. Can you spot the mistakes and correct them?

As Wallace (2010) explains that Stefan Smith was initially very successful after taking over HSF and managed to grow the business steadily. He subsequently realized the threat of dependence on the big supermarkets and a very cost-sensitive product, and recognizing that the company had to change. According to Wallace, she says that Smith took decisive action and commissioned a '*strategic marketing analysis*' *(p. 6)* from external consultants as part of a comprehensive strategic rethink.

I Write a paragraph for a university lecturer, summarizing the strengths of Stefan Smith as a manager based on Wallace's case study. Decide whether you should quote or paraphrase the material from the text.

96

Soft Fruit Growers, 09.2010

Hadford Soft Fruits *A case study in good strategic management* **By Hillary Wallace**

Stefan Smith bought Hadford Soft Fruits (HSF) in the early 1990s after selling his Birmingham property development business. He invested a substantial amount of his own money in developing the business, gaining market share in the local soft fruit market with products that were mostly produced intensively, under glass. Over the next ten years, Smith grew the business steadily, adding an import/export division and borrowing £1 million to invest in new buildings and equipment to improve efficiency and productivity. Smith was the first Hadford producer to secure a contract with a major supermarket, which at that time was keen to exploit HSF's offer to supply soft fruits from March to November.

Within five years, HSF was turning over in excess of £6 million and showing healthy year-on-year growth and profits of £300,000. In order to do this, it was employing 80 people on its farms, in its transport business and in its new headquarters in Birmingham city centre. Smith was clear about the reasons for their success. His own commitment to hard work, service and fair dealing was mirrored throughout the company. His products met stringent quality standards at a time when there wasn't the interest in them in this sector that there is today.

However, in the middle of the 1990s, things began to change. The supermarkets, which had been the mainstay of the business, providing more than 70% of the HSF profits, began to use their buying power to squeeze prices, forcing suppliers into long-term, exclusive contracts. Smith realized the threat of dependence on the big supermarkets, a narrow customer base and a very cost-sensitive product. Although HSF was successful, with its prize-winning branded soft fruit, it had taken out large loans to fund growth, after which it faced substantial debt servicing and repayment costs, so these especially meant change was needed.

First of all, HSF commissioned a strategic marketing analysis from a firm of consultants specializing in the agribusiness sector. The report, including a detailed SWOT analysis, endorsed Smith's concerns, but it also identified new opportunities. Consequently, with this, he was able to go to work with his management team to produce a radical new strategy for the company. This highlighted three key areas for action: product diversification, branding and market segmentation. To implement the strategy, HSF initiated a process of re-engineering the business, starting with consensus building around the vision of healthy, high-quality, environmentally friendly products. While going organic proved too big a risk, the company committed to a pesticide-free policy. It increased the product range through the acquisition of ten acres of rare variety apple and pear orchards and it began producing its own-brand honey 'Brilliant Fruits' label.

Despite the fact that income from soft fruit sales fell, HSF's profitability had never been better. It secured significant grant funding to support its work encouraging biodiversity in its orchards, and it used its enhanced reputation to build a retail/farm shop spin-off. By focusing on the premium market, it attracted high-end supermarket business at the same time as using its membership of 'Regional Food and Drink', its industry-led body, to build a substantial new network of niche retail customers. There is growing competition from local producers, some of whom have learned from HSF, but with a commitment to becoming carbon neutral within three years, the strategy is to move from radical, breakthrough changes to securing sustained success by a process of continuous improvement (CIP) through a series of incremental steps.

6

97

General note

Read the *Vocabulary bank* and *Skills bank* at the end of the Course Book unit. Decide when, if at all, to refer students to them. The best time is probably at the very end of the lesson or the beginning of the next lesson, as a summary/revision.

Lesson aims

- understand rhetorical markers in writing (*but* and *so* categories)
- use direct quotations from other writers:
  common mistakes
  missing words
  fitting to the grammar of the sentence
  adding emphasis to a quote
  continuing to quote from the same source

Further practice in:

- indirect quotations/paraphrases/summaries
- summarizing with a series of topic sentences
- rhetorical markers (adding points)
- deictic reference and relative pronouns

Introduction

Revise the main SWOT concepts from Lesson 12.1. What does each letter stand for?

To prepare students for the lesson's theme, ask them to think of a journey they have made recently and to describe the transport system, the route, cost, value for money, time taken, the good things and the bad things about the journey.

Exercise A

Set for pairwork or class discussion. Accept any reasonable suggestions.

Answers

Possible answers:

1 Supermarkets expect the supplier to provide an excellent product that meets agreed quality standards, they expect them to deliver on time (i.e., to meet service standards) and they expect staff to be polite and cooperative and work with them to help improve the supermarket's performance.

2 There are no guarantees of success, but a company that knows what its customers want, and that invests in its products and people to ensure that they consistently offer better value and quality than the competition while maintaining good financial systems, creates the framework necessary for success.

Exercise B

Remind students about surveying a text (skim-reading to get an approximate idea of the text contents by looking at the title, looking at the beginning few lines and the final few lines of the text, and by looking at the first sentence of each paragraph).

Set for individual work and pairwork discussion. Each pair should agree three questions. Feed back with the whole class. Write some questions on the board.

Exercise C

Set for individual work followed by pairwork discussion. Feed back with the whole class. Ask whether the questions you have put on the board have been answered in the text.

Exercise D

Set for individual work and pairwork checking. This activity could also be done using Resource 12C in the additional resources section. Photocopy and cut up the sentences and hand them out in a jumbled order. Tell students to put them in the correct order.

Answers

4	Stefan Smith called in consultants.
7	HSF secured grants for biodiversity.
2	HSF achieved turnover of more than £6 million.
1	Stefan Smith sold his property business.
3	Stefan Smith recognized the threat of dependence on a narrow customer base.
6	HSF underwent re-engineering.
8	HSF committed to becoming carbon neutral.
5	HSF received a strategic marketing analysis report.

Exercise E

Set for individual work and pairwork discussion. The topic sentences should suggest a suitable title.

Answers

Possible answers:

	Topic sentence	Para title
Para 1	Stefan Smith bought Hadford Soft Fruits (HSF) in the early 1990s …	The purchase of Hadford Soft Fruits
Para 2	Within five years, HSF was turning over in excess of £6 million …	HSF achieves early success
Para 3	However, in the middle of the 1990s, things began to change.	HSF encounters problems
Para 4	First of all, HSF commissioned a strategic marketing analysis from a firm of consultants …	Stefan Smith calls in the experts
Para 5	Despite the fact that income from soft fruit sales fell, its profitability had never been better …	The future looks bright for HSF

Exercise F

Set for individual work and pairwork checking.

Answers

Word	Line	Refers to
that	14	early 1990s
this	19	turnover more than £6m, showing good growth and healthy profits
them	26	stringent quality standards
its	37	HSF
which	39	securing loans
these	40	debt servicing and repayment costs
this	47	the report (could refer to endorsement and ideas)
whom	74	local producers

Exercise G

1 Refer students to the highlighted words. Elicit that they are all linking words and phrases.

2 With the whole class, elicit from the students some linking words or phrases that can be used for:

- contrast and concession (i.e., words which have a *but* meaning)
- result and reason (i.e., words which have a *so* or *for* meaning)

Build the table in the Answers section on the board, reminding students of the difference between between- and within-sentence linking words (refer to Unit 11 *Vocabulary bank*).

3 Set for individual work. Encourage students to rewrite the sentences using a different type of linking word from the original (i.e., swapping between- and within-sentence linkers).

Answers

Possible answers:

2 See table below.

3 *But*, in the middle of the 1990s, things began to change.

In spite of the fact that/Despite the fact that HSF was successful, with its prize-winning branded soft fruit, it had taken out large loans to fund growth, after which it faced substantial debt servicing and repayment costs. *As a result/Consequently/ Therefore,* these especially meant change was needed.

As a result/So/Therefore, with this, he was able to go to work with his management team to produce a radical new strategy for the company.

Although going organic proved too big a risk, the company committed to a pesticide-free policy.

Although/In spite of the fact that income from soft fruit sales fell, HSF's profitability had never been better.

There is growing competition from local producers, some of whom have learned from HSF. *However/Nevertheless*, with a commitment to becoming carbon neutral within three years, the strategy is to move from radical, breakthrough changes to securing sustained success by a process of continuous improvement (CIP) through a series of incremental steps.

Language note

The first of these sentences is the beginning of a paragraph and the linking word links to the previous paragraph. Therefore, a between-sentence linking word must be used.

Exercise H

Set for individual work and pairwork checking. Feed back with the whole class.

Answers

Model answers: see table on the next page.

	Between-sentence linking words/phrases	Within-sentence linking words/phrases
Addition *(and)*	Furthermore, … In addition, … Additionally, …	… and … … too … also … … along with … … together with …
Contrast (*but*) used when comparing	However, … In/By contrast, … On the other hand, …	… but … … whereas … … while …
Concession (*but*) used to concede/accept a point which simultaneously contrasts with the main point of a sentence or paragraph	However, … At the same time, … Nevertheless, … Despite/In spite of (*this*/noun), … Yet …	… although … … despite/in spite of the fact that …
Result (*so*)	So, … As a result, … Consequently, … Therefore, …	…, so … … so that … … with the result that …
Reason (*for*)	Because of (*this*/noun), … Owing to (this/noun), … Due to (*this*/noun), …	… because … … since … … as … … due to/owing to the fact that …

Corrected version	Comments
As Wallace (2010) explains,	Note the grammar here: either *As Wallace explains* or *Wallace explains that*, but not both. This is a common mistake.
Stefan Smith was initially very successful after taking over HSF and managed to 'grow the business steadily' (p. 6).	1. The words which are the same as the original need quotation marks. 2. The page number should be given in brackets at the end of the quote.
He subsequently 'realized the threat of dependence on the big supermarkets ... and a very cost-sensitive product' and recognized that the company had to change (p. 6).	1. It is important to make a quotation fit the grammar of a sentence. Failing to do this properly is a common mistake. 2. The words which are the same as the original need quotation marks. 3. Some words have been left out. Where this happens, three dots are used to signify an omission. It is important that a quote is exactly the same as the original. Any changes (such as omitting words) need to be clearly shown. 4. The page number should be given in brackets at the end of the quote.
According to Wallace (ibid.),	1. You do not need a verb of saying with *According to*. 2. When referring to the same place in the same source, use *ibid.* instead of the full source reference. If it is the same publication (but not the same place in the text), use *op. cit.*
Smith took decisive action and 'commissioned a *strategic marketing analysis'* [emphasis added] from a firm of consultants' as part of a comprehensive strategic rethink.	1. The quotation marks must be added to the words which are the same as the source. 2. If you want to emphasize a part of a quote, use *italics* and then put [emphasis added] after the quote.

Exercise I

Set for individual work, possibly for homework. Alternatively, set for pair or small group work. Students can write the paragraph on an OHT or other visual medium, which you can display and give feedback on with the whole class.

Answers

Possible answers:

In the case study, Hillary Wallace (2010) clearly believes that Stefan Smith is a special type of manager, with a range of the qualities needed for success. He appears to combine foresight and the courage of his convictions. As Wallace indicates, he was prepared to invest '... a substantial amount of his own money in developing the business ...' (p. 6). He is also a go-getter, a man of action: we are told that he was the first local supplier to secure a deal with a major supermarket. Interestingly, Wallace (ibid.) also indicates that Smith is a self-aware individual, who recognizes that his personal qualities – a commitment to working hard, good service and dealing fairly with people – are key to his success. Smith has other strengths too, as is made plain in the study. Perhaps the most impressive is his ability to create a successful company, recognize that all was not well and provide the *leadership* needed to take HSF through a very substantial change process to renewed success.

Closure

Ask students to discuss these questions.

1 What are the qualities that a good manager needs?

2 Imagine you work in HSF's marketing department. What areas of the business would you like to research in order to ensure sustained success?

Accept any reasonable suggestions.

12.3 Extending skills

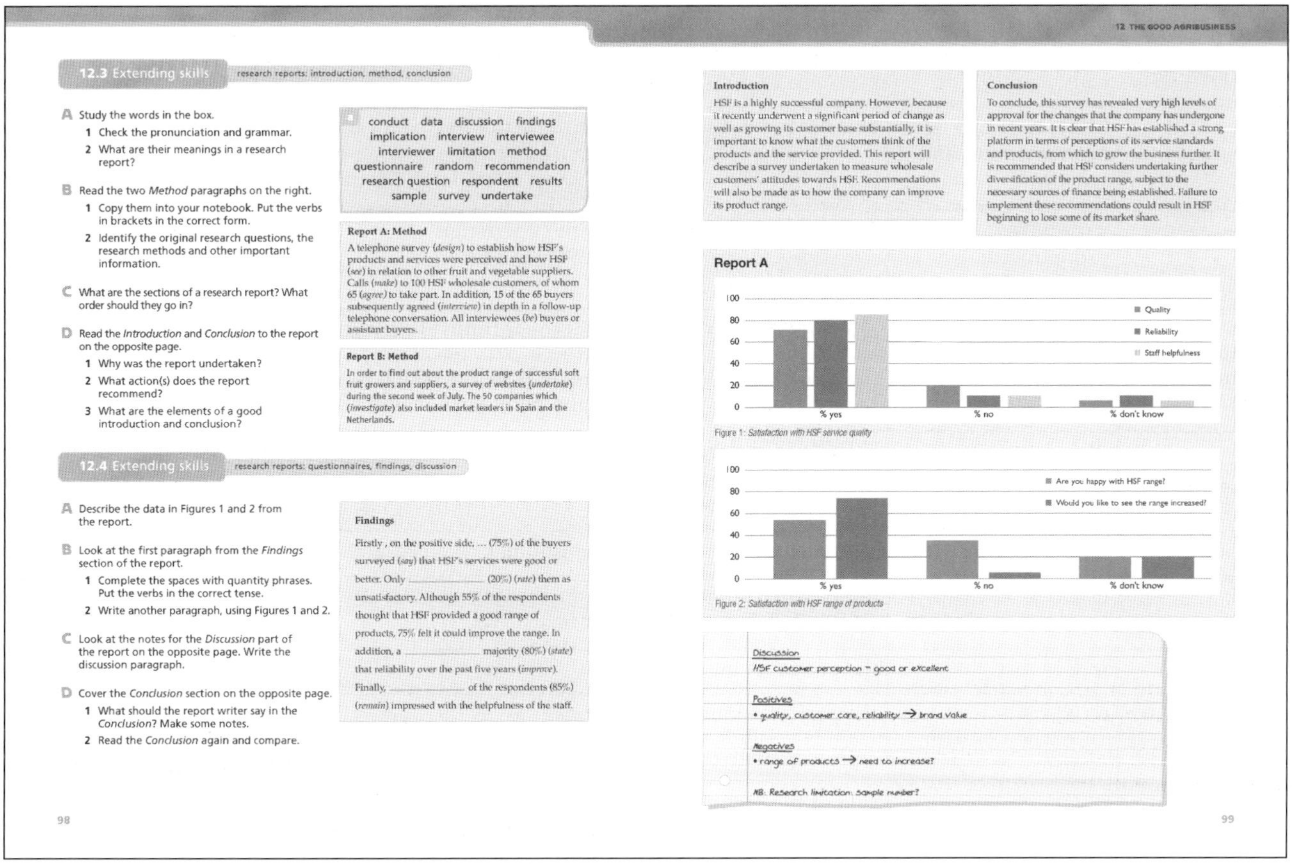

12.3 Extending skills research reports: introduction, method, conclusion

A Study the words in the box.
1. Check the pronunciation and grammar.
2. What are their meanings in a research report?

conduct data discussion findings implication interview interviewee interviewer limitation method questionnaire random recommendation research question respondent results sample survey undertake

B Read the two *Method* paragraphs on the right.
1. Copy them into your notebook. Put the verbs in brackets in the correct form.
2. Identify the original research questions, the research methods and other important information.

C What are the sections of a research report? What order should they go in?

D Read the *Introduction* and *Conclusion* to the report on the opposite page.
1. Why was the report undertaken?
2. What action(s) does the report recommend?
3. What are the elements of a good introduction and conclusion?

Report A: Method

A telephone survey (*design*) to establish how HSF's products and services were perceived and how HSF (*see*) in relation to other fruit and vegetable suppliers. Calls (*make*) to 100 HSF wholesale customers, of whom 65 (*agree*) to take part. In addition, 15 of the 65 buyers subsequently agreed (*interview*) in depth in a follow-up telephone conversation. All interviewees (*be*) buyers or assistant buyers.

Report B: Method

In order to find out about the product range of successful soft fruit growers and suppliers, a survey of websites (*undertake*) during the second week of July. The 50 companies which (*investigate*) also included market leaders in Spain and the Netherlands.

12.4 Extending skills research reports: questionnaires, findings, discussion

A Describe the data in Figures 1 and 2 from the report.

B Look at the first paragraph from the *Findings* section of the report.
1. Complete the spaces with quantity phrases. Put the verbs in the correct tense.
2. Write another paragraph, using Figures 1 and 2.

C Look at the notes for the *Discussion* part of the report on the opposite page. Write the discussion paragraph.

D Cover the *Conclusion* section on the opposite page.
1. What should the report writer say in the *Conclusion*? Make some notes.
2. Read the *Conclusion* again and compare.

Findings

Firstly, on the positive side, ... (75%) of the buyers surveyed (*say*) that HSF's services were good or better. Only ________ (20%) (*rate*) them as unsatisfactory. Although 55% of the respondents thought that HSF provided a good range of products, 75% felt it could improve the range. In addition, a ________ majority (80%) (*state*) that reliability over the past five years (*improve*). Finally, ________ of the respondents (85%) (*remain*) impressed with the helpfulness of the staff.

98

Introduction

HSF is a highly successful company. However, because it recently underwent a significant period of change as well as growing its customer base substantially, it is important to know what the customers think of the products and the service provided. This report will describe a survey undertaken to measure wholesale customers' attitudes towards HSF. Recommendations will also be made as to how the company can improve its product range.

Conclusion

To conclude, this survey has revealed very high levels of approval for the changes that the company has undergone in recent years. It is clear that HSF has established a strong platform in terms of perceptions of its service standards and products, from which to grow the business further. It is recommended that HSF considers undertaking further diversification of the product range, subject to the necessary sources of finance being established. Failure to implement these recommendations could result in HSF beginning to lose some of its market share.

Report A

Figure 1: *Satisfaction with HSF service quality*

Figure 2: *Satisfaction with HSF range of products*

Discussion
HSF customer perception = good or excellent

Positives
• quality, customer care, reliability → brand value

Negatives
• range of products → need to increase?

NB: Research limitation: sample number?

99

Lesson aims

- structure a research report:
  introduction
  method
  findings
  conclusion

Further practice in:
- essay structure
- research methods

Introduction

In preparation for looking at the structure of a research report, revise the sections for an essay: introduction, body, conclusion.

Ask students what should go in each section. Elicit ideas for introductions and conclusions. Do not correct at this point.

Remind students about the methods for doing research (see Unit 5). Ask students what kinds of research would be appropriate if you want to find out what customers think of a company's products or services. (Primary sources are best: survey, questionnaire, interview, quantitative and qualitative methods.)

Ask students what kinds of research would be appropriate if you want to find out what business activities companies in a particular sector are involved in. (Secondary sources are the easiest, e.g., Internet research, company reports, trade magazines.)

Tell students that the next two lessons will focus on writing up research in reports. Ask for suggestions for suitable sections of a research report. Do not correct at this point.

Subject note

In the model presented here, the report is executed at a very simple level. For instance, in a real academic research report, there will be a literature review section before the methods section, and the research questions will be linked with this review. There are also different models for reports. For example, a business report (as opposed to an academic research report) may put conclusions and recommendations near the beginning and the findings as the final section. For the complete text of the report, see Resource 12D in the additional resources section.

Exercise A

Set for individual work and pairwork checking. Feed back with the whole class.

Answers

Model answers:

Word	Notes on pronunciation and grammar	Meaning in a research report
con'duct	v (noun is pronounced: 'conduct)	do (some research, a survey, an experiment)
'data	pl n	information; can be numerical (quantitative) or verbal (qualitative)
dis'cussion	n (U/C)	the title of the section in a research report which discusses the findings. Sometimes the discussion is included in the *Findings/Results* section
'findings	pl n	the title of the section in a research report which details what has been found out; each finding should be linked with a research question. The title *Results* can also be used for this section
impli'cation	n (C)	possible effect or result of the findings
'interview	n (C), v	noun: when someone is asked questions in a survey verb: ask someone questions in a survey
interview'ee	n (C)	the person being questioned
'interviewer	n (C)	the person asking the questions
limi'tation	n (C)	a problem with the research methods; an aspect which the research could not address
'method	n (C)	title of the section in a research report which explains how the research was carried out. In the plural, it refers to the research methods used
questio'nnaire	n (C)	a written set of questions
'random	adj	in no fixed order; with no organizing principle
recommen'dation	n (C)	suggestion for action as a result of the findings of the research
re'search question	n (C)	what the researcher wants to find out
re'spondent	n (C)	a person taking part in a questionnaire survey
re'sults	pl n	Same as *Findings*. Used more or less interchangeably
'sample	n (C), v	noun: the group of people taking part in the research verb: ask research questions to a selected group of people
'survey	n (C), v	noun: a type of research in which the researcher sets out to describe a situation or set of ideas or behaviours, by reading a variety of documents or asking people questions verb: conduct this type of research
under'take	v	do (some research, a survey)

Exercise B

Explain to the students that these paragraphs are examples of the *Method* section of a research report.

1 Set for individual work. Feed back with the whole class, drawing students' attention to the use of the past tense when reporting methods of research, as well as the use of the passive.

2 Set for individual work and pairwork checking. Tell students that they should transform the research questions into real, direct questions. Feed back with the whole group, pointing out that the information given in the *Method* section should include these types of details.

Answers

Possible answers:

	Research questions	Research Method	Other important information
Method (A) A telephone survey (*design*) was designed to establish how HSF's products and services were perceived and how HSF (*see*) was seen in relation to other fruit and vegetable suppliers. Calls (*make*) were made to 100 HSF wholesale customers, of whom 65 (*agree*) agreed to take part. In addition, 15 of the 65 buyers subsequently agreed (*interview*) to be interviewed in depth in a follow-up telephone conversation. All interviewees (*be*) were buyers or assistant buyers.	1. What are wholesale customers' views on HSF's service? 2. How does HSF compare with other suppliers?	telephone survey	100 HSF customers contacted = buyers/assistant buyers; 65 telephone surveys completed
		in-depth telephone interview	15 (of the 65) follow-up telephone interviews conducted
Method (B) In order to find out about the product range of successful soft fruit growers and suppliers, a survey of websites (*undertake*) was undertaken during the second week of July. The 50 companies which (*investigate*) were investigated also included market leaders in Spain and the Netherlands.	What range of products do successful soft growers provide?	survey of websites	• done in second week of July • 50 companies/websites = market leaders in Spain the Netherlands

Exercise C

Use this to confirm that students understand the organization of a research report. Elicit the answers from the whole class.

Answers

Model answers:

Section	Order in a research report
Introduction	1
Method	2
Findings/Results	3
Discussion	4
Conclusion	5

Subject note

Different disciplines and reports for varying purposes may have different section names or organization. The model suggested here is a rather general one, and is a pattern commonly adopted in an academic context, though there are variations depending on the level of the writing (whether, for example, it is a Master's or PhD dissertation). If students are going to write about 500 words only, you may wish to include *Discussion* with *Findings/Results* or with the *Conclusion*.

Exercise D

Explain to the students that these are examples of a typical introduction and conclusion. The introduction explains why the report was undertaken.

Set for pairwork discussion. Feed back with the whole class. Bring the class's attention to the tenses that are used here (present perfect, present simple, future) as well as the use of the passive.

Answers

Model answers:

1 The report was undertaken because HSF has recently been through a period of change, and it was necessary to investigate what the customer base thought of the company's service and products.

2 The recommended action is further diversification of the product range subject to finding the necessary funding.

3 See tables on next page.

Good introduction	Example sentence(s)
Introduce the topic. Give some background information.	HSF is a highly successful company.
Say why the topic is important.	However, because it has recently undergone a significant period of change, as well as growing its customer base substantially, it is important to know what the customers think of the products and the service provided.
Say what you will do in the report. Give a general statement of the purpose of the research.	This report will describe a survey undertaken to measure wholesale customers' attitudes towards HSF. Recommendations will also be made as to how the company can improve its product range.

Good conclusion	Example sentence(s)
Give a general summary/restatement of findings.	To conclude, this survey has revealed very high levels of approval for the changes that the company has undergone in recent years. It is clear that HSF has established a strong platform in terms of perceptions of its service standards and products, from which to grow the business further.
Say what your recommendations are.	It is recommended that HSF considers undertaking further diversification of the product range, subject to the necessary sources of finance being established.
Set out the implications of not taking action.	Failure to implement these recommendations could result in HSF beginning to lose some of its market share.

Language note

The impersonal use of the passive for research reports is not absolutely required. It is often possible to find students' work (assignments, dissertations) which contains the use of the first person singular. However, in more formal writing, such as in journal articles, the passive is usually used.

Closure

1 Refer students to the *Skills bank* to consolidate their understanding of the sections of a research report and their contents.

2 Ask students to choose a company into which they would like to carry out some customer research. They should think about aspects such as quality, service, reliability, customer perceptions, etc. What topics would they ask customers about in a questionnaire?

12.4 Extending skills

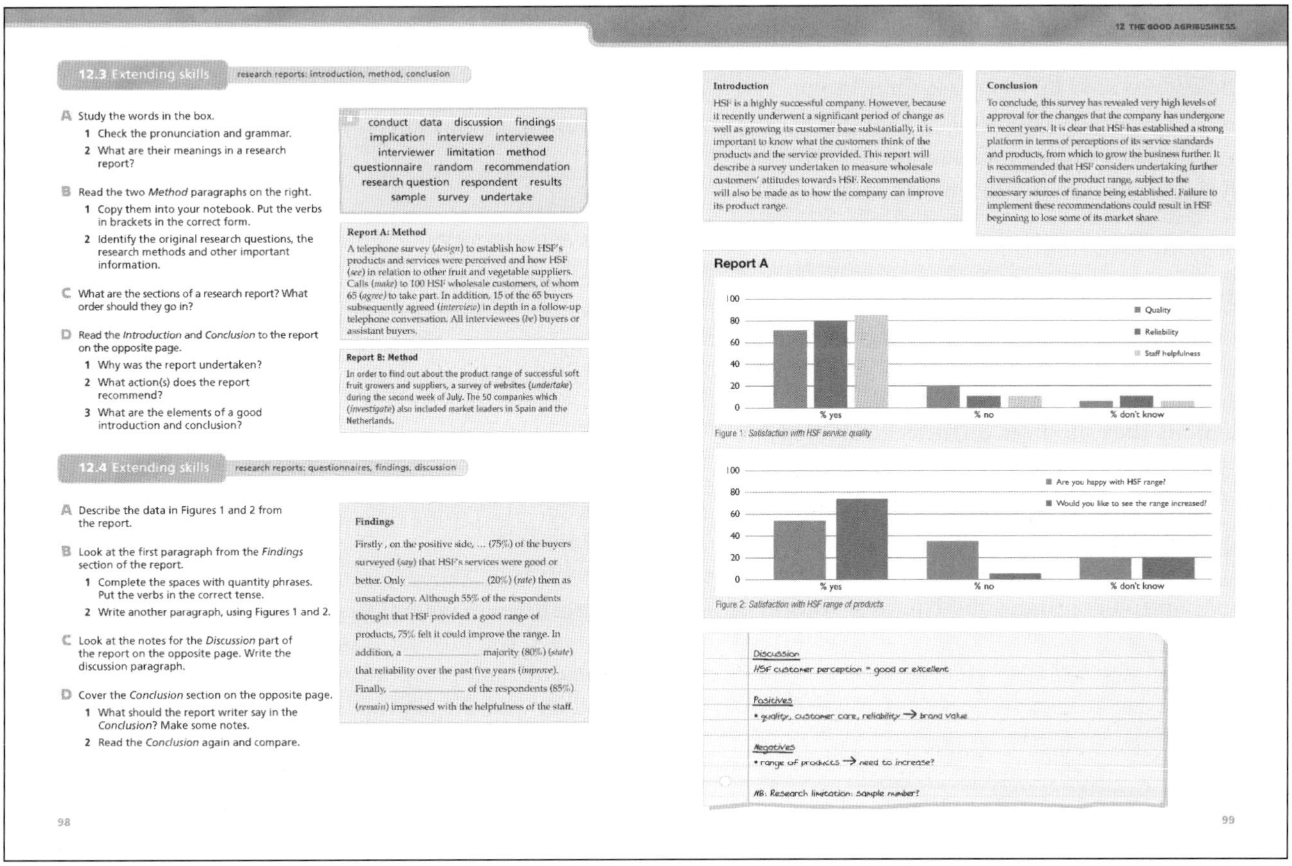

Lesson aims

- write part of a research report: *Findings* and *Discussion*
- analyze and use research data and information

Further practice in:

- talking about numbers and quantities

Introduction

Write up the table below on the board. Give some example phrases and ask students to say approximately what percentage they represent, e.g., a large majority = 80% approximately?

A/An	overwhelming large significant slight small insignificant tiny	majority minority number	(*of* + noun)
Over		half a quarter a third	
More	than		
Less		x %	

Note that *of* is needed if the category for the numbers is given: *A slight minority of respondents said that …* but *A slight minority said that …*

Ask students: What is the difference between *many and most?*

Exercise A

Set students to work in pairs to talk about the key elements of the numbers shown in the charts. If you wish, ask students to write some sentences. Feed back with the whole class, writing some example sentences on the board. Ask the class what these results show about HSF (some of its strengths).

Answers

Possible answers:

Figure 1

A substantial majority (70%) of respondents said that HSF provided a quality service. Unusually, given these were industry buyers, 5–10% of answers for each question were 'don't knows'.

A remarkable percentage (80%) of respondents said HSF was reliable.

Most said that the staff were helpful, with only a minority (10%) saying they weren't.

Figure 2

Although 55% of respondents said they were happy with the range HSF offered, a significant minority (35%) didn't agree with them.

An overwhelming majority of respondents (75%) said they would like to see HSF increase their range.

Again, it was surprising that when asked the question 'would you like to see the range increase?', 20% of those surveyed gave a 'don't know' response to the question.

Exercise B

1 Set for individual work and pairwork checking. Tell students that each space may be for more than one word. They will also need to practise the expressions they used for quantity in Exercise A.

Feed back with the whole class, pointing out the use of past tenses when reporting findings.

2 Set for individual work. Remind students to use linking words and to begin with a topic sentence. This paragraph continues the *Findings* section of the report.

Answers

Possible answers:

Findings

1 Firstly, on the positive side, <u>a large majority</u> (75%) of the buyers surveyed (*say*) <u>said</u> that HSF's services were good or better. Only <u>a minority</u> (20%) (*rate*) <u>rated</u> them as unsatisfactory. Although 55% of the respondents thought that HSF provided a good range of products, 75% felt it could improve the range. In addition, a <u>significant</u> majority (80%) (*state*) <u>stated</u> that reliability over the past five years (*improve*) <u>had improved</u>. Finally, <u>most</u> of the respondents (85%) (*remain*) <u>remained</u> impressed with the helpfulness of the staff.

2 The survey revealed many positive aspects for HSF. Firstly, a majority of the respondents (70%), said the company provided a quality service, although a significant minority indicated 'don't know' to all questions, indicating a problem with the question structure. The overwhelming majority confirmed that HSF are seen as reliable, with a reliable and helpful service (80% and 85% respectively). Moreover, although only half of interviewees felt that the HSF product range was currently adequate, there was strong endorsement overall for the company, with a big majority (75%) indicating that they wish to see HSF offering a wider range of products.

Exercise C

Tell students to look at the notes on the right-hand page. Ask checking questions about the information in the notes, e.g.:

Why does the writer think that HSF's customers' perception of their service is good or 'excellent'?

What things stand out as positive about HSF?

What does the writer identify as negative in the feedback?

What is the possible limitation of the research?

Tell students that the *Discussion* section of a report is where they interpret findings and therefore can give their opinions on the findings. They should write a paragraph using the ideas in the notes. Set for individual work.

Answers

Possible answers:

Discussion

It is clear from the results of this survey that HSF has a very positive image in the perception of its target customers, the buyers. It is not all good news though: on the one hand, it scores strongly on its service, reliability and the attitude of its staff, suggesting a strong brand value among buyers. On the other hand, the results indicate that while its products are perceived as good, there is a view that the range is limited. The survey involved a relatively small cohort, although the return rate was high enough to be valid.

Exercise D

1 Get students to cover the *Conclusion*. They may well remember what it said, but, even if they don't, they can work out what it *should* say based on the *Findings* and the *Discussion* and, of course, on the *Introduction*. Set for individual work and pairwork checking. Do not confirm or correct.

2 Refer students to the *Conclusion* to check their ideas.

Closure

1 Ask students to work out the original questions used in the HSF customer survey.

First, suggest some question types for questionnaires. Elicit the following:

- yes/no
- multiple choice
- open-ended

Tell students to concentrate on the *yes/no* or *multiple choice* types (open-ended questions will elicit qualitative information which is often hard to analyze) and to look at the data in Figures 1 and 2 and the sample *Findings* paragraph. They should try to formulate the actual questions given in the customer survey questionnaire.

Set for pairwork. Feed back with the whole class, writing examples of good questions up on the board. Refer to the model questionnaire in the additional resources section (Resource 12E).

2 Set a research report based on a questionnaire survey for homework. Students can use the ideas they have already discussed in this unit. They should write questionnaires, carry out the research amongst a suitable group of customers (20–40 respondents is fine) and then write up the report. See extra activity 3 for suggestions for topics.

Alternatively, students could choose to find out about the business activities of several global companies in the same sector.

3 If you wish, students could investigate to see what information they can find about business strategy in the soft fruits market. They could look at the following websites (URLs correct at the time of writing):

www.eurofrutta.co.uk

www.hdc.org.uk

www.dole.com

Extra activities

1 Work through the *Vocabulary bank* and *Skills bank* if you have not already done so, or as revision of previous study.

2 Use the *Activity bank* (Teacher's Book additional resources section, Resource 12A).

A Set the wordsearch for individual work (including homework) or pairwork.

Answers

A	R	S	U	B	S	T	A	N	T	I	A	L	I	O	U	Y
Q	Y	G	R	V	G	M	A	J	O	R	I	T	Y	I	R	K
H	O	K	J	S	U	P	P	L	I	E	R	M	F	H	J	N
C	O	N	T	E	N	D	I	N	C	E	N	T	I	V	E	S
U	W	N	I	Q	U	E	S	T	I	O	N	N	A	I	R	E
I	N	C	R	E	M	E	N	T	A	L	W	K	T	H	C	V
Y	Q	N	E	T	W	O	R	K	H	B	K	W	O	A	O	T
E	L	E	G	I	S	L	A	T	I	O	N	D	Q	N	N	M
B	D	C	O	N	C	E	D	E	P	I	N	C	U	E	S	I
P	H	Y	D	R	O	P	O	N	I	C	K	Z	D	D	E	N
A	R	P	E	S	T	I	C	I	D	E	Z	N	N	U	R	O
P	U	E	Q	J	I	C	K	E	B	G	O	E	J	R	V	R
A	I	I	M	B	C	H	B	Q	K	P	R	Z	A	A	A	I
F	I	N	D	I	N	G	S	F	S	T	A	S	Q	N	T	T
D	C	W	O	R	S	U	U	E	J	D	N	Y	O	D	I	Y
P	V	O	S	N	L	E	R	F	X	M	K	G	M	O	O	N
S	U	R	V	E	Y	T	S	D	I	K	K	X	Z	M	N	D

B Set for individual work and pairwork checking.

Answers

95%	the great majority
70%	a significant majority
53%	just over half
50%	half
48%	slightly less than half
10%	a small proportion
2%	a tiny minority

3 Ask students to practise making questionnaires for customer surveys. They could choose from the following topics (or other appropriate topics):

'beauty without cruelty' cosmetics

organic produce (versus non-organic)

a particular brand of organic soft drink

Activity bank

A Solve the crossword.

Down

1 A carbon … is a term used to describe the impact of a person or agribusiness on the environment.

3 A person with an interest in a business, including shareholders, customers and local communities, for example, is called a … .

4 Most new agribusiness product ideas have advantages and … .

6 Industrial farming tends to be efficient at producing food … , but doesn't always deliver food quality.

12 … is the money left over from selling a product when all the costs involved in making it have been paid.

13 A … agribusiness is a business with several departments.

Across

2 A … approach is the same as a complete or total approach.

5 An … farm operates by Soil Association rules.

7 A … process is a process that gives us what we need now and in the future as well.

8 An … farm puts profits and production first.

9 Primary businesses tend to have … organizational structures.

10 A building where things are manufactured is called a … .

11 The difference between the price a shop pays for a product and the price it sells the product for is known as … .

14 An agribusiness cooperative involves several agribusinesses working together to help each other in … .

15 You use a … analysis to analyze the internal workings of an agribusiness.

16 We call a cow a … animal; a tiger is a wild animal.

B Play noughts and crosses. Use the words in context or explain what they mean.

extractive	recycling	processed
industrial	genetic	unskilled
non-organic	overproduction	stewardship

ethics	human resources	philosophy
welfare	opportunity	free-range
weakness	secondary production	battery

Activity bank

A Find 20 words from this unit in the wordsearch.

- Copy the words into your notebook.
- Check the definition of any words you can't remember.

R	O	T	A	T	I	N	G	Q	K	N	L	U	R	N	H	L
Z	E	I	R	R	I	G	A	T	I	O	N	E	A	E	J	N
C	P	R	P	I	O	N	E	E	R	R	Z	W	G	F	W	S
M	U	T	O	S	Q	L	D	L	K	I	S	S	R	F	V	R
Z	U	L	M	T	A	P	J	F	L	S	E	E	I	I	T	V
R	Y	N	T	I	A	D	B	I	E	T	F	L	C	C	E	J
D	X	A	E	I	L	T	T	D	A	C	C	E	U	I	C	E
N	N	T	E	E	V	R	I	L	W	H	O	C	L	E	H	N
B	F	R	I	Q	E	A	P	O	C	B	M	T	T	N	N	C
J	Z	Y	S	F	W	M	T	T	N	O	B	I	U	T	O	L
P	I	A	K	W	E	J	I	I	H	M	I	V	R	E	L	O
R	P	F	O	T	U	D	F	Z	O	L	N	E	E	R	O	S
T	J	S	O	E	N	T	K	F	K	N	E	K	J	K	G	U
P	R	O	P	A	G	A	N	D	I	S	T	Z	D	P	I	R
D	R	I	L	L	I	N	N	O	V	A	T	I	O	N	C	E
O	A	R	E	V	O	L	U	T	I	O	N	U	S	V	A	M
A	L	G	H	L	Z	P	L	O	U	G	H	K	A	R	L	O

B Do the quiz.

1. Who invented the combine harvester?
2. When was the combine harvester invented?
3. Arthur Young was one of the propagandists mentioned in this unit. Who was the other?
4. Who invented the horse-powered seed drill?
5. What was Lord Townshend's nickname? How did he get this nickname?
6. What innovation did Bakewell pioneer?
7. What was the name of Bakewell's new breed of cattle?
8. Who developed the Durham Shorthorn cattle?
9. In the Norfolk four-course crop rotation, which crop was used for cleansing?
10. In the Norfolk four-course crop rotation, which crop replaced nutrients?

In the 1750s, a number of factors began to change agriculture in England.	One of these factors was the urgent need to feed a growing population. People disagree about the changes that took place. Some believe that they were completely new. Others argue that they were a continuation of a process that had started much earlier. However, there are two reasons for saying that events in this period comprised a revolution in agriculture.
Firstly, from 1750, there were growing population levels sustained by self-sufficiency in food.	The population at this point was approximately 6 million. This was a level probably achieved twice before in England's history. However, in previous periods, growth in population to this level had been followed by a corresponding decline. This was largely because agriculture at the time could not sustain the extra numbers. But after 1750, the population continued to grow, reaching nearly 17 million by 1850. This time, food production grew at a sufficient rate to maintain this level of population.
Secondly, productivity rose after 1750, even though the proportion of the population employed in agriculture dropped consistently.	There were fewer people working on the land, but they were producing more food. This was due to increases in the amount of land under cultivation and the intensification of its use. In 1850, with the population at record levels, agriculture employed only 22 per cent of the workforce. This is the lowest percentage in English history, and a lower percentage than any of England's European competitors.
This period of rapid change is called the Agrarian or Agricultural Revolution and it came about for different reasons.	Social changes were one factor in starting and sustaining the Revolution. For example, the Enclosure Laws led to many ordinary people losing their right to use common land for grazing their animals. Landowners used the laws to increase their access to the land for their own use. The Enclosures, as they are known, were bitterly resented by many and are still contentious today.
The Enclosures were important in encouraging the growth of larger, more efficient farms.	Of course, many ordinary people were forced to leave rural areas as a result. But there is no doubt that life for ordinary people was very hard anyway during this period. And the cities must have offered many poor farm labourers and their families an attractive alternative, with new types of jobs and better wages.
Some, but not all, of the revolutionary ideas originated in England.	However, some were imported from Holland and Denmark and developed in East Anglia. They fall roughly into four categories. The first relates to the science of agriculture and includes innovations such as crop rotation and improvements to soil quality. The second comprises the technological developments and new machines that began to spread quickly. Thirdly, word of mouth was important to assure the rapid spread of new practices from East Anglia to the rest of the country. Finally, new stock breeding techniques were significant. These involved the use of only the best animals. They led to the availability of better blood lines, higher productivity and survival rates, and greater profits. This, in turn, led to more investment in farming.
There is general agreement that the century from 1750 to 1850 was one of *revolution*, not *evolution*.	There is substantial evidence that the period was a turning point. Agriculture in England moved from a low-intensity system to a high-intensity system, which was based on arable crops. The template for our modern agricultural industry was established.

From the 1750s onwards, farming productivity was improved in a number of ways.

A small number of key individuals, who became known as 'agriculturalists', started many of the changes.

Lord 'Turnip' Townshend introduced the Norfolk four-course crop rotation system to England.

Jethro Tull is well known today, though many of the innovations he proposed were not implemented in his lifetime.

Robert Bakewell pioneered selective breeding, using simple and effective methods.

Innovation was crucial to the Agricultural Revolution, but people were also needed to spread the word.

It is difficult to identify the most influential innovation.

From the 1750s onwards, farming productivity was improved in a number of ways.

A small number of key individuals, who became known as 'agriculturalists', started many of the changes.

Lord 'Turnip' Townshend introduced the Norfolk four-course crop rotation system to England.

Jethro Tull is well known today, though many of the innovations he proposed were not implemented in his lifetime.

Robert Bakewell pioneered selective breeding, using simple and effective methods.

Innovation was crucial to the Agricultural Revolution, but people were also needed to spread the word.

It is difficult to identify the most influential innovation.

From the 1750s onwards, farming productivity was improved in a number of ways.	These improvements were numerous developments in the techniques which farmers employed to increase productivity. They had an impact on the lives of the rural workforce and on the wider economy and social life of England.
A small number of key individuals, who became known as 'agriculturalists', started many of the changes.	There is some argument over which of them contributed most. The group brought new expertise, and shared an interest in the development of agriculture and a commitment to change. They tend to fall into three categories. First, there were men like Robert Blakewell who developed the science of farming. Jethro Tull was typical of a second group who improved the technology. Finally, Arthur Young pioneered the use of publicity.
Lord 'Turnip' Townshend introduced the Norfolk four-course crop rotation system to England.	Under his system, farmers rotated wheat, turnips, barley and clover annually. The turnips were used as a cleansing crop. Clover replaced the nutrients in the soil. Turnips and clover were used for animal fodder, enabling farmers to feed animals over the winter. This meant fresh meat all year. Townshend also introduced a method called 'marling' to increase productivity where topsoil is mixed with sandy soil to give better cropping.
Jethro Tull is well known today, though many of the innovations he proposed were not implemented in his lifetime.	His contributions include a horse-powered seed drill. This replaced a system where workers distributed seeds by hand, with a machine that used a rotating cylinder. The seed drill distributed the seeds into ploughed furrows, and closed the furrows.
Robert Bakewell pioneered selective breeding, using simple and effective methods.	He bred from the best animal. His methods created the Leicester Longwool sheep and the Dishley Longhorn cow. He used record keeping to ensure good health and this became the template for breeders. The Collings brothers used his ideas as a basis to breed Durham Shorthorn cattle, which combined high milk yields and lean meat.
Innovation was crucial to the Agricultural Revolution, but people were also needed to spread the word.	Arthur Young and Thomas Coke were examples of early agricultural propagandists. They both recognized the need to establish good channels of communication but they used contrasting methods. Young edited a journal and was Secretary of the Board of Agriculture. He spread the message. Coke was a Norfolk landowner who used his own farm to demonstrate the benefits of change. His tenant farmers were offered leases of 20 to 40 years and were encouraged to try new methods. He organised regular events on his estate to demonstrate these methods.
It is difficult to identify the most influential innovation. Periods of rapid and dramatic change tend to be characterized by a range of events and influences.	The period from 1750 to 1850 is a good example.

Activity bank

A Solve the crossword.

Down

1 What we call the weather conditions that exist in a particular area or country.
2 Disagreements between two or more parties.
3 Clearing land of trees, often to make way for other uses of the land such as cattle farming.
5 An official document which proves that someone has achieved something.
7 The collective name we give plants grown to produce food.
9 A raw material such as rice or wheat.
10 Another word for the surroundings in which we live.
11 Harmony or agreement.
13 An increase.

Across

4 A place where items are built or processed for sale.
6 To improve something, for example adding organic matter to arable land to ensure good crops.
8 A measurement of land roughly twice the size of an acre.
12 Operating in many countries.
14 Historically, the staple food of Ireland.

B Play bingo.

- Think of words for each of the categories and write them on card 1.
- Each student says one of their words. Cross the squares on card 2 when you hear a word from that category.

1

arable ______	organic ______	developing nation ______
developed nation ______	dairy product ______	20th century technology ______

2

arable	organic	developing nation
developed nation	dairy product	20th century technology

Activity bank

A Find 20 verbs from the first four units in the wordsearch.

- Copy the verbs into your notebook.
- Write the noun for each verb.

A	V	V	E	N	P	R	O	T	E	C	T	Z	P	X	N
H	L	A	C	I	N	F	E	C	T	C	W	Z	C	T	C
P	R	O	M	O	T	E	G	B	W	O	D	A	B	K	O
R	B	I	N	N	O	V	A	T	E	N	E	W	I	C	O
E	S	Q	W	J	W	E	M	I	T	T	F	R	W	C	R
D	E	G	R	A	D	E	Z	M	A	A	O	O	X	O	D
C	P	F	M	A	G	G	H	V	S	M	L	T	T	O	I
O	A	J	I	O	M	R	I	N	E	I	I	A	R	P	N
N	R	I	O	I	T	T	O	C	L	N	A	T	A	E	A
S	A	V	Z	J	L	I	P	W	E	A	T	E	N	R	T
E	T	N	J	U	U	Z	V	X	C	T	E	C	S	A	E
R	E	I	C	C	Z	I	S	A	T	E	K	U	F	T	K
V	Q	G	K	O	P	E	R	A	T	E	K	L	O	E	U
E	F	B	C	K	D	I	S	G	B	E	Z	T	R	I	V
E	Z	T	E	J	B	G	K	P	H	T	O	A	M	Q	K
I	R	R	I	G	A	T	E	V	M	B	A	N	T	Y	Y

B Play noughts and crosses. You must say the abbreviation or acronym and give the original words to place your symbol in a square.

FAQ	PPT	HTTP
MP3	RAM	CAL
ROM	PC	HTML

DVD	LAN	URL
WWW	LCD	ISP
CPU	RTF	PDF

Although we live in a time of rapid technological change, new technologies have been having an impact on agribusinesses and agriculture for centuries.

New technologies have also always brought with them both opportunities and threats.

Another example of technology impacting on agribusiness is industrial salmon farming.

The revolution of the 20th and 21st centuries has been in information technology.

The use of computers is widespread among agribusinesses.

Although we live in a time of rapid technological change, new technologies have been having an impact on agribusinesses and agriculture for centuries.

New technologies have also always brought with them both opportunities and threats.

Another example of technology impacting on agribusiness is industrial salmon farming.

The revolution of the 20th and 21st centuries has been in information technology.

The use of computers is widespread among agribusinesses.

CAL	DVD
HTML	HTTP
ISP	LCD
PIN	ROM
URL	USB
WAN	WWW

Activity bank

A Solve the synonyms crossword. Find words with the same meaning as the clues.

Across

1 retail outlet
5 task
7 in fact
8 known as
10 principal
11 conclude
13 promote
14 requirements
17 fall sharply
18 firm (n)
20 decline
21 talk about

Down

2 naturally (2, 6)
3 purchase
4 overview
6 fundamentally
8 client
9 type
11 crucial
12 means
15 rise sharply
16 aim
19 satisfy

B Play opposites bingo.

- Choose six words from the box and write one word in each square of your bingo card.
- Your teacher will call out some words. If you have the **opposite** word on your card, cross it out.
- The first person to cross out all the words on their card is the winner.

careful child easy elderly fall large low luxury male poor professional rigid sharply single slightly slow traditional unskilled

Describing trends

Verbs	Nouns	Adverbs	Adjectives
rise		gradually	
increase		sharply	
grow		slightly	
improve		markedly	
fall		significantly	
decrease		rapidly	
drop		steeply	
decline		steadily	

Poor contributions	**Student A**	**Student B**	**Student C**
disagrees rudely			
doesn't explain how the point is relevant			
doesn't understand an idiom			
dominates the discussion			
gets angry when someone disagrees with them			
interrupts			
is negative			
mumbles or whispers			
says something irrelevant			
shouts			
sits quietly and says nothing			
starts a side conversation			
other:			

Good contributions	**Student A**	**Student B**	**Student C**
allows others to speak			
asks for clarification			
asks politely for information			
brings in another speaker			
builds on points made by other speakers			
contributes to the discussion			
explains the point clearly			
gives specific examples to help explain a point			
is constructive			
links correctly with previous speakers			
listens carefully to what others say			
makes clear how the point is relevant			
paraphrases to check understanding			
says when they agree with someone			
speaks clearly			
tries to use correct language			
other:			

Student A

Previous research: primary research

Your research task:

Find out about **secondary research** and report back to your group. Say what this kind of research is, its advantages and disadvantages.

Your discussion task:

Contribute to the discussions on primary, qualitative and quantitative research, for example by asking for information or clarification, agreeing or disagreeing. If you have also researched any of these other types of research, say where you found your information.

Student B

Previous research: secondary research

Your research task:

Find out about **primary research** and report back to your group. Say what this kind of research is, its advantages and disadvantages.

Your discussion task:

Contribute to the discussions on secondary, qualitative and quantitative research, for example by asking for information or clarification, agreeing or disagreeing. If you have also researched any of these other types of research, say where you found your information.

Student C

Previous research: qualitative research

Your research task:

Find out about **quantitative research** and report back to your group. Say what this kind of research is, its advantages and disadvantages.

Your discussion task:

Contribute to the discussions on qualitative, primary and secondary research, for example by asking for information or clarification, agreeing or disagreeing. If you have also researched any of these other types of research, say where you found your information.

Student D

Previous research: quantitative research

Your research task:

Find out about **qualitative research** and report back to your group. Say what this kind of research is, its advantages and disadvantages.

Your discussion task:

Contribute to the discussions on quantitative, primary and secondary research, for example by asking for information or clarification, agreeing or disagreeing. If you have also researched any of these other types of research, say where you found your information.

1 <u>Secondary</u>

= info from sources, e.g., books, Internet, trade mags, reports, etc. (i.e., already exists)

\+ cheap; good overview of market; based on real sales statistics; relatively fast

\- reports → sometimes expensive; poss. out of date

2 <u>Primary</u>

= new info: from (1) people, e.g., customers, retailers; (2) observation, e.g., of shoppers

\+ info = recent; can ask specific questions (good method for psych. research)

\- expensive; time-consuming

3 <u>Quantitative</u>

= statistical info, usually thro' questionnaires

\+ good for factual info; overview of trends ∵ large nos.

\- sample <u>must</u> be v. big; people lie ∴ results may ≠ reliable; low response rate for questionnaires

4 <u>Qualitative</u>

≠ numbers; usually verbal info; used to find out attitudes, beliefs, etc. Methods inc. interviews, focus groups, etc.

\+ reveal unknown probs; basis for quant. methods

\- in groups, opinions easily led by one person; only small numbers ∴ difficult to generalize

Activity bank

A Find 20 verbs from this unit in the wordsearch.

- Copy the words into your notebook.
- Write the noun for each verb.

V	V	S	Y	K	J	D	L	A	U	N	C	H	Y	G	W
E	R	K	U	C	E	S	E	G	M	E	N	T	V	O	A
Q	E	D	I	S	P	L	A	Y	S	E	Y	L	Y	R	C
P	T	E	O	Q	R	W	A	X	K	E	T	Q	Z	E	I
J	A	V	A	D	L	C	U	S	A	T	U	R	A	T	E
I	I	A	U	E	D	L	U	B	I	N	C	L	U	D	E
E	N	L	G	S	O	X	U	Q	W	A	S	S	E	S	S
X	O	U	M	I	B	I	N	T	E	G	R	A	T	E	W
T	Q	A	E	G	R	E	P	A	C	K	A	G	E	A	U
E	Z	T	N	N	J	A	G	V	O	N	Q	H	R	Q	E
N	D	E	T	S	P	R	O	D	U	C	E	D	Y	V	V
D	V	Y	W	B	E	N	E	F	I	T	H	U	E	N	W
I	N	D	I	C	A	T	E	I	W	T	P	I	Q	H	U
M	S	H	D	J	N	E	B	H	I	F	H	Z	G	Q	X
X	X	S	T	A	G	E	V	W	Q	C	B	O	R	Y	G
R	M	I	Y	K	K	F	T	L	A	U	P	D	A	T	E

B Think of a word or words that can go in front of each of the words below to make a phrase from agribusiness. Explain the meaning.

Example: market = *mature market, emerging market, target market, niche market*

______________ assurance
______________ benefits
______________ cycle
______________ edge
______________ flow
______________ habits
______________ market

______________ phase
______________ period
______________ practice
______________ prices
______________ segments
______________ services
______________ span

______________ standards
______________ strategy
______________ strategy
______________ strategy
______________ tools
______________ value

1 Cash flow is ______________________ during the development phase.

2 In the ______________________ phase, unit costs are high.

3 Cash flow becomes positive in the ______________________ phase.

4 ______________________ sales mean that unit costs are reduced.

5 Sales are at their peak once the market has ______________________ .

6 Sales may start to ______________________ when the market reaches saturation.

1	Cash flow is negative during the development phase.
2	In the introduction phase, unit costs are high.
3	Cash flow becomes positive in the maturity phase.
4	Increasing sales mean that unit costs are reduced.
5	Sales are at their peak once the market has stabilized/matured.
6	Sales may start to fall/drop/decline when the market reaches saturation.

a	It is not until the maturity phase is reached that cash flow is no longer negative.
b	Peak sales figures are achieved at the maturity phase.
c	While the product is being designed and trialled, there are many expenses but no income.
d	There may be a reduction in business if there are too many competitors.
e	It costs a lot to produce each unit early in the cycle.
f	Once the products start to sell, the company can make them more cheaply.

Original sentence	Student A	Student B
Applecart has proven commercial potential, therefore.	Applecart has therefore demonstrated its commercial promise.	Thus, there is no doubt that Applecart's business has the potential to turn a profit.
	Completely unsatisfactory paraphrase: the student has reworded the sentence merely by making lexical replacements and moving the word 'therefore'.	*Acceptable paraphrase:* ● *the student has used the idiomatic phrase 'there is no doubt' to replace 'proven'* ● *the student has altered the sentence structure by defining the nature of Applecart's 'potential'* ● *the student has used the idiomatic phrase 'to turn a profit'.*
But agritourism is relatively new and Applecart must not become complacent or make assumptions based on traditional tourism.	However, agritourism is fairly new and Applecart must not make judgements based on conventional tourism.	It would be dangerous for Applecart to base business decisions solely on examples from conventional tourism, given that agribusiness is quite a new field.
	Unsatisfactory paraphrase: the student's lexical changes (replacing 'but' with 'however', 'relatively' with 'fairly', 'assumptions' with 'judgements' and 'traditional' with 'conventional') are not enough on their own. The student has not altered the sentence structure. This is patch-writing.	*Acceptable paraphrase:* ● *the student has interpreted original sentence to conclude that 'It would be dangerous …'* ● *the student has reversed the order of the ideas 'agri-tourism is new' and 'Applecart shouldn't base its decisions on conventional businesses'.*

Original sentence	Student A	Student B
Fortunately, it assessed its situation and took a proactive approach.	Luckily, Applecart weighed up the situation and took a positive approach.	Applecart decided to meet the challenges of the market head on, after considering all the options.
	Unacceptable paraphrase: again, the student has merely made a few minor lexical changes ('assessed' → 'weighed up', 'proactive' → 'positive').	*Acceptable paraphrase:* ● *the student has reversed the order of the information 'assessed its situation' and 'Applecart took a proactive approach'.* ● *the student has made significant lexical changes, introducing new phrases 'meet the challenges of the market head on' and 'after considering all the options'.*
Philippe and Anna called in a marketing consulted who, from the start, introduced them to new ways of analyzing their situation.	Philippe and Anna consulted a marketing consultant who, from the beginning, showed them new methods of analyzing their situation.	A marketing specialist introduced Philippe and Anna to many new ways of looking at the market.
	Completely unacceptable paraphrase: once again, the student has merely made some minor lexical changes and made no effort to reorder any ideas or restructure the sentence.	*Acceptable paraphrase:* ● *the student has made a lexical change (consultant > specialist)* ● *the student has reworded the final idea significantly* ● *the student has changed the sentence structure by making the marketing specialist the subject.*

Central to the marketing mix is the agriproduct itself.

Applecart Farm is a traditional agribusiness that has diversified into agri-tourism products.

Applecart depends on the Internet for its marketing.

Applecart has proven commercial potential, therefore.

An Ansoff Matrix assumes four strategies for achieving growth.

Central to the marketing mix is the agriproduct itself.

Applecart Farm is a traditional agribusiness that has diversified into agri-tourism products.

Applecart depends on the Internet for its marketing.

Applecart has proven commercial potential, therefore.

An Ansoff Matrix assumes four strategies for achieving growth.

	Main subject	Main verb	Main object/ complement	Other verbs + their subjects + objects/complements
A	the name (of the farm cereal)	was changed	(to) 'Toasted Fruit'.	which was called 'Fruitflakes'
B	Three of the many ways	will be described		in which a product may be evaluated
C	You	can eat	products like this	1. … which is very convenient 2. that has played a major part in the expansion of the cereal market.
D	an agribusiness	must be	aware (of the qualities of its products.)	As well as understanding its target markets, …
E	the company	has seen	excellent results.	Having taken these steps as part of a strategy for achieving growth, …

A added	A analytical	A buying
A cash	A competitive	A customer
A decline	A discounted	A extension
A functional	A good	A life
A market	A marketing	A mature
A portfolio	A quality	A quality
A shoulder	B assurance	B benefits
B cycle	B edge	B flow
B habits	B market	B period
B phase	B practice	B prices
B segments	B services	B standards
B strategy	B strategy	B strategy
B tools	B value	

Activity bank

A Solve the crossword.

Across

1 Storage workers use a … truck to lift cartons and boxes onto lorries.

3 A …-market product is the opposite of a niche market product.

6 Another word for an Ishikawa diagram is a … diagram.

8 … is the process by which a company gets the product to the customer.

10 A … fuel is a type of renewable energy source.

12 … management is a type of management that is flexible and constantly responsive to external events.

13 … is the opposite of maximize.

14 In the UK, tomatoes are normally grown under glass in a … .

Down

1 Coal and oil are examples of … fuels.

2 Boxes of, e.g., chutney, are normally put on a … before being put on a lorry.

4 … is the management technique where you try and rise above the situation to get an overview of it.

5 Businesses often … or 'kick around' a lot of different ideas with a group before they decide on a new product.

7 In a … business, the customers also own the business.

9 A Critical Path Analysis is a way of planning a … of events or activities.

11 A …-priced product is a high-priced or high quality product.

	Fixed phrase	Followed by …	Actual information (suggested answers)
1	An important concept (is) …	a new idea or topic that the lecturer wants to discuss	
2	What do I mean by … ?	an explanation of a word or phrase	
3	say …	a hypothetical example	
4	… the point is …	a key statement or idea	
5	… in financial terms, …	there is a financial point to be made	
6	Looking at it another way, …	a different way to think about the topic	
7	What I'm referring to is …	connected point(s)	
8	apart from …	a new topic or point	
9	… as you can see, …	a comment about a 'given' or a subject already in discussion	
10	… in this way …	a concluding comment giving a result of something	

7.4 Eco-communities

- part of UK government vision
- small communities around 20,000 homes
- good infrastructure and public transport
- low-carbon technology
- low-cost housing to encourage young families and first-time homebuyers
- BBC news reports on eco-towns: news.bbc.co.uk/1/hi/uk/8152985.stm

7.4 Biomass fuels

- anything that can either burn or decompose
- other terms are biomass energy or biofuel
- 'bio' refers to life
- more popular due to the rising costs of fossil fuels
- develop own fuels 'at home'
- reduces pollution, helps control carbon dioxide emissions
- examples: http://wiki.answers.com/Q/Three_examples_of_biomass_fuels

7.4 Co-housing projects

- group living, shared common aims and activities + own self-contained accommodation and personal space
- set up and run by members for mutual benefit
- common space facilitates shared amenities like laundries, heating systems, guest rooms, transport
- UK network: http://www.cohousing.org.uk/?q=Map

7.4 Cooperatives

- different business organizations have different legal status and rules
- for mutual benefit: members control and benefit from the operation
- serve members' interests, rather than investors'
- based around the concepts of self-help, self-responsibility and self-organization
- more stable, caring and responsible employers
- greater job satisfaction and variety, encourage a strong work commitment
- more responsible to the customer and the community in which they operate
- innovative and very competitive businesses, e.g., John Lewis, partnership spirit: www.johnlewispartnership.co.uk

Activity bank

A Find 15 words from this unit in the wordsearch. All the words are uncountable nouns in the texts in this unit.

- Copy the words into your notebook.
- Check the definition of any words you can't remember.

Q	S	J	L	T	U	I	I	L	T	Y	Y	G	B	J	L
W	I	F	C	M	A	I	N	T	E	N	A	N	C	E	E
Q	G	L	W	O	W	Z	J	G	Q	L	V	F	H	S	A
F	N	E	V	T	E	L	Q	R	O	G	V	O	P	S	R
A	I	X	R	L	X	M	G	R	D	Q	R	O	A	N	N
R	F	I	H	T	M	H	T	U	Z	D	Q	O	O	I	I
M	I	B	A	R	C	N	A	T	U	R	E	I	W	K	N
I	C	I	P	J	O	L	M	H	F	Z	T	O	A	T	G
N	A	L	P	C	L	U	U	K	C	C	F	W	Q	L	H
G	N	I	I	A	D	A	P	T	A	B	I	L	I	T	Y
P	C	T	N	D	T	J	B	F	O	C	T	P	P	K	D
S	E	Y	E	R	S	R	S	U	C	G	B	R	V	N	H
X	S	U	S	T	A	I	N	A	B	I	L	I	T	Y	W
C	H	F	S	U	T	D	W	E	A	L	T	H	R	E	R
S	D	W	K	A	X	I	M	P	O	R	T	A	N	C	E
X	E	P	S	E	C	O	N	O	M	I	C	S	I	A	H

B Rearrange the letters in the words to form a correctly spelt word from this unit.

Jumbled word	Correct spelling
sngiiro	
ngbeni	
nogrica	
nactfisngii	
kratem	
iterflirez	
toilxep	
lolbag	
rocam	
mehp	

Activity bank

A Solve the crossword.

Down

2 The amount someone is paid is called their … , and it's usually paid monthly.
3 In business, a … is a plan for achieving something for the business.
4 A … owns a share or shares in a company.
8 If you give money to a company in the hope of a future profit, it's called an … .
9 The total business revenue during a defined period of time is called its … .
10 When we break down the essential features of a problem, we call it a problem … .

Across

1 The opposite of profit is a … .
5 A ... is a plan for income and expenses during a future period.
6 A ... is an action in which money or goods are exchanged.
7 … is another word to describe a method of doing something.
11 An ... is someone who invests in a company.
12 … is another word for what a company spends.
13 A company's ... is the difference between its market value and what it owes.
14 The ... of a company is another word to describe its assets.
15 … is the practice of auditing and analyzing the financial circumstances of a business.

B Are the nouns countable or uncountable? Use a dictionary to check.

Noun	Countable or uncountable?	Notes
assets		
cash		
cost		
economy		
liabilities		
loss		
margin		
profit		
sale		
share		
taxation		

Review reduce + recite + review	**Notes** record
Here you write only important words and questions; this column is completed *after* the lecture. Later this column becomes your study or revision notes. You can use it by covering the right-hand column and using the cue words and questions here to remember the contents on the right.	This column contains your notes. You should underline headings and indent main ideas. After the lecture or reading, you need to identify the key points and write them in the review column as questions or cue words.
Summary reflect + recite + review After the class, you can use this space to summarize the main points of the notes on the page.	

Cornell page layout: answer key

Review	**Notes**
2 types of accounting are...?	Accounting in companies: management accounting (MA) and financial accounting (FA)
Importance of finance means ...?	Finance is *important*: standard tools/methods companies use both *FA* + *MA*
	<u>MA</u>
Main purpose of ...?	1) *Forward* planning
Information supplied by ...?	• information given in annual *budgets*
Aims of MA? (5 specific aims)	• aim is to help company: a) decide how to *spend* its money b) decide whether to *borrow funds* (e.g., to invest in *new machinery*, to take on more staff) c) monitor *cash flows* & value of *investments* d) predict *profits*/losses e) help monitor performance (incl. staff)
Required by law?	2) Not a *legal* requirement
Management tool	3) Internal management mechanism
(advantages: why used? by whom?)	• good *practice* • powerful → *decisions* are based on the data • not useful for *outsiders*

Summary

There are two types of accounting: management and financial. Management accounting is a useful management tool, used internally within a company. It's not a legal requirement, unlike financial accounting.

	Jan 06	Jan 05	Jan 04
Assets			
	6,414	5,488	5,199
Net receivables	2,662	1,715	1,254
	32,191	29,447	26,612
	79,290	68,567	58,530
Other current assets	2,557	1,841	1,356
Total assets	123,114	107,058	92,951
Liabilities			
	40,178	35,107	31,051
Short-term debt	8,648	7,781	6,367
	30,171	23,669	20,099
Other liabilities	6,019	4,270	3,772
Total liabilities	85,016	70,827	61,289
Shares outstanding	4,165	4,234	4,296

Activity bank

A Find 20 words from this unit in the wordsearch..

- Copy the words into your notebook.
- Check the definition of any words you can't remember.

B	P	O	D	L	U	C	O	N	T	A	M	I	N	A	T	E	U	D
C	W	U	A	U	N	M	R	I	A	F	R	E	E	Z	E	G	D	U
E	G	T	F	X	M	N	M	M	L	Y	W	N	L	O	R	M	J	S
B	J	B	E	L	F	I	C	O	D	L	I	X	U	B	E	P	H	R
I	P	R	W	N	L	M	H	A	D	V	O	V	S	L	B	V	O	T
F	M	E	J	C	T	F	F	G	O	K	R	A	B	Z	U	N	E	R
D	P	A	X	O	D	E	C	B	G	B	X	O	N	K	Y	O	V	A
M	S	K	C	Q	U	Y	R	O	R	M	R	W	C	U	E	X	A	I
L	V	E	S	H	T	T	A	P	M	P	S	L	O	F	R	F	J	N
R	E	K	N	I	T	I	X	I	R	P	T	V	P	K	K	D	G	I
L	A	G	L	X	R	A	X	O	A	I	L	J	J	G	F	O	N	N
R	Q	A	E	E	Y	K	S	X	L	V	S	Y	T	D	Q	L	D	G
L	U	I	T	T	D	G	U	Q	W	A	S	E	N	W	H	H	S	E
Q	Q	C	E	B	S	T	A	N	D	A	R	D	E	L	B	J	X	B
R	A	F	S	A	L	M	O	N	E	L	L	A	U	Z	A	L	L	M
B	A	O	Z	A	V	I	R	U	S	N	I	N	S	P	E	C	T	M
S	H	A	Z	A	R	D	D	L	P	L	I	N	V	E	S	T	S	R

B Think of a word or words that can go after each of the words or phrases below to make a phrase from agribusiness and agriculture. Explain the meaning.

Example: *food = food hygiene, food safety, food poisoning, food-borne*

cross ______________________________

environmental ______________________________

food ______________________________

hazard ______________________________

hygiene ______________________________

imported ______________________________

local ______________________________

notifiable ______________________________

smoking ______________________________

spoilable ______________________________

Introduction		**Examples of ideas**
introduce the topic area give the outline of the essay		
Body	**Para 1:** situation/problems (general)	
	Para 2: problems (specific examples)	
	Para 3: solutions	
	Para 4: evaluations of solutions	
Conclusion		

Early MP & Adams CR. (2004) Principles of horticulture *(Fourth Ed)* Oxford: Elsevier/Butterworth-Heinemann

HG Baker (1974) The Evolution of Weeds *Annual Review of Ecology & Systematics 5* pp.1-24

Bond W, Turner RJ (2005) *A review of mechanical weed control*. HDRA http://www.gardenorganic.org.uk/organicweeds/downloads/mechanical.pdf (accessed 14/9/07)

Mortimer M and Roger Cousens. *Dynamics of Weed Populations* Cambridge: CU 1995

Batish D, Kohli K & Singh HP: Handbook of Sustainable Weed Management 2006. Birmingham: Harworth Press

John Morrison & John Williams JB, (1987) *ADAS colour atlas Of Weed seedlings.* Wolfe Publishing Ltd, Netherlands

Early MP & Adams CR. (2004) Principles of horticulture *(Fourth Ed)* Oxford: Elsevier/Butterworth-Heinemann

HG Baker (1974) The Evolution of Weeds *Annual Review of Ecology & Systematics 5* pp.1-24

Bond W, Turner RJ (2005) *A review of mechanical weed control*. HDRA http://www.gardenorganic.org.uk/organicweeds/downloads/mechanical.pdf (accessed 14/9/07)

Mortimer M and Roger Cousens. *Dynamics of Weed Populations* Cambridge: CU 1995

Batish D, Kohli K & Singh HP: Handbook of Sustainable Weed Management 2006. Birmingham: Harworth Press

John Morrison & John Williams JB, (1987) *ADAS colour atlas Of Weed seedlings.* Wolfe Publishing Ltd, Netherlands

Activity bank

A Solve the crossword.

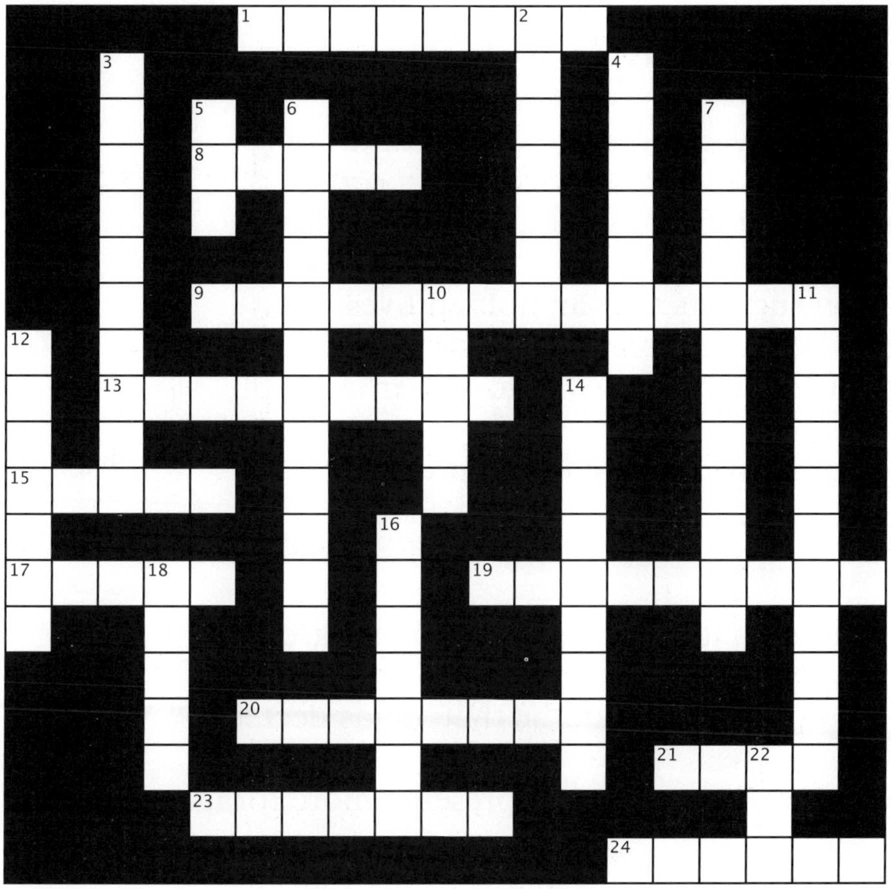

Across

1 Greenpeace is an incredibly effective … group.
8 Fuel … inflation is a regular feature of the global economy.
9 People who study the weather are called … .
13 Rearing chickens in cages or … will soon be a thing of the past.
15 Fossil … like coal and gas are unsustainable.
17 Governments don't always tighten regulations: they sometimes … them.
19 There are … people who simply don't believe the facts.
20 Wildlife isn't confined to the country; there are natural … in cities too.
21 Periods of rapid growth tend to be followed by a slowing … in the economy.
23 Three years without rain is guaranteed to cause … .
24 The 2008/2009 economic crash is usually referred to as the credit … .

Down

2 We don't want a conventional approach to the problem; we want a … one.
3 Today, the not-for-profit and … sector are competing with the for-profit sector.
4 Carbon … is a form of carbon dioxide removal.
5 The acronym … refers to a satellite-based technology.
6 … is important: we need as many different species as possible.
7 The damage caused by the tsunami was appalling, quite … .
10 The … is for more and more people to want to visit the countryside.
11 When people buy commodities because they believe they'll increase in price, we call this … .
12 Most people care about animal … and want animals to be well looked after.
14 Pressure groups work by raising … of issues.
16 Tree … should only be carried out in renewable forests.
18 … schemes like the 'Good Egg' can be helpful.
22 When farms improve their environmental impact and their profits, it's a …-win situation.

B Match a word from the first box with a word from the second box to make a two-word phrase. Make sure you know what they mean.

carbon consultation drive economic environmentally global greenhouse habitat raise renewable

awareness destruction dioxide downturn energy exercise friendly gases up warming

According to a recent scientific report, farms in the UK need to start adapting to cope with reduced water due to higher temperatures and summers with lower rainfall. Moreover, there is a clear risk of more frequent droughts as well as flooding, both of which could lead to falling outputs. The information suggests that an upland livestock farmer might need to consider bringing animals indoors earlier in the season because of the rainier autumns in this more unpredictable environment. In addition, farmers need to look at the consequences for their livestock. The report suggests that there may be a longer growing period for forage crops. One result of this could be that yields decrease after the early season, meaning farms may need to preserve more forage for winter feeding. Not only this, but there is also a warning that for arable farms in certain elevated or vulnerable areas, if there are increased temperatures and a drier summer, as a result, there could be problems establishing crops because of a lack of moisture. Another point is that lower precipitation in the summer could mean wetter autumns, which could in turn mean difficulties for farmers accessing their land. The report advises that this might have a knock-on effect firstly on farm labourers, and secondly on machinery use. What is more, it could encourage more spring sowing.

Activity bank

A Find 20 words from this unit in the wordsearch.

- Copy the words into your notebook.
- Check the definition of any words you can't remember.

A	R	S	U	B	S	T	A	N	T	I	A	L	I	O	U	Y
Q	Y	G	R	V	G	M	A	J	O	R	I	T	Y	I	R	K
H	O	K	J	S	U	P	P	L	I	E	R	M	F	H	J	N
C	O	N	T	E	N	D	I	N	C	E	N	T	I	V	E	S
U	W	N	I	Q	U	E	S	T	I	O	N	N	A	I	R	E
I	N	C	R	E	M	E	N	T	A	L	W	K	T	H	C	V
Y	Q	N	E	T	W	O	R	K	H	B	K	W	O	A	O	T
E	L	E	G	I	S	L	A	T	I	O	N	D	Q	N	N	M
B	D	C	O	N	C	E	D	E	P	I	N	C	U	E	S	I
P	H	Y	D	R	O	P	O	N	I	C	K	Z	D	D	E	N
A	R	P	E	S	T	I	C	I	D	E	Z	N	N	U	R	O
P	U	E	Q	J	I	C	K	E	B	G	O	E	J	R	V	R
A	I	I	M	B	C	H	B	Q	K	P	R	Z	A	A	A	I
F	I	N	D	I	N	G	S	F	S	T	A	S	Q	N	T	T
D	C	W	O	R	S	U	U	E	J	D	N	Y	O	D	I	Y
P	V	O	S	N	L	E	R	F	X	M	K	G	M	O	O	N
S	U	R	V	E	Y	T	S	D	I	K	K	X	Z	M	N	D

B Match the percentages with a suitable phrase to describe numbers of respondents.

95%	______	a significant majority
70%	______	a small proportion
53%	______	a tiny minority
50%	______	half
48%	______	just over half
10%	______	slightly less than half
2%	______	the great majority

SWOT analysis

When a company's performance is looking poor, it may need to make some changes. However, before the company can do this, it should establish where it is now through a strategic analysis. A commonly used technique for this is the SWOT analysis. In this method, managers carry out an 'external audit' in which they examine their business and economic environment, as well as the market conditions they face, in order to understand the opportunities and threats to the company. Secondly, the organization needs to complete an 'internal audit' in which its strengths and weaknesses are compared with those of the competitors. This means that managers should look at all the functional areas: finance, HR, marketing and operations. The results of such audits are presented in a four-box summary of the business's current strengths and weaknesses, and the opportunities and threats which will affect its future development.

Stefan Smith called in consultants.

HSF secured grants for biodiversity.

HSF achieved turnover of more than £6 million.

Stefan Smith sold his property business.

Stefan Smith recognized the threat of dependence on a narrow customer base.

HSF underwent re-engineering.

HSF committed to becoming carbon neutral.

HSF received a strategic marketing analysis report.

Stefan Smith called in consultants.

HSF secured grants for biodiversity.

HSF achieved turnover of more than £6 million.

Stefan Smith sold his property business.

Stefan Smith recognized the threat of dependence on a narrow customer base.

HSF underwent re-engineering.

HSF committed to becoming carbon neutral.

HSF received a strategic marketing analysis report.

Questionnaire survey of HSF customers

Introduction

HSF is a highly successful company. However, because it has recently undergone a significant period of change as well as growing its customer base substantially, it is important to know what the customers think of the products and the service provided. This report will describe a survey undertaken to measure wholesale customers' attitudes towards HSF. Recommendations will also be made as to how the company can improve its product range.

Method

A telephone survey was designed to establish how HSF's products and services were perceived and how HSF was seen in relation to other fruit and vegetable suppliers. Calls were made to 100 HSF wholesale customers, of whom 65 agreed to take part. In addition, 15 of the 65 buyers subsequently agreed to be interviewed in depth in a follow-up telephone conversation. All interviewees were buyers or assistant buyers.

Findings

The survey revealed many positive aspects for HSF. Firstly, a majority of the respondents (70%), said the company provided a quality service, although a significant minority indicated 'don't know' to all questions, indicating a problem with the question structure. The overwhelming majority confirmed that HSF are seen as reliable, with a reliable and helpful service (80% and 85% respectively). Moreover, although only half of interviewees felt that the HSF product range was currently adequate, there was strong endorsement overall for the company, with a big majority (75%) indicating that they wish to see HSF offering a wider range of products.

Discussion

It is clear from the results of this survey that HSF has a very positive image in the perception of its target customers, the buyers. It is not all good news though: on the one hand, it scores strongly on its service, reliability and the attitude of its staff, suggesting a strong brand value among buyers. On the other hand, the results indicate that while its products are perceived as good, there is a view that the range is limited. The survey involved a relatively small cohort, although the return rate was high enough to be valid.

Conclusion

To conclude, this survey has revealed very high levels of approval for the changes that the company has undergone in recent years. It is clear that HSF have established a strong platform in terms of perceptions of its service standards and products, from which to grow the business further. It is recommended that HSF considers undertaking further diversification of the product range, subject to the necessary sources of finance being established. Failure to implement these recommendations could result in HSF beginning to lose some of its market share.

Hadford Soft Fruits (HSF) Survey

We are ringing you today as a valued customer to ask you some questions about HSF. We will read you the question and then ask you to tell us which one of our answers most closely reflects your opinion. Please feel free to give us your own comments. Any information you give us will be used to help us further improve our service to you.

1 *Are you satisfied with the quality of HSF's service?*
A yes *B no* *C don't know*

2 *Do you consider HSF to be a reliable supplier?*
A yes *B no* *C don't know*

3 *Are the staff at HSF helpful?*
A yes *B no* *C don't know*

4 *Are you happy with the range of HSF's products?*
A yes *B no* *C don't know*

5 *Would you like to see the range increased?*
A yes *B no* *C don't know*

6 *Any other comments?*

Thank you for your help!